ENVIRONMENTAL AND RESOURCE VALUATION
WITH REVEALED PREFERENCES

THE ECONOMICS OF NON-MARKET GOODS AND RESOURCES

VOLUME 7

Series Editor: Dr. Ian J. Bateman

Dr. Ian J. Bateman is Professor of Environmental Economics at the School of Environmental Sciences, University of East Anglia (UEA) and directs the research theme Innovation in Decision Support (Tools and Methods) within the Programme on Environmental Decision Making (PEDM) at the Centre for Social and Economic Research on the Global Environment (CSERGE), UEA. The PEDM is funded by the UK Economic and Social Research Council. Professor Bateman is also a member of the Centre for the Economic and Behavioural Analysis of Risk and Decision (CEBARD) at UEA and Executive Editor of Environmental and Resource Economics, an international journal published in cooperation with the European Association of Environmental and Resource Economists. (EAERE).

Aims and Scope

The volumes which comprise *The Economics of Non-Market Goods and Resources* series have been specially commissioned to bring a new perspective to the greatest economic challenge facing society in the 21st Century; the successful incorporation of non-market goods within economic decision making. Only by addressing the complexity of the underlying issues raised by such a task can society hope to redirect global economies onto paths of sustainable development. To this end the series combines and contrasts perspectives from environmental, ecological and resource economics and contains a variety of volumes which will appeal to students, researchers, and decision makers at a range of expertise levels. The series will initially address two themes, the first examining the ways in which economists assess the value of non-market goods, the second looking at approaches to the sustainable use and management of such goods. These will be supplemented with further texts examining the fundamental theoretical and applied problems raised by public good decision making.

For further information about the series and how to order, please visit our Website
http:/www.wkap.nl/series.htm/ENGO

Environmental and Resource Valuation with Revealed Preferences

A Theoretical Guide to Empirical Models

by

Nancy E. Bockstael

Department of Agricultural and Resource Economics,
University of Maryland, U.S.A.

and

Kenneth E. McConnell

Department of Agricultural and Resource Economics,
University of Maryland, U.S.A.

Springer

A C.I.P. Catalogue record for this book is available from the Library of Congress.

ISBN-10 0-7923-6501-1 (HB)
ISBN-13 978-0-7923-6501-3 (HB)
ISBN-10 1-4020-5318-5 (e-book)
ISBN-13 978-1-4020-5318-4 (e-book)

Published by Springer,
P.O. Box 17, 3300 AA Dordrecht, The Netherlands.

www.springer.com

Printed on acid-free paper

Cover images taken before and after the 1989 Exxon Valdez oil spill in Prince William Sound, Alaska.

Dedication

To Ivar and my mom, both of whom have put up with a lot over the years.

N.E.B.

For Ginny, for helping in many ways.

K.E.M.

Contents

List of Figures

Preface

The goal of this book is to lay out the basic principles of using observations on individual agents' behavior for valuing changes in public goods, especially environmental amenities. We have tackled revealed preference methods because that is our chief interest. We waste no time arguing that behavioral methods are superior to stated choice methods. That methodological battle is long over.

With this book, we hope to provide a sense of the major conceptual issues in the literature. We also draw attention to some important but neglected issues. No time is spent arguing for more refined estimation procedures or better data. Instead we ask what can be said about welfare effects under the typical circumstances that researchers encounter.

This book got its start with an EPA project in the 1980's that investigated behavioral methods. This research involved Ivar Strand, Michael Hanemann, Maureen Cropper, Tim Phipps and others. It spawned research in recreation demand studies and hedonic models. We were fortunate to have Cathy Kling, Doug Larson, Bruce Madariaga, and Terry Smith working with us as graduate students.

Many people have helped us in the long process of writing this book. These include, in particular, Ray Palmquist, Laura Taylor, Kerry Smith, Cathy Kling and Joe Herriges. Countless others have added to our knowledge and appreciation of the issues. Numerous graduate students, including Sarah Adelman, Dave Herberich, Sonia Jarvis and Constant Tra, have read chapters and helped.

The Department of Agricultural and Resource Economics, our academic home for the past 25 years, has been conducive to research and has fostered the writing of this book. The need to apply research tools is an unexpected benefit of working in a land grant department, and the liberal attitude towards what constitutes good applied research has been instrumental. We thank our colleagues here. These include Richard Just and John Horowitz, and especially Ivar Strand, with whom much of the earlier work was done. Especial credit is due to the many graduate students who, over the years, have suffered through the Applied Welfare Economics course, challenging our reasoning and continually contributing to our understanding.

Chapter 1

Setting the Stage

1.1 Oil Spills and Valuation

When the tanker Exxon Valdez collided with Bligh Reef on March 24, 1989, it released 11 million gallons of oil into the pristine waters of Prince William Sound and subsequently induced a torrent of valuation studies on the economic damages of the spill. The potential magnitude of Exxon's liability unleashed a conflict between oil companies and resource trustees over measurement of these damages that has had a lasting influence on environmental valuation.[1] The conflict led to the investment of a vast quantity of resources in valuation methods by both sides. In the struggle to establish the magnitude of economic damages something of greater significance emerged: a universal admission that resource damages represent real losses to people. Although defendants challenged the use of stated choice methods in measuring *'non-use'* values associated with toxic spills, both plaintiffs and defendants accepted the concept of measuring damages for injury to public resources by the amount of compensation individuals in society would need in order to restore their well-being. This acceptance, characteristic of both the Exxon Valdez oil spill and other important but less spectacular cases, implicitly ratified the economic model of individual choices as the basis for economic value.

The willingness to use economic principles to establish the economic damages from the Exxon Valdez oil spill is part of the steady growth in the application of economics to public resource allocation. Whether the goal is to estimate the economic damages from injury to natural resources or to assess government

[1] Exxon eventually settled with the state of Alaska for damages of about $3 billion. Details of the oil spill and the settlement can be found on the website http://www.evostc.state.ak.us/.

regulatory analysis, economists cannot proceed without knowing the benefits and costs. Benefit-cost analysis has been applied to the evaluation of an astonishingly wide spectrum of issues. These include childhood reading programs, transport facility design, prevention of recidivism among convicted felons, a wide variety of health care initiatives and the measurement of the benefits of pollution reduction which is the chief concern of this book. The demand for good benefit estimates for non-marketed goods seems insatiable.

Two basic approaches to valuation have emerged: methods based on questioning that directly elicits the values that economists seek and methods based on observing behavior from which economists can deduce these values. In the development of benefit measures, the earliest methods were crude examples of the latter approach. Economists believed fervently that the only reliable evidence about how much a person would pay for a good or service would come from a situation where the person actually paid.[2] This belief has proved difficult to live by. Often benefit-cost analysis requires valuing services that have never been purchased nor are related in any way to observable behavior, making the revelation of values based on behavioral decisions impossible.

In the struggle over the size of the damages from the Exxon Valdez oil spill, both the plaintiffs and the defendants invested large sums of money in research on valuation techniques, but the focus of the research was rather narrow. Essentially the chief question was whether researchers could infer '*non-use*' values, the economic losses from injury to resources in their natural state that people would never experience directly. The nature of this valuation task required researchers to use direct interview techniques–stated preferences–rather than rely on observations of behavior. Consequently, most of the research focused on stated preference techniques, leading to substantial advances in the approach.

In this research environment, revealed preference approaches were relatively neglected. Compared with stated preference, revealed preference approaches seem unwieldy, not nearly as accessible or satisfying as a simple graph of consumer surplus would suggest. When one admits errors of measurement or specification, even the simplest of revealed preference applications seems fraught with difficulty. But experience has taught us that good stated preference analysis may be no less challenging. Meeting the NOAA panel recommendations (Arrow *et al.*, 1993) is no easy task. Both revealed and stated preference analyses pose difficulties, and while some valuation problems simply

---

[2]This attitude was equally true for firm behavior and is best represented by Friedman (1953) who argued that comparing the implications of models is a better methodological approach than interviewing businesses to get them to reveal how they made decisions–in particular whether they are maximizing profits. It may be true that competition in the asset market will force firms to maximize profits. Nothing, however, prevents households from behaving irrationally.

can not be addressed using revealed preference, others are vastly improved by having information on behavior and therefore the ability to predict how related markets may adjust.

The purpose of the book is to clarify and extend the economic theory and economic models that provide the foundations for behavioral approaches to valuation. It is written in the long shadow of the Exxon Valdez oil spill, with the view of filling in gaps and bringing new life to revealed preference approaches.

1.2 Where Do We Begin?

Anyone who has attempted to communicate the ideas of non-market valuation to non-economists or students in introductory classes will have encountered the many misconceptions of the basic idea of economic welfare. It has been variously construed as measured by GDP or changes in GDP, value added, expenditures (i.e. sales revenue) or changes in expenditures, etc. The latter is the most common mistake–that 'economic impacts' (measures of sales revenues) are synonymous with welfare. To state and local governments, impact analysis may be of interest as it measures economic activity in a locality and looks like an increase in income to local inhabitants. In fact, a project that increases sales revenue and tax receipts in one location may lead to an increase in welfare in that locality, but it may do so at the expense of sales and tax receipts in another locality and therefore may be neutral at the national level. More fundamentally, though, economic activity is simply not the same thing as social welfare. Admittedly, an increase in an individual's real income, all other things equal, represents an increase in that person's well-being. But increases in local economic activity involve much more than simply increases in income. Perhaps the most convincing argument against the interpretation of impacts as welfare comes in the form of an extreme example. A major hurricane will increase local expenditures dramatically both in terms of expenditures made to protect property *a priori* and expenditures made *ex post* for replacements and repairs. These show up as increases in revenues to construction and materials supply firms. Yet no one would argue that social welfare is enhanced by a hurricane.

Throughout the book we take care in defining the concept of economic welfare. A broadly accepted definition of individual welfare is the amount of potential compensation–paid or received–that leaves the individual with the same level of utility after a well-defined change in the individual's circumstances as before it. This definition stems from early work by Hicks (1939), Kaldor (1939), and Scitovsky (1941) who struggled with defining meaningful welfare measures more broadly useful than the restrictive Pareto criterion. Measures conceived as compensation are especially relevant for damage assessment because they satisfy the legal idea that individuals not be made worse off than they were before the injury to the natural resources. This definition of economic welfare, based on the Hicks-Kaldor compensation principle, is central to all we do in this book.

Weitzman (1976) has shown that the welfare significance of net national product (NNP) lies in its equivalence with the present discounted value of consumption, so that policies that maximize the former also maximize the latter. While having welfare significance, these aggregate social accounting measures serve a very different purpose from the compensation measures we will be interested in. Compensation measures are useful for the analysis of projects, policies, rule making and damage assessment while NNP and similar measures are useful for longer run social planning.

Throughout the book we emphasize the importance of posing well-formed welfare questions. For a welfare question to be well-formed, it must relate to some clearly defined change in exogenous circumstances. The change may, however, induce a chain reaction of further effects, all of which are ultimately attributable to the initial 'shock'. This will often include behavioral responses by individuals, and may lead to market effects if sufficient numbers of individuals alter their behavior. The fact that welfare measures are defined on exogenous changes, but that the ultimate welfare effects might include further induced changes, is important. Revealed preference methods have the advantage over stated preference methods in their capacity to account for these induced reactions, because they are based on knowledge of behavior.

Relative to stated preference methods, revealed preference methods are admittedly at a disadvantage for other reasons. Our concept of 'economic welfare' is a compensatory one, but behavior rarely reveals compensation-type measures. At least in the case of price changes, economists have known since the work of Willig (1976) that in most cases one can approximate these compensation measures quite well using information from ordinary demand curves. In subsequent chapters we will investigate how well this result holds up when measuring the economic welfare effects of changes in parameters other than prices.

1.3 The Purpose and Approach of the Book

The goal of this book is to provide a resource for economists doing applied welfare economics. The purpose is chiefly to clarify and occasionally to contribute to understanding models of individual behavior that can be used with empirical observations to estimate monetary welfare measures. Further it brings together the many conceptual ideas that may be found in the literature in a single place. The book has no particular conceptual outlook to foster on readers. Rather the idea is simply to try to understand how to use observations on behavior to infer economic values.

The concept of revealed preference is straightforward when applied to price changes. The essence of this book is figuring out what sorts of behavior will reveal welfare measures when the levels of environmental goods and services change. Another way of looking at the problem is to ask what kinds of restrictions on preferences will allow a given type of behavior to reveal these values. Of course, it is always possible to 'make assumptions' about preferences that will enable researchers to capture welfare measures from a given behavioral observation. The challenge is not so much the theoretical issue of whether there exists an assumption that will work. Rather, are there plausible and intuitively attractive stories about preferences that provide restrictions leading to the appropriate welfare measures? Some intuitively plausible restrictions are quite useful while other quite useful restrictions lack credibility.

The organization of our investigations differs somewhat from many studies of non-market valuation. Rather than beginning with the particular kind of empirical model, such as on-site travel cost for example, we begin with the structure of the model. From particular model structures, we derive welfare measures for changes in public goods, and then see how behavior can be used to measure the welfare effects. Chapters 3, 4 and 5 explore various revealed preference approaches that can be used when a public good enters directly into an individual's preference function but the good is not sold on the market. These cases cover the travel cost models as well as other approaches. Chapters 6 and 7 devise methods of valuation when goods with environmental quality dimensions are bought and sold on the market. This initial theoretical construction leads logically to both housing and wage hedonic applications. Chapters 8 and 9 investigate valuation approaches that are relevant when public goods or bads (e.g. pollution) enter, together with marketed goods, into some sort of production framework. Beginning with this structure lets us evaluate directly the usefulness of models of defensive expenditures and costs of illness as welfare measures. It also takes us into the domain of the firm.

1.4 The Maintained Assumptions

This book is about measuring welfare using individual choices. The behavior of individuals gives us the data. The models come from ideas about the motives that induce the behavior. We start with the assumption that individuals try to make the most of what they have–that is they attempt to maximize their own welfare. The arguments of the preference function and the nature of constraints may change, but throughout the book we work with models that have a similar set of maintained assumptions. Our most important maintained assumptions are fairly standard, but we state them explicitly here.

- **Coincidence of the individual and the household**

In our models, the individual and the household are the same, and we use the terms largely interchangeably. In effect we are assuming a unified model of the household and do not recognize differences among different members of a household. Such intellectual 'sloppiness' in accounting for the appropriate agent is admittedly troubling, yet it reflects the rather substantial gap in the literature. Closing this gap is a significant research issue in valuation, but one we have not undertaken in this book.

- **Well behaved preference functions**

Throughout the book we assume that preferences have the usual properties. Households make choices that reflect preferences for more of a good rather than less, their choices are transitive and they do not become satiated. Although we proceed with these assumptions about basic preference structures, they do not always imply straightforward indirect utility or expenditure functions. Two kinds of choice problems create some unusual duality results. When the budget constraint is non-linear, the resultant indirect utility and expenditure functions may not be well-behaved and Marshallian demands may not be well-defined. Further, when the choice problem involves a set of mutually exclusive alternatives, dual functions and demand functions will be composed of many disjoint parts, each depending on relative prices, income and the quantities of public goods.

- **Rational consumers**

In developing models for calculating benefits of public programs and policies, we have little choice but to assume that consumers always make rational choices. Since our models are static, we do not ask for the kinds of extreme rationality that dynamic optimization assumes. We do consider the consequences of imperfect information, but we do not allow the kinds of anomalies in revealed preferences that behavioral economics has uncovered in other settings. Indeed, it seems only a matter of time before behavior inherent in revealed preference models is shown to be contaminated by myopia and reference dependence.[3] Welfare questions when consumers fail to follow the basic axioms of preferences will prove especially challenging.

---

[3]See the book edited by Camerer, Loewenstein and Rabin (2004) for numerous examples.

1.5 What the Book Omits

Readers may pick up this book looking for topics and issues that we fail to address. The inclusion of some topics and exclusion of others has most to do with the extent of coverage elsewhere, the natural constraints of space and time, and our own comparative advantage–or lack of it. Some omissions warrant a brief explanation, although most are obvious from the nature of the book.

- **Stated preferences**

This is a book about revealed preferences. Several books on stated preferences are available.[4] There are similarities with stated preferences, in the sense that the approaches might begin with similar statements of welfare. Further, there are often implied behavioral responses in stated preference approaches. For example, contingent behavior analyses (including conjoint analyses) might offer a household the choice among a set of discrete alternatives much like the choice problems we investigate in Chapter 5, although the responses remain hypothetical. The differences are slight in those cases, but substantial in other applications, with the key distinction being the degree to which behavioral adjustments make revealed preference approaches more complicated. There are obviously other differences. Typically, revealed preference approaches require a longer chain of analysis between response data and welfare measurement. Observed behavior hardly ever reveals welfare measures without a struggle.

It would be natural to cover the combining of stated and revealed preference approaches, but the issues here are principally econometric and our coverage of econometric issues is slight.

- **Econometrics**

For the most part we suppress most econometric issues for the simple reason that we have our hands full with conceptual and empirical ones.[5] The two exceptions involve discrete choice models and hedonic wage analysis. With discrete choice models, there is little distinction between the model structure and econometric specification. Hence we introduce stochastic specification in the chapter on discrete choices. When we investigate welfare measurement using hedonic wage analysis, and especially the use of hedonic wage analysis to

[4]Among the many books on the topic, Louviere, Hensher and Swait (2000) provides a complete treatise on stated choice methods of valuation. A manual of stated preference procedures can be found in Bateman *et al.* (2002). Freeman's (2003) well-known book on environmental valuation includes chapters on stated as well as revealed preference methods, and a forthcoming book by Carson offers a history and complete bibliography.

[5]Haab and McConnell (2002) offer a treatment of the econometric issues that typically arise in both revealed and stated preference analysis.

infer the 'value of a statistical life', we confront econometric issues. Hedonic wage analysis is different from other valuation approaches because almost all researchers working in the area draw on the same datasets for wages and risk information. Finally much of the controversy in estimating the 'value of a statistical life' involves questions of data and/or model specification in estimating hedonic wage equations.

- **General equilibrium**

Our investigations are largely restricted to partial equilibrium models. We investigate models for valuation in three settings: individuals interact with public goods with no intervening markets; individuals purchase private goods on the market but government actions or public goods influence the quality of the private goods; public goods or bads enter households' and firms' production processes. In the first setting, where markets are peripheral to the chief problem, the analysis does not even warrant being called partial equilibrium, because there are no market effects at all. In the second setting, it would be feasible to incorporate more general equilibrium effects. Smith, Sieg, Banzhaf and Walsh (2004) have developed a locational equilibrium model of the housing market for example. When some salient external circumstances change differentially across local areas, a new equilibrium is likely to occur. This requires a general equilibrium model, which they have developed. In the production case, changes in environmental inputs can easily induce further market effects and lead naturally to the sorts of multiple-market welfare analyses developed by Just, Hueth and Schmitz (2004). We discuss some consequences of these models but refrain from venturing into general equilibrium modeling.

- **Dynamic analysis**

In all cases, we limit the analysis to a single period. This is not because we believe that all environmental valuation issues really are static, but because the more important aspect of most problems concerns the structure of household decisions rather than the dynamic element. Yet some of the most compelling environmental and natural resource problems are inherently dynamic and inter-generational–such as climate change and long term resource depletion. For such problems welfare analysis requires accounting for the stream of benefits and costs over time, raising important questions of sustainability and intergenerational equity. Proper treatment of these issues would involve adding considerably to the concepts and tools used in this book, as well as adding enormously to the book's length.

The applicability of the book is therefore limited to the many environmental problems that have rather more immediate consequences. There are plenty of these; finding examples in the literature has not been difficult. For these types of problems, we feel that static models do no serious injury to the analysis and that no important conclusions would be reversed by incorporating dynamic considerations.

1.6 A Look Ahead

Throughout this book we begin each analysis with the ultimate goal of estimating measures of compensation. It is usually quite easy to *express* the precise welfare concept in terms of indirect utility functions or expenditures for each change in exogenous circumstances considered. Obtaining empirical measures of the concept, based on observed behavior, is always the real challenge. In the case of price changes, the connection between the welfare concept and empirical measurement is well-worked out and widely understood. However, when valuing changes to public goods, the roadmap connecting theory to measurement is not so familiar.

There is no one answer to this question that applies in all settings. The need to value a wide array of changes in public policies, public goods, regulatory rules, and acute pollution events has led naturally to many different kinds of non-market valuation approaches. Economists have been creative. In the following chapters we try to provide the rationale for measuring welfare in many of these circumstances.

Chapter 2

Welfare Economics for Price Changes

2.1 Introduction

The phrase *environmental valuation* has come to be applied to the practice of evaluating the social gains and losses from environmental degradation or improvement. Economists practice valuation by applying welfare economics to environmental outcomes. There is, of course, a good deal of debate as to what is meant by valuation, particularly among the broader science community. Evaluation of benefits and costs often evokes strong objections, even when applied in the well-defined context of welfare economics. Because the objections have an even greater propensity to emerge in environmental applications, the principles of applied welfare economics deserve a quick reminder. This chapter reviews the theory of welfare measurement, but as with the entire book, the ultimate empirical application remains foremost in our minds. The methods for recovering the welfare measures we seek (or good approximations of them) are indirect and will depend on careful reasoning and sound econometrics. Although the importance of the econometric details can not be underestimated, in this book we focus on the logic that connects behavior with estimation.[1]

Ultimately welfare economics is concerned with social choice–whether some states of the world are better than others. But because welfare economics is individualistic, in the sense that the well being of society is based on the well being of its individuals, we start with how we determine individual well being.

[1] Haab and McConnell (2002) address many of the econometric issues that arise in using revealed and stated preference techniques in environmental valuation.

Here economics is unequivocal: an individual is the arbiter of his own welfare. When an individual faces two alternatives A and B and chooses A, we recognize that an individual is better off with A. We say he has 'revealed his preference' for A. Given the emphasis on revealed preferences, it may seem puzzling that all models begin with the individual maximizing an unobservable index we call utility. This puzzle is solved by showing that when individual behavior satisfies a set of plausible axioms of choice, the individual has a preference ordering that can be represented as a utility function.[2]

The puzzle of whether society is better off in different states has plagued economists for centuries. The concept of 'utility' that forms the basis of theoretical models of individual behavior cannot, on its own, provide what we need to answer the larger question of comparing social states. Utility is not measurable by researchers, nor is it likely to be measurable in a cardinal way even by individuals themselves; and it can not be compared across people. Further, Arrow (1951) and Samuelson (1956) have shown the impossibility of using a social welfare function to aggregate individual levels of utility or preferences to order different states. The familiar Pareto criterion (Pareto, 1896) avoids such meaningless pursuits and forms the logic underlying economists' attempts to make normative statements about alternative 'states of the world'. This criterion has become the basis for defining economic efficiency. An economically efficient solution is one in which no one can be made better off without making someone else worse off—the very statement of Pareto optimality. The Pareto criterion is appealing because only ordinal rankings of states by each individual is required. No cardinal measurement of preferences is needed, nor interpersonal utility comparisons. However the Pareto principle has at least two drawbacks. For one thing, it provides a very incomplete ordering of states. The Pareto criterion is almost never helpful in practice because most changes improve the well being of some while making others worse off. As a consequence, moves from the current situation (the *status quo*) are usually ruled out by this principle. The second and related drawback is that the Pareto criterion attributes to the *status quo* a preferred status that may be unwarranted. All is compared to the *status quo*, and no movement from this position is allowed if it harms anyone, even if that individual loses very little and is among the most well-off in society. In fact, the set of Pareto efficient solutions *depends* on the *status quo* (i.e. where we start from) and therefore depends on the distribution of initial endowments across individuals in society–a distribution that may or may not be considered 'fair' by society.[3] Change those endowments and the

---

[2]These well known axioms are discussed in Deaton and Muellbauer (1980), chapter 2, or Cornes (1992), chapter 2.

[3]Endowments are best interpreted as wealth, natural ability, access to education, etc., rather than income which is effectively endogenous.

set of efficient solutions changes. Pareto efficiency can be defined only with reference to a given initial distribution.

To overcome the shortcomings of the Pareto criterion, economists have developed the methods of applied welfare economics on the foundation of the compensation principle proposed by Hicks (1939) and Kaldor (1939). A policy, project or event passes the Hicks-Kaldor compensation test if it is *possible* for those who benefit by the change to compensate those who lose. In other words, *potential* Pareto improvements are allowed, not just actual ones. The compensation principle helps but does not entirely do away with either of the drawbacks of the Pareto criterion. For one thing Scitovsky's (1941) reversal paradox suggests that even when using the compensation principle not all states are necessarily comparable. It may be possible for gainers to compensate losers when a move from state A to state B takes place, but if compensation is not actually paid then it is sometimes possible, once in state B, for the move back to state A also to pass the compensation test. If one requires the more stringent test that the same state must be found preferable irrespective of the starting point, then some states will be non-comparable according to the compensation principle. This sort of reversal rarely arises in real policy problems for reasons we will get to later in the book. When it does arise, the choice among 'states' must be made on other grounds.[4]

The compensation principle would do nothing to reduce the overpowering role of the *status quo* if actual compensation were required. But the principle inherently leaves the decision of whether compensation will take place to another public process. The fact that the compensation principle does not require compensation to be paid goes only part way in reducing the influence of the *status quo,* however. Although it is now possible for a policy that helps the poor more than it hurts the rich to pass the 'benefit-cost' test even without compensating the rich, it is also a certainty that a policy that makes the rich better off by more than it makes the poor worse off will also pass the test—and compensation need not be paid, so that policies that widen the gap in relative economic status are not necessarily discouraged.

The importance of initial endowments must be kept in mind even when using the compensation principle. If we could chose among states by adding up and comparing the utilities of winners and losers, then the endowments problem would not exist. But we can not. In fact, we are really answering something different. We are really answering the questions: how much would losers from a change need to be compensated so that if the change were enacted they would be just as well off after the change as they were in the initial situation?

[4]The choice may be dictated legally if either winners or losers from the change have the legal property rights.

And, how much could the winners afford to pay in compensation, so that if the change were enacted they are at least as well off as they were in the initial situation? This does not measure utility change because utility is returned to its initial level. It generates measures of compensation—the right amount of compensation, received or paid, in some units of a 'numeraire' to make the losers and winners indifferent.

Economists normally expect utility to be increasing at a decreasing rate in its arguments. As long as this is true of the numeraire used to calculate compensation then, all else equal, a person endowed with more of the numeraire will need more compensation (or be willing to pay more compensation) than an individual with exactly the same preferences but less endowment. This is simply because a given increment in the numeraire means less to someone endowed with more of the numeraire to begin with. The numeraire most frequently used to define compensation is money, because it is the most fungible and can be easily redistributed. Because the marginal utility of money is expected to fall with wealth, an individual's compensation measured in money will be greater the wealthier he is. From a distributional perspective, this result is disturbing because it means that those with more money implicitly have greater weight in our calculations. While this fact does not always have a distorting influence on the analysis, it will be most troubling when the changes being analyzed break down in such a way that winners and losers from the proposed change differ in terms of their wealth and do so dramatically.

Even if one is indifferent to the distributional implications, an interesting conceptual dilemma can arise because of the role of the numeraire. Alternative numeraires can lead to different answers to the compensation test, if the alternative numeraires are not completely fungible and if the relative endowments of the two numeraires differ substantially across sub-groups of the population. Bockstael and Strand (1985) consider two numeraires—money and time, and two alternative sportfishing policies—one that reduces the monetary cost of sportfishing and one that reduces time costs of access. They show that if two subgroups of anglers differ in their endowments of the two 'compensation numeraires' in that one has more discretionary time but less discretionary income than the other, then the outcome of a benefit-cost analysis that compares the two policies will be reversed depending on the numeraire used to calculate the compensation measures.

Even without differences in initial endowments, the numeraire choice will matter if preferences differ, as long as the two numeraires are not completely fungible. Market equilibrium conditions induce equality of marginal values for each market good for all consumers. For public goods, heterogeneity among consumers can make the equilibrium marginal values differ and, as Brekke (1997) illustrates, will lead to ambiguity in benefit cost analysis. Brekke's example is one of a materialist and an environmentalist with different preferences over money and the environment, but identical initial endowments. The environmentalist has higher marginal utility for the environment and lower marginal utility for income relative to the materialist. Because the materialist values money highly relative to environmental quality, his bid for any given project will be smaller if made in units of money than if made in 'environmental units' (Brekke's alternative 'compensation currency'). The reverse will be true for the environmentalist. Now suppose an environmental improvement project is proposed. Given that the environmentalist has stronger preferences for the project than the materialist, the project will more likely pass the benefit-cost test if the 'currency' used to evaluate it is money rather than environmental units. The environmentalist who feels strongly about the project also values money less and can give up more of it for any utility gain. If the benefit-cost analysis is conducted using environmental units as a currency, the reverse will be true. So the final answer *may* depend on the 'currency' or numeraire when the good is a public good and individuals have different marginal rates of substitution between public and private goods.

There is no satisfactory resolution to this dilemma. It stems from the fact that we can not compare preferences across people directly. Reversals such as the ones described above are only likely to take place in extreme cases where winners and losers differ dramatically in their marginal values of money and when proposed changes are large. All we can do is recognize the limitations of this criterion and resist embracing it wholeheartedly irrespective of context. At the heart of the problem is something we must keep in mind throughout, despite our constant reliance on the concept of individual compensating variation. Compensating variation is *not* a measure of welfare or utility change, although for convenience we will often refer to it as a 'welfare measure'. Instead, it is a measure in some numeraire—generally money—of what is required to return an individual to a given utility level, and therefore it can not measure a *change* in utility. It makes sense that compensating variation should not be a measure of utility change, since such measures could never be comparable across individuals. It is the act of translating utility into some 'currency' comparable over individuals that causes the endowment problem described above. It is also important to remember that when summed over all individuals affected by a policy, project or event, the resulting measure can not tell us whether that policy, project or event has produced a more efficient outcome unless it includes the actual payment of compensation. Whether economic efficiency is even desirable, however, depends on the circumstances of the change and the distributional objectives of society.

2.2 Compensation Measures

The compensation measures inherent in the application of the compensation principle are defined in the context of individuals' choices. These measures are dependent not only on the preferences of the individual and his endowments but also on the alternative actions available to the individual and his ability to adjust his behavior. If an environmental hazard such as a serious air pollution event were to occur, the amount of compensation necessary to return the individual to his original utility level would be smaller if he were able to acquire equipment (such as an air filter) that would mitigate the danger at a smaller cost. In many contexts, the behavior of an individual faced with alternative choices can tell us something about the compensation he would need to accept or be willing to pay to avoid externally imposed changes. When choices or actions of the individual provide critical information about compensation measures, revealed preference methods for calculating or approximating these measures are possible.

A framework for describing and analyzing behavior is a necessary starting point for using revealed preference methods. In constructing this framework, we begin where most economists begin–with the individual's constrained utility maximization problem. Of course, even this can be challenged and *is* challenged by those who do not view individuals as rational or informed decision makers. The issue of information will come up again and again as we deal with the practical concerns that face us in empirically modeling decisions.

The framework is also challenged by those who object to the use of the representative individual. Admittedly this has two drawbacks in practice, one that can be easily overcome and one that involves some unresolved conceptual issues. The *application* of welfare measurement necessarily involves heterogeneous consumers, and in practice this heterogeneity can be accommodated either by measuring the impact of individual characteristics on behavior and welfare or by allowing unobserved, random differences. Consequently, there is considerable latitude for generalizing the 'representative' part of the representative agent.

A more serious challenge to the analysis of the representative individual stems from the fact that individuals are members of households and that many decisions are made jointly within households (and in some cultures within communities or villages). The allocation of household resources has been modeled in a variety of ways. Household decisions can be viewed as the consequence of joint maximization of a household preference function. This is the unitary model of household decision making, essentially the Becker (1981) model of households. This provides a means of obtaining results quite analogous to standard models, although in practice even the Becker model entails some revisions to the individual decision model. There are two competing models: the bargaining model, in which households use threat points to secure household resources (e.g. McElroy and Horney, 1981; Manser and Brown, 1980), and the collective model. In the collective model (e.g. Browning and Chiappori, 1998), individuals within the household have their own preferences, and joint household outcomes are the result of collective decisions that are efficient in a Pareto sense.

Economists have found it difficult to incorporate these new theories into empirical work. To our knowledge, the sole application to environmental valuation work is provided by Smith and Van Houtven (2004).[5] The paucity of this type of work is somewhat surprising given that much of environmental valuation depends on behavior which can be characterized as household production and that depends on the allocation of time. Nonetheless, we will persist in framing the problem in terms of the unitary model–a model that treats the many-person household as having one set of objectives and a single maximization process so that the mathematical formulation is identical to the individual. And, we will use the terms 'individual' and 'household' somewhat interchangeably in later chapters, if for no other reason than to remind the reader that the household is sometimes the more appropriate decision making unit. The consequences of using the unitary model rather than another model of household decision making are not well worked out in the literature. These are issues that deserve attention that we will not be able to provide here.[6]

The simplest representation of the decision process is one of utility maximization subject to an exogenous and linear budget constraint. The indirect utility function is the solution of that problem:

$$v(\mathbf{p}, y) = \max_{\mathbf{z}} \{u(\mathbf{z}) | \mathbf{p} \cdot \mathbf{z} \leq y, \mathbf{z} \geq 0\} \tag{2.1}$$

where $u(\mathbf{z})$ is the quasi-concave utility function, $\mathbf{z}$ is an n-dimensional vector of the consumption levels of goods, $\mathbf{p}$ is a corresponding price vector, and y is income, the latter being treated as exogenous for now. The maximization problem yields the ordinary or Marshallian demand functions

$$\mathbf{z} = \mathbf{f}(\mathbf{p}, y),$$

which characterize the behavior that is typically observable. The non-negativity restriction on the commodity bundle is unremarkable; it appears in the typical

---

[5] Finding consistency between the individual and household choices is likely to be an important research topic for valuation for a variety of reasons, not the least of which is the effort to value the well being of children.

[6] The issue arises with particular severity in stated preferences, where the researcher cannot tell whether the respondent speaks for the individual or the household. Munro (2005) has shown that when individuals pool income, the respondent provides compensating variation for the household.

statement of the consumer's problem, but it is especially important for many valuation problems. In later chapters we encounter circumstances in which the consumer chooses positive quantities of some commodities and zero quantities of others.

In the framework of (2.1), we could define how much compensation an individual would need to be paid (or would need to give up) if the price vector were to change from $\mathbf{p}^0$ to a final vector, $\mathbf{p}^1$. The compensating variation of such a change can be defined either implicitly or explicitly. Using the indirect utility function, Hicksian compensating variation (CV) is given by

$$v(\mathbf{p}^0, y^0) = v(\mathbf{p}^1, y^0 - CV). \tag{2.2}$$

In this expression CV is defined in money terms as the change in exogenous income necessary to return the individual to the utility level that he experienced before the change in the price vector. The initial utility level equals $v(\mathbf{p}^0, y^0)$. Therefore CV is indeed a measure of the level of compensation, but the signing of this measure can be confusing, a circumstance exacerbated by inconsistencies in the literature. Some papers (e.g. Willig, 1976; Hausman, 1981) use the more literal convention, defining CV as the compensation (positive or negative) that must be *paid* to return the individual to his initial utility level. Therefore, if the price change makes the individual better off, the 'compensation' necessary to return him to his initial utility level must be negative. Just, Hueth and Schmitz (1982, 2004) adopt the convention that the CV associated with a change that is improving should itself be positive. This merely requires a slight semantic change in definition. CV is now the amount (positive or negative) that must be taken away from the individual to return him to his initial utility level. In the case of a welfare improvement, where $v(\mathbf{p}^1, y^0) > v(\mathbf{p}^0, y^0)$, it is necessary that $CV > 0$, for $v(\mathbf{p}^0, y^0) = v(\mathbf{p}^1, y^0 - CV)$ to hold. Here CV would be a payment—the maximum amount the individual would pay rather than forego the price decrease. We employ the latter definition, as it seems most useful to match the sign of the welfare measure with the direction of the welfare change.

The companion concept, equivalent variation (EV), is defined implicitly using the indirect utility function as:

$$v(\mathbf{p}^1, y^0) = v(\mathbf{p}^0, y^0 + EV) \tag{2.3}$$

The equivalent variation measure takes as the baseline the new level of utility that would be possible were the price change put into effect. Equivalent variation is the adjustment of income necessary to achieve this new level of utility but without the price change. Thus EV is a compensation measure with a different benchmark–the utility level achieved after the change. The welfare measure is the change in income necessary to attain that utility level under the initial price circumstances. Again, our signing convention forces the 'welfare' measure to have the same sign as the welfare effect.

Why do we bother with EV when CV appears sufficient for our purposes? The origins of EV go back to the Scitovsky reversal paradox mentioned earlier in the chapter. In comparing two second best alternative states, it is possible to encounter a reversal paradox such that *state B* looks preferred to *state A* from the vantage point of *state A*, but *state A* looks preferred to *state B* from the vantage point of *state B*. This is the same as saying that even if the sum of compensating variations associated with a potential move from *state A* to *state B* is positive, once in *state B* the sum of compensating variations associated with a potential move back to *state A* could also be positive. Scitovsky's criteria required that there be an unequivocal gain in moving from A to B in order to pass the compensation test. He required that the CV of the move be positive and the CV of the reverse move be negative. Equivalent variation is a useful concept in this light. The EV of a move from A to B is defined to be exactly equal to $-CV$ of a move from B to A. Thus the Scitovsky test requires both ΣCV and ΣEV to be positive.[7]

While more intuitive when expressed as in equations (2.2) and (2.3), CV and EV can be represented in terms of the expenditure function. The expenditure function is the solution to the cost minimization problem

$$m(\mathbf{p}, \bar{u}) = \min_{\mathbf{z}} \{\mathbf{p} \cdot \mathbf{z} | u(\mathbf{z}) \geq \bar{u}, \ \mathbf{z} \geq 0\} \tag{2.4}$$

which incorporates the same information as the utility maximization problem. This minimization problem yields utility-constant or Hicksian demand functions

$$\mathbf{z} = \mathbf{z}^h(\mathbf{p}, \bar{u})$$

which describe a type of behavior that is rarely observable but useful as an abstract construct. Initial income, y^0, in equation (2.1) equals the $m(\mathbf{p}, \bar{u})$ of equation (2.4) evaluated at initial prices and utility, and $\bar{u}$ of equation (2.4) equals $v(\mathbf{p}, y)$ of equation (2.1) evaluated at initial prices and income. The expenditure function is itself measured in terms of units of income—specifically the amount of income required to reach a specified level of well-being, given the

[7] As early as Hause (1975) some have argued that equivalent variation is the only acceptable measure because, unlike compensating variation, it is ordinally consistent with utility rankings. As an example, start with an initial situation and measure the CV and EV associated with a) an income change and b) a price change, both of which cause the individual to end up on the same new level of utility. The EV measures of the price and income change will be identical but the CV measures will not. Just, Hueth and Schmitz (2004) (Appendix 6C) convincingly argue that this is not a sufficient basis to choose between the measures, since once a change takes place, the EV test for reversing the change is equivalent to the CV test for making it, and policy should not be based on unstable criteria that can be reversed. The only sensible thing is to consider both CV and EV if they are expected to differ substantially. However, substantial differences are not often encountered.

price vector. As a consequence, the difference in expenditure functions, properly conditioned on exogenous variables, equals the compensating variation:

$$CV = m(\mathbf{p}^0, u^0) - m(\mathbf{p}^1, u^0). \tag{2.5}$$

Likewise, equivalent variation is

$$EV = m(\mathbf{p}^0, u^1) - m(\mathbf{p}^1, u^1). \tag{2.6}$$

Previous signing conventions are preserved; the measures are positive or negative, according to whether the change is welfare enhancing or welfare reducing. Once again, the distinction between equations (2.5) and (2.6) is the reference level of utility. Utility is at u^0 before the person experiences the change. After the change, and after the person maximizes utility constrained by the new parameters and the budget, utility is at u^1. If the sum of the CV measures over all affected individuals is positive, then the policy, project or event passes the test set forth by Hicks and Kaldor. If the sum of the CV measures and the sum of the EV measures are both positive, then there is no reversal paradox and the change passes Scitovsky's more stringent test.

2.2.1 Willingness to Pay and Willingness to Accept

We have discussed CV and EV as if they are quite different in magnitude. One of the chief debates in valuation over the last 20 years is whether they are likely to be very different. The question most frequently debated is whether willingness to pay (WTP) is different from willingness to accept compensation (WTA), but we can connect these concepts with CV and EV, using the baseline and new utility levels, $v(\mathbf{p}^0, y^0)$ and $v(\mathbf{p}^1, y^0)$. Given the construction of the concepts in equations (2.2) and (2.3), CV will be equal in magnitude to WTP for situations where the utility increases and to WTA for situations where utility decreases. In all cases, WTP and WTA are nonnegative. EV is equal in magnitude to WTA for utility increases and to WTP for utility decreases. For improvements in utility, CV is the maximum amount an individual would pay rather than go without the improvement. Hence when $v(\mathbf{p}^0, y^0) < v(\mathbf{p}^1, y^0)$ it makes sense to write

$$v(\mathbf{p}^0, y^0) = v(\mathbf{p}^1, y^0 - WTP).$$

On the other hand, if $v(\mathbf{p}^0, y^0) > v(\mathbf{p}^1, y^0)$ an individual will need to receive a payment to overcome the decline in utility, so that WTA is the minimum amount an individual would accept and we can write

$$v(\mathbf{p}^0, y^0) = v(\mathbf{p}^1, y^0 + WTA).$$

Thus CV will be the same as WTP for improvements and will be equal to minus WTA for deteriorations.

There is now a well-developed literature on the disparity between WTA and WTP based on experimental and stated preference studies. This disparity was originally observed in early contingent valuation studies of natural resources. Hammack and Brown (1974) were the first to find that respondents claimed to require a great deal more compensation to give up a resource than they were willing to pay to obtain it. This disparity has persisted in various sorts of stated preference studies. It has in fact led to a systematic study of responses to survey questions, which some have claimed undermines neoclassical theory of preferences. For the purposes of this book however, the importance of the debate is lessened by the paucity of evidence from behavioral models of a disparity between WTA and WTP.[8]

2.3 From Behavior to Welfare Measures

This book explains revealed preference methods for estimating gains and losses people experience when environmental goods and services are altered. Because of the focus on *revealed* preference, it must be possible to deduce gains and losses from observations on the behavior of individuals. Collecting data on individuals' behavior will rarely provide a direct means of calculating CV or EV for the cases of most interest in this book. However, there are indirect means for estimating CV and EV or bounded approximations of these concepts, and it these that we will pursue.

The price change case in equations (2.2) and (2.3) is a special case in welfare economics, and for this special case there is a link between the compensating measures and behavior. This connection is made through Shephard's lemma, from which we know that

$$\frac{\partial m(\mathbf{p}, u)}{\partial p_i} = z_i^h(\mathbf{p}, u) \tag{2.7}$$

where $z_i^h(\mathbf{p}, u)$ is the Hicksian or compensated demand function for good i. Suppose our interest centers on the welfare effects of a change in the price of commodity 1 from p_1^0 to p_1^1. Let $\mathbf{p}_{-1}$ and $\mathbf{z}_{-1}$ represent the price and commodity vectors that exclude commodity 1. Given Shephard's Lemma, one can

[8] An exception might arise in the context of housing hedonic applications because housing represents such a large portion of an individual's expenditures. Chattopadhyay (2002) attempts identification of underlying preferences in a hedonic context and recovers CV and EV meausures that differ by discernible amounts. Brookshire, Thayer, Tschirhart, and Schulze's 1985 study of earthquake threats is a context in which such a divergence might be expected.

obtain the CV measure of a change in the price of commodity 1 by integrating over this compensated demand function between the initial and final prices:

$$CV(p_1^0, p_1^1, u^0) = -\int_{p_1^0}^{p_1^1} \frac{\partial m(\mathbf{p}, u^0)}{\partial p_1} dp_1 = -\int_{p_1^0}^{p_1^1} z_1^h(p_1, \mathbf{p}_{-1}, u^0) dp_1 \qquad (2.8)$$

where the minus sign is added to reflect the fact that the order of integration is reversed from what is implied by equation (2.5).

In equation (2.8) the welfare measure, CV, is related to a behavioral function $z_1^h(p_1, \mathbf{p}_{-1}, u^0)$, but this is compensated demand and is not directly observable. The problem of course is that we need expenditure minimization behavior to obtain our welfare measures, but utility maximization is a more appropriate assumption for the data generating process. The description of how behavior actually changes as p_1 changes is best captured by the set of Marshallian or income-constant demand curves, $\mathbf{z} = \mathbf{z}^m(p_1, \mathbf{p}_{-1}, y)$, where the vector, $\mathbf{z}$, is the set of observable arguments that solves the maximization problem in equation (2.1). Although we do not observe behavior that stems from expenditure minimization, duality results from consumer theory assure us of a well-defined correspondence between the two. The initial choice of z_1^0 at price, p_1^0, is a solution common to both compensated and uncompensated demands. Once price diverges from p_1^0, the two functions will diverge except in special circumstances. The new utility maximization solution at z_1^1, after price has changed to p_1^1, represents another point on the same Marshallian demand function and is common to a second compensated demand function—one conditioned on u^1. Since this demand function $z_1^h(\mathbf{p}, u^1)$ must relate to the expenditure function $m(\mathbf{p}, u^1)$ according to Shephard's lemma, it must be true that the integral over this new function is the measure of equivalent variation:

$$EV(p_1^0, p_1^1, u^1) = -\int_{p_1^0}^{p_1^1} \frac{\partial m(\mathbf{p}, u^1)}{\partial p_1} dp_1 = -\int_{p_1^0}^{p_1^1} z_1^h(p_1, \mathbf{p}_{-1}, u^1) dp_1. \qquad (2.9)$$

The relationship among these compensated and uncompensated demand functions is depicted in Figure 2.1 and is described by the Slutsky equation. The initial optimum is characterized by $z_1^h(\mathbf{p}^0, u^0) = z_1^m(\mathbf{p}^0, m(\mathbf{p}^0, u^0))$, where $z_1^m(\mathbf{p}^0, m(\mathbf{p}^0, u^0))$ is the Marshallian demand function evaluated at initial prices and income. Divergences from the initial price level are described by the Slutsky equation:

$$\frac{\partial z_1^h(\mathbf{p}, u^0)}{\partial p_1} = \frac{\partial z_1^h(\mathbf{p}, y)}{\partial p_1} + \frac{\partial z_1^m(\mathbf{p}, y)}{\partial y} \frac{\partial m(\mathbf{p}, u^0)}{\partial p_1}. \qquad (2.10)$$

Since $\partial m(\mathbf{p}, u^0)/\partial p_1 = z_1 \geq 0$, and since $\partial z_1^m(\mathbf{p}, y)/\partial y$ is the Marshallian income effect which will be positive for normal goods, the negative demand

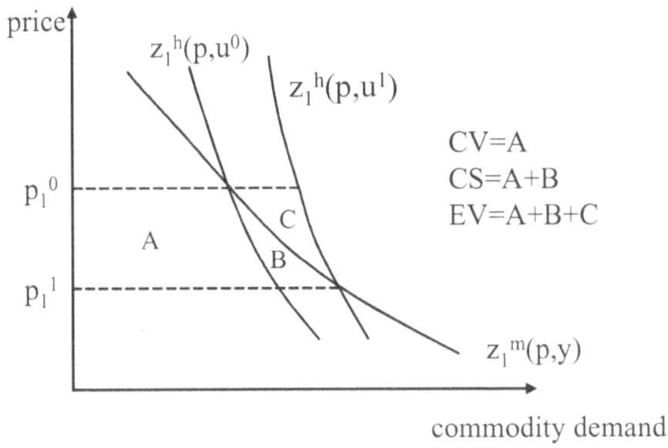

FIGURE 2.1. Relationship among CV, EV, and CS: Price Decrease for Consumption Good

response to a change in price must be greater for uncompensated behavior than for compensated behavior. Drawn with price on the vertical axis, this implies a flatter uncompensated than compensated demand function.[9]

The Slutsky equation makes clear that the area behind the compensated demand function between two prices will not equal the corresponding area behind the ordinary demand function, unless there are no income effects. Figure 2.1 illustrates the differences. CV and EV will be measured as the areas between the two prices behind the compensated demands conditioned on u^0 and u^1 respectively. The area behind the ordinary demand function is, of course, the well-known concept of consumer surplus (CS), first drawn by Marshall. Both the graph and the Slutsky equation helps us to order the three measures. For a consumer good's price change, $CV \leq CS \leq EV$, *given our signing convention.* For a price increase, this implies that CV will be larger in absolute value than CS, which will be larger in absolute value than EV, but all will be negative.

2.3.1 So What is Wrong with Consumer Surplus?

Alfred Marshall (1930) defined consumer surplus as a means of evaluating price changes. Why do we now prefer CV and EV measures instead? Consumer surplus seems preferable on practical grounds: it is defined in terms of observable behavior and does not require a choice of baseline utility. At first blush, CS also seems preferable to the compensating measures, even on theoretical grounds. To see the basis of this appealing but misleading statement, write

[9]For inferior goods–goods with negative income effects–the reverse will be true. In extreme cases, when income effects are large and negative, it is at least feasible for Marshallian demands to be upward sloping.

out consumer surplus in mathematical form:

$$CS = -\int_{p_1^0}^{p_1^1} z_1^m(p_1, \mathbf{p}_{-1}, y)dp_1. \tag{2.11}$$

Using Roy's identity, which equates $z_1^m(p_1, \mathbf{p}_{-1}, y)$ to $-v_1/v_y$, consumer surplus can be restated as:

$$CS = \int_{p_1^0}^{p_1^1} \frac{v_1(p_1, \mathbf{p}_{-1}, y)}{\lambda(p_1, \mathbf{p}_{-1}, y)} dp_1 \tag{2.12}$$

where λ is the Lagrangian multiplier implied in the constrained maximization problem of (2.1) and equals v_y, the marginal utility of income. Subscripts on the indirect utility function indicate partial derivatives. If λ could be expected to remain constant over the price change, then CS could be written as

$$CS \stackrel{?}{=} \frac{1}{\lambda} \int_{p_1^0}^{p_1^1} \frac{\partial v(p_1, \mathbf{p}_{-1}, y)}{\partial p_1} dp_1 = \frac{1}{\lambda}[v(p_1^1, \mathbf{p}_{-1}, y) - v(p_1^0, \mathbf{p}_{-1}, y)]. \tag{2.13}$$

This expression has great appeal, as it is simply a rescaling of the *utility change*. The change in utility brought about by the change in price is converted to money using the marginal utility of income.

However, a number of early papers, including a famous one by Samuelson (1942), criticized the use of consumer surplus. Samuelson argued that λ is not constant over the parameter change and therefore can not be moved outside of the integral sign, leaving us with equation (2.12), which has no meaningful definition.

Samuelson's second and related criticism of consumer surplus, that of non-uniqueness, stems from the fact that the line integral of ordinary demand functions over multiple price changes is not path independent. To understand what this statement means, note that to write the consumer surplus of a multiple price change, one must express it as the line integral

$$CS = -\int_L \sum_{j=1}^{J} z_j^m(\mathbf{p}, y)dp_j \tag{2.14}$$

where L is a path for integrating over the J price changes and $z_j^m(\mathbf{p}, y)$ is the ordinary demand for the jth good whose price changes. A line integral is an integral of a sum of interindependent multivariate functions along a specific path between the initial and final values of the multiple variables of integration. In this case, the interdependent functions are the J demands each of which conceivably depends on all J prices that are changing. A line integral can be evaluated as the sum of *ordinary* definite integrals along some path of integration. For example, if there were three price changes and the path of these price changes was given by: $(p_1^0, p_2^0, p_3^0) \longrightarrow (p_1^1, p_2^0, p_3^0) \longrightarrow (p_1^1, p_2^0, p_3^1) \longrightarrow (p_1^1, p_2^1, p_3^1)$, then the line integral could be evaluated as

$$\begin{aligned} CS &= -[\int_{p_1^0}^{p_1^1} z_1^m(p_1, p_2^0, p_3^0, y^0)dp_1 + \\ &\int_{p_3^0}^{p_3^1} z_3^m(p_1^1, p_2^0, p_3, y^0)dp_3 + \int_{p_2^0}^{p_2^1} z_2^m(p_1^1, p_2, p_3^1, y^0)dp_2]. \end{aligned} \tag{2.15}$$

If the line integral we began with in (2.14) is not 'path independent', then evaluating it as a sequence of definite integrals as in (2.15) will yield different answers depending on the path of integration. In other words, reordering the sequence of multiple price changes will generally produce different answers to this line integral. A condition for a line integral to be path independent is that the integrand be an exact differential of some function. But the integrand of (2.14) is not an exact differential of any function. If the marginal utility of income (λ) could be extracted from the integrand, then the remainder *would* be an exact differential of the indirect utility function, but it cannot–again because it does not remain constant with price changes. So the problem remains that consumer surplus is in principle neither meaningful nor unique (for a more careful and complete treatment of this, see Just, Hueth, and Schmitz, 2004, Chapter 5). If this had been the end of the story, applied welfare economics might have died an early death.

2.4 From Ordinary Demands to Welfare

In the early 80's, empirical welfare measurement received a boost from several economics papers that set out means for moving between uncompensated and the corresponding compensated functions. Hausman (1981) is perhaps the best known although Hanemann (1980) demonstrated the result at about the same time. Hausman shows that a Marshallian demand function, stated in terms of the good's price, income, and an implicit price of 1 for a numeraire good, can be integrated back to the indirect utility function or expenditure function. From this, CV and EV measures can easily be calculated. As long as the Marshallian function is consistent with utility maximization, integration is *in theory* possible. For this reason, the approach is clearly more appropriate for individuals' demand functions, although Hausman motivates the procedure in the context of aggregate market demands.

Integrating back analytically works well for some forms of demand functions but analytical solutions are sometimes not possible. Vartia (1983) demonstrated a numerical integration algorithm that one can employ even if analytical integration is infeasible because of the functional form. Vartia's technique is adaptable to systems of equations as well.

Applying any of these techniques produces answers only as good as the initial input–the Marshallian demand function specification. With sufficient observations on $\mathbf{z}$, $\mathbf{p}$, and y, and parsimonious assumptions about the parameters and functional form of $\mathbf{z}^m(\mathbf{p}, y)$, we can often estimate a sufficient portion of preferences to allow *behavior* to be predicted in the neighborhood of the observed data. But perfect information about the entire ordinary demand function is far more elusive. For one thing, there often are large portions of the ordinary demand function for which we have no observable data, and assuming any explicit functional form will be perilously close to guess-work on our parts.

Despite the emphasis given to the divergence between compensated and ordinary demands in the theoretical literature, the problem has received less attention in the empirical literature. Researchers have occasionally calculated both CV and EV by integrating back from observable demands, but these calculations often reveal little difference between the two measures. This is entirely consistent with a paper by Willig (1976) that is often alluded to by those wishing to forego Hausman/Vartia type procedures. Willig derived the following bounds on the error from using consumer surplus to approximate the compensating variation of a price change, where we restate his expression to reflect our notation and *our signing convention* ($CV, CS, EV > 0$ for improvements):

$$\begin{aligned}\frac{[1+(\eta_L-1)\frac{CS}{y^0}]^{\frac{1}{1-\eta_L}}-1+\frac{CS}{y^0}}{\frac{|CS|}{y^0}} &\leq \frac{CS-CV}{|CS|} \\ &\leq \frac{[1+(\eta_U-1)\frac{CS}{y^0}]^{\frac{1}{1-\eta_U}}-1+\frac{CS}{y^0}}{\frac{|CS|}{y^0}}.\end{aligned} \qquad (2.16)$$

CS and CV are consumer surplus and compensating variation, respectively, and y^0 is income. The only additional information needed are the upper (η_U) and lower (η_L) bounds on the income elasticity of demand for the good whose price has changed.

Willig's paper is perhaps best known, however, for his demonstration that for often-encountered ranges of elasticities and for goods for which the consumer surplus is not exceptionally large relative to income, errors from using CS rather than CV or EV will be quite small. To give more precision to this statement, for cases when the consumer surplus associated with consumption of the good is less than 90% of total income and when the relevant income elasticities times this ratio are less than 0.1, the following rules of thumb will hold:

$$\frac{\eta_L|\frac{CS}{y^0}|}{2} \leq \frac{CS - CV}{|CS|} \leq \frac{\eta_U|\frac{CS}{y^0}|}{2} \text{ and}$$
$$\frac{\eta_L|\frac{CS}{y^0}|}{2} \leq \frac{EV - CS}{|CS|} \leq \frac{\eta_U|\frac{CS}{y^0}|}{2}. \tag{2.17}$$

Simplifying further, if income elasticity doesn't change significantly over the range of the price change, then compensating and equivalent variation can be approximated by the following:[10]

$$CV \approx CS[1 - \frac{\eta CS}{2y^o}] \; and \; EV \approx CS[1 + \frac{\eta CS}{2y^o}], \tag{2.18}$$

where $\eta \approx \eta_L \approx \eta_U$.

These bounding expressions can be used to show that for many measured consumer surpluses, the equivalent and compensating variations are quite close, and given the types of errors one can expect to encounter in demand function estimation, provide little impetus to calculate CV or EV instead of consumer surplus. Consider an example in which the estimate of consumer surplus is \$1000, while income is \$20,000, and income elasticity equals 2. This example is far more favorable to a disparity between EV and CV than one usually finds in practice. Consumer surpluses are usually smaller and a much smaller proportion of income, and income elasticities are usually much lower, making the bounds much tighter. Although our hypothetical conditions are more likely than usual cases to create divergences, they imply estimates of CV and EV of \$950 and \$1050 which do not differ dramatically from the CS estimate of \$1000.

Willig argues that the errors from using CS instead of CV or EV 'will often be overshadowed by the errors involved in estimating the demand curve' (p 589). Even when not approximately zero, it is likely that estimates of CV and EV lie well within a reasonable confidence interval constructed for CS, given the randomness in parameters of estimated demand functions. Kling (1992) derives the expression for the variance of Willig's difference between CS and CV, where these two measures are obtained by estimating a demand function. She shows the more stringent requirements for the error to be significantly different from zero once the variance of the error is taken into account.

---

[10] Remember that our versions of these results will look slightly different from the way Willig presents them, for we use a different signing convention for CS, CV, and EV.

2.4.1 Multiple Price Changes

Results for the single price change case, with some modification, apply to cases in which more than one price changes. First, information about compensated demands is still sufficient to recover compensating and equivalent variation. The only added complication is that the multiple price changes must be sequenced. Equations (2.8) and (2.9) must now be written in line integral form. The compensated variation associated with J price changes is given by the line integral:

$$CV = -\int_L \sum_{j=1}^{J} \frac{\partial m}{\partial p_j} dp_j = -\int_L \sum_{j=1}^{J} z_j^h(\mathbf{p}, u^o) dp_j \tag{2.19}$$

where L denotes a path of price changes, such as the one implied by:

$$\begin{aligned} CV &= -[\int_{p_1^0}^{p_1^1} z_1^h(p_1, p_2^0, p_3^0, u^0) dp_1 + \\ & \int_{p_3^0}^{p_3^1} z_3^h(p_1^1, p_2^0, p_3, u^0) dp_3 + \int_{p_2^0}^{p_2^1} z_2^h(p_1^1, p_2, p_3^1, u^0) dp_2]. \end{aligned} \tag{2.20}$$

The actual path of price changes is typically not known, but CV and EV calculated from Hicksian demands have the desirable property that they are invariant to the path chosen for evaluation of (2.19). Path independence holds for any line integral whose integrand is an exact differential of a function and from equation (2.19) we see that the integrand of $\sum_{j=1}^{J} \frac{\partial m}{\partial p_j} dp_j$ is an exact differential of the expenditure function. While it does not matter *which* path is chosen to evaluation CV and EV, *some* path must be chosen and the ordinary integrals correspondingly sequenced. Equation (2.20) offers only one example.

Second, Just, Hueth and Schmitz show that Willig's bounds can be extended to the multiple price change case. Consider price changes for goods $z_j, j = 1, ..., J$ and define $\ddot{\eta}_L = \min(\eta_{L_1}, \eta_{L_2}, ..., \eta_{L_J})$ where η_{L_j} is the lower bound income elasticity for good j and $\ddot{\eta}_U = \max(\eta_{U_1}, \eta_{U_2}, ..., \eta_{U_J})$, where η_{U_j} is the upper bound on good j's income elasticity. Then

$$\begin{aligned} \frac{\ddot{\eta}_L |\frac{CS}{y^0}|}{2} &\leq \frac{CS - CV}{|CS|} \leq \frac{\ddot{\eta}_U |\frac{CS}{y^0}|}{2} \quad and \\ \frac{\ddot{\eta}_L |\frac{CS}{y^0}|}{2} &\leq \frac{EV - CS}{|CS|} \leq \frac{\ddot{\eta}_U |\frac{\mathbf{CS}}{y^0}|}{2}. \end{aligned} \tag{2.21}$$

CV and EV are the compensating and equivalent variations of the multiple price changes as defined in (2.20), and CS is an analogous expression calculated using ordinary demands. It is the sum of *sequenced* individual consumer surpluses associated with the J goods whose prices have changed, where 'sequencing' implies conditioning the ordinary demands on sequentially changing prices. For example, CS could be calculated as in (2.15).

We know from the last section that, unlike CV and EV, CS is not unique but will depend on the chosen path of price changes. What (2.21) shows us is that different paths may produce different CS answers, but all are bounded by the unique CV and the unique EV in the way described above. Just, Hueth and Schmitz give some empirical examples for multiple price changes. These suggest that the difference between CS and CV or EV for commonly encountered income elasticities and modest CS measures will not be sufficiently great to justify calculating the more exact measures.

Finally, we saw that in the single good case it is possible to integrate back from a Marshallian demand function for some z_i and obtain the indirect utility function (or equivalently the expenditure function) up to a constant of integration that cannot include z_i's price (Hausman, 1981). Likewise, it is in concept possible to recover from a *system of demands* all the information embedded in the indirect utility function (or expenditure function) up to a constant of integration that varies with absolute utility but not prices (LaFrance, 1990; LaFrance and Hanemann, 1989). All the information we need to recover the indirect utility or expenditure function is embodied in the complete demand system $\mathbf{z}^m = \mathbf{f}(\mathbf{p},y)$ as long as the following integrability conditions hold:

- the Marshallian demands are homogeneous of degree zero in prices and income;
- the demands are non-negative;
- total expenditure equals income;
- the Slutsky matrix of substitution terms is symmetric and negative semi-definite, where a typical element of this substitution matrix is given by $\partial z_i^h / \partial p_j + z_j \, \partial z_i^m / \partial y$.

These conditions are equivalent to the existence of the expenditure function or an indirect utility function with the usual properties. When the integrability conditions hold, we can in principle recover all the information we need to measure welfare effects of multiple price changes.

In practice researchers work in a data-sparse and econometrically difficult setting and so only limited parts of the preference function can be recovered. Justifications for doing so rely on either partial or incomplete demand systems. (See the discussion by Hanemann and Morey, 1992.) To illustrate the two, we begin with the following utility function:

$$u(\mathbf{z}) = u(\mathbf{z}_A, \mathbf{z}_B)$$

and correspondingly partition the price vector as $\mathbf{p} = (\mathbf{p}_A, \mathbf{p}_B)$. A partial system results from the maintained assumption that the group of demands of interest, denoted $\mathbf{z}_A$, is weakly separable from the rest of the demands. This implies a separable utility function that can be written as:

$$u(\mathbf{z}_A, \mathbf{z}_B) = \hat{u}(\varphi(\mathbf{z}_A), \mathbf{z}_B).$$

The demand functions of interest can be estimated in the form:

$$\mathbf{z}_A = \mathbf{z}_A^m(\mathbf{p}_A, y_A) \tag{2.22}$$

where y_A is the income allocated to the consumption of $\mathbf{z}_A$. An intertemporal context could be the setting for a partial demand system. Estimating the demand for current goods and services as a function of the prices of those goods and services, as well as current income, would constitute a partial system, requiring intertemporal separability. Hanemann and Morey (1992) show that for full recovery of the relevant portion of preferences, one needs non-negative commodity demands, a sub-budget y_A that is strictly less than y, homogeneity of degree zero in prices and the sub-budget y_A, and symmetry and negative semi-definiteness of the Slutsky matrix associated with $\mathbf{z}_A$. Welfare analysis on partial systems can yield welfare measures of price changes in the partial systems, conditional on the income devoted to the partial group.

In contrast to a partial system, an incomplete system includes information on all prices and income, at least in concept, and can be written as:

$$\mathbf{z}_A = \mathbf{z}_A^m(\mathbf{p}_A, \mathbf{p}_B, y), \tag{2.23}$$

although one estimates behavior only for the subset of demand functions in group A. The incomplete system with some prices assumed constant is frequently what one estimates. If prices $\mathbf{p}_B$ were constant at $\bar{\mathbf{p}}_B$ across observations, then we could rewrite equation (2.23) as

$$\mathbf{z}_A = \tilde{\mathbf{z}}_A^m(\mathbf{p}_A, y) = \mathbf{f}_A(\mathbf{p}_A, \bar{\mathbf{p}}_B, y) \tag{2.24}$$

where the fixed prices $\bar{\mathbf{p}}_B$ are implicitly incorporated into the parametric structure of the demand curve $\tilde{\mathbf{z}}_A^m(\mathbf{p}_A, y)$. An incomplete system allows welfare analysis of price changes in the group of observed prices, conditional on total income. The incomplete demand system appears similar to the partial demand system except that full income is the appropriate argument in the former and a sub-budget is appropriate in the latter. They are not, however, the same functions.

The specifications most frequently used in the applied literature often look like equation (2.24) but the assumption that the $\mathbf{p}_B$ vector is constant may often be violated and in a systematic way. This will be true, for example, when important substitutes are omitted from the incomplete system but their prices vary over the sample. This is likely to happen in situations where goods (including substitutes) are not purchased at constant market prices but are produced by households facing different cost structures. Examples of this sort include travel cost models, which will figure centrally in Chapters 4 and 5, and averting behavior models, which will be discussed in Chapter 8.

2.5 Income and Welfare Effects

There is one more complication to add to the story. Typically the characterization of the individual's decision problem takes income as an exogenous constraint. As described in equation (2.1), both prices and income are parameters to the individual, but policies, public projects, or events can lead to changes in income as well as prices. This is obvious when wage rates change, but also happens any time a resource owner experiences an increase in the price of the resource. Unlike price changes, *exogenous* income changes are simple to evaluate. The CV of an income change equals the EV of that change, and both are simply equal to the exogenous income change. It is easy to see that this must be true by referring back to the definition of compensating and equivalent variation. We have in fact defined the compensatory payments in terms of exogenous income payments. For an income change, CV (and EV) will be the amount of exogenous income necessary to compensate the individual for the exogenous income change. The only question is how to sign this magnitude. Following our convention, we sign the compensating and equivalent variation measures according to the welfare effect. If the policy increases exogenous income, then the CV and EV measures will be positive.

This logic can be shown to be consistent with our previous definitions of CV and EV. It is easiest to show this by using the definitions presented in equations (2.2) and (2.3)—the ones based on the indirect utility function. The CV of an exogenous income change from y^0 to y^1 is

$$v(\mathbf{p}^0, y^0) = v(\mathbf{p}^0, y^1 - CV),$$

which by definition must be equal to $y^1 - y^0$. Likewise, the EV measure is given by:

$$v(\mathbf{p}^0, y^1) = v(\mathbf{p}^0, y^0 + EV),$$

so that $EV = y^1 - y^0$ as well.

Exogenous income changes pose no particular difficulties for path independence in Hicksian terms. If changes occur in exogenous income *and* prices, then the CV measure is simply the change in income plus the CV expression for the multiple price changes, as in equation (2.20). The Marshallian approximation is only slightly more complicated in that it must take into account the fact that ordinary demands are functions of income. In the Marshallian case, the income change must be sequenced along with the price changes. To make this clear, consider circumstances that lead to a change from (p_1^0, p_2^0, y^0) to (p_1^1, p_2^1, y^1). There will be multiple paths along which the relevant line integral might be evaluated. Whatever path is chosen, it must be a legitimate (i.e. sequenced) path such that subsequent functions are evaluated conditional on previously changed parameters. One such arbitrary path, chosen for illustration, involves the calculation:

$$-\int_{p_1^0}^{p_1^1} z_1^m(p_1, p_2^0, y^0)dp_1 - \int_{p_2^0}^{p_2^1} z_2^m(p_1^1, p_2, y^0)dp_2 + (y^1 - y^0). \tag{2.25}$$

Unfortunately, as we have seen earlier, the consumer surplus measure will be different depending on which path of price and income changes is chosen. Consumer surplus is not path independent. This is especially obvious for changes in exogenous income. An alternative path to that in (2.25) is:

$$(y^1 - y^0) - \int_{p_1^0}^{p_1^1} z_1^m(p_1, p_2^0, y^1)dp_1 - \int_{p_2^0}^{p_2^1} z_2^m(p_1^1, p_2, y^1)dp_2, \tag{2.26}$$

which will obviously be different from (2.25) if income effects are not zero.

2.5.1 Endogenous Income

In practice, individuals often have some control over the income they receive. In the long run, an individual can train for different types of work that will yield greater returns. In the shorter run, he can choose whether or not to be employed and how many hours he will work. There are often discontinuities and constraints on the number of work hours associated with a given job, but individuals may be able to work overtime, take leave, or work at secondary jobs. When the researcher is interested in consumption decisions for goods that do not have large money prices nor require significant time allocations by the household, treating the problem as though income were exogenous does not compromise the results. In environmental valuation, though, numerous circumstances arise in which time allocation is the behavior that helps reveal values for amenities. This is true for recreation decisions when individuals choose among recreational sites to experience different levels of environmental amenities, and it holds for averting behavior when people try to mitigate or avoid health risks. Also common are situations in which households' income earning decisions are tied to levels of environmental services, such as farm production decisions of households in developing countries where environmental circumstances affect production. Considering the time as well as the money constraint provides important insights and raises interesting and additional welfare issues.

Here we establish the framework for the analysis. Consider the individual's (or household's) decision problem when income is the result of a choice process. The individual may have some exogenous income, which we will label $\bar{y}$. This is income derived from return on investments, perhaps, or from entitlements such as retirement pensions, social security, etc. The endogenous income accrues from selling some resource that he owns, either his time or the rights to his real property. For some people, $\bar{y}$ may be zero, but in any case it represents an 'account' into which exogenous compensation could in theory be paid. We treat the factor supply choice as a simple and unconstrained one, although the labor literature (e.g. Killingsworth, 1983) illustrates a variety of institutional restrictions that can make the labor supply function more complex. Doubtless other complications apply to the sale or rental of other factors.

Measuring welfare changes when an individual has an endowment of goods, some of which can be consumed and some sold, means that the purchase and sale activity influence income. We provide the analysis for the case of labor supply only.[11] Denote the individual's labor supply by x, and the wage rate by w. Incorporating returns to factors begins by rewriting the budget constraint as $\bar{y} + xw = \mathbf{p} \cdot \mathbf{z}$. However, the individual has an endowment of time, denoted T, that can be used as labor or leisure. It makes sense to measure this endowment of time net of activities that are not fungible, such as sleeping, eating and other biological essentials. Time not sold on the labor market is time that can be used for household production or leisure. This leads to the well known trade-off between labor and leisure.

With this framework, the indirect utility function is the result of the decision problem:

$$v(\mathbf{p}, w, \bar{y}, T) = \max_{\mathbf{z},x}\{u(\mathbf{z}, T - x) | \mathbf{p} \cdot \mathbf{z} \leq \bar{y} + wx,\ \mathbf{z} \geq 0, T \geq x \geq 0\}, \tag{2.27}$$

and the expenditure function is the result of the dual expenditure minimization problem:

$$m(\mathbf{p}, w, \bar{u}, T) = \min_{\mathbf{z},x}\{\mathbf{p} \cdot \mathbf{z} - wx | u(\mathbf{z}, T - x) \geq \bar{u},\ \mathbf{z} \geq 0, T \geq x \geq 0\}. \tag{2.28}$$

The ordinary labor supply function, $x^m(\mathbf{p}, w, \bar{y}, T)$, and compensated labor supply function, $x^h(\mathbf{p}, w, \bar{u}, T)$, derive from these optimization problems.

In equations (2.27) and (2.28), labor supply is the focus of attention, but *own-consumption* of the time is leisure. We rewrite equation (2.27) in terms

---

[11] Just Hueth and Schmitz (2004) review the welfare analysis for endowments in the general case.

of leisure, which we denote s, noting that it is defined as $s = T - x$. Written in these terms, expressions analogous to equations (2.27) and (2.28) are

$$v(\mathbf{p}, w, \bar{y}, T) = \max_{\mathbf{z},s}\{u(\mathbf{z},s)|\mathbf{p}\cdot\mathbf{z}+ws \leq \bar{y} + wT,\ \mathbf{z} \geq 0, T \geq s \geq 0\} \quad (2.29)$$

and

$$m(\mathbf{p},w,\bar{u},T) = \min_{\mathbf{z},s}\{\mathbf{p}\cdot\mathbf{z}+ws - wT|u(\mathbf{z},\mathbf{s}) \geq \bar{u},\ \mathbf{z} \geq 0, T \geq s \geq 0\}. \quad (2.30)$$

Now suppose a policy, project, or event induces a change in the wage rate. What would be the welfare consequences for this individual? As always, the compensating variation can be expressed as the difference in the expenditure function, conditioned on the initial and final levels of the parameter that changes. Signing appropriately, we have

$$CV = m(\mathbf{p},w^0,\bar{u},T) - m(\mathbf{p},w^1,\bar{u},T). \quad (2.31)$$

Applying Shephard's Lemma to equation (2.28), $\partial m/\partial w = -x^h(\mathbf{p},w,\bar{u},T)$, where x^h is the Hicksian supply of labor. Once again, Shephard's Lemma provides a conceptual means of relating the desired welfare measure and a behavioral function. Not surprisingly we find that the compensating variation of a wage rate change can be measured as the area to the left of the labor supply function between the initial and final wages:

$$CV = \int_{w^0}^{w^1} x^h(\mathbf{p},w,u^0,T)dw, \quad (2.32)$$

This is equal to the amount A in Figure 2.2 (where we suppress inessential arguments). The EV measure could, of course, be found using the compensated labor supply function conditioned on the new utility level, u^1.

It would also be possible to obtain the CV (or EV) measure using the Hicksian demand function for the consumption of leisure. But in doing so, one must be careful to take note of the explicit form of the expenditure function in equation (2.30). CV will *not* be measurable simply as the loss of the area to the left of the demand for leisure, between the initial and final price, because this is not the full effect of the wage change. The increase in the wage makes leisure more expensive, leading to the *loss* of the area depicted by C in Figure 2.2, but the wage increase also leads to a *gain* equal to the income change, $T(w^1 - w^0)$, depicted as *area* $(B + C)$. The sum of the two effects is *area* B. Note that the term, $T(w^1 - w^0)$, is the increase in Becker's full income–a concept that we revisit in Chapter 4.

This result emerges from the application of Shephard's Lemma to equation (2.30):

$$\partial m/\partial w = -T + s(\mathbf{p},w,,u^0,T).$$

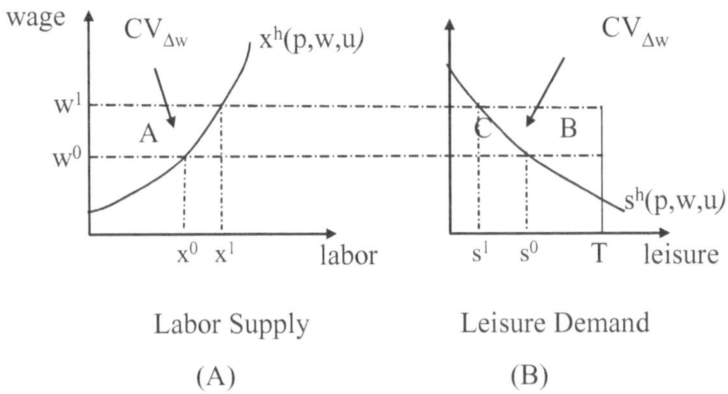

FIGURE 2.2. Compensating Variation of Wage Change Using Labor Supply or Leisure Demand

The compensating variation of a factor price change, in terms of the own-consumption of the factor, is measured as

$$CV = -\int_{w^0}^{w^1} \frac{\partial m}{\partial w} dw = T(w^1 - w^0) - \int_{w^0}^{w^1} s(\mathbf{p},w,,u^0,T)dw. \tag{2.33}$$

This is clearly a mathematical statement of *area B* in Figure 2.2.

For equations (2.32) and (2.33), Marshallian analogs, conditioned on $(\mathbf{p},w,\bar{y},T)$, result from the substitution of ordinary for compensated functions. We examine the relationship between the ordinary and compensated functions using the Slutsky equation and find it differs somewhat from the standard consumption good case. Define $x^m(\mathbf{p},w,\bar{y},T)$ as the Marshallian labor supply function and $x^h(\mathbf{p},w,T,u^0)$ as the compensated labor supply function. Then

$$\frac{\partial x^h(\mathbf{p},w,T,u^0)}{\partial w} = \frac{\partial x^m(\mathbf{p},w,\bar{y},T)}{\partial w} + \frac{\partial x^m(\mathbf{p},w,\bar{y},T)}{\partial \bar{y}} \frac{\partial m(\mathbf{p},w,T,u^0)}{\partial w}. \tag{2.34}$$

The first term on the left is the Hicksian labor supply response which, from the properties of the expenditure function, we know to be positive.[12] The last term is easy to sign, as $\partial m/\partial w = -x^h < 0$, and $\partial x^m/\partial \bar{y}$ will be negative if the own-consumption good (i.e. leisure) is a normal good. This means that the Hicksian effect of factor price on labor supply is more positive than the Marshallian effect. There is no other information to sign the slope of the Marshallian supply function, which could in fact be either positive or negative.

[12] To be more precise, we would say that the Hicksian labor supply response to a wage increase is non-negative.

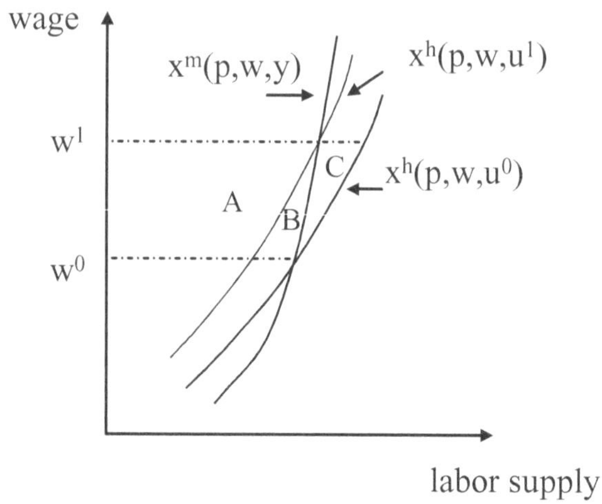

FIGURE 2.3. CV, EV, and CS for Factor Supply Functions: Price Increase for Factor

For large levels of labor supply and/or large income effects (which are quite realistic for a good like leisure), the ordinary Marshallian factor supply function could be downward sloping. Such circumstances could produce the oft-talked of, and occasionally documented, backward bending labor supply curve, for example. But we need not venture into these difficult waters to encounter welfare measurement complications.

Let's assume that the ordinary labor supply function for x is increasing in w. Drawn with the wage on the vertical axis this produces an ordinary supply function that is steeper than its compensated counterpart as in Figure 2.3 (with inessential arguments suppressed). The compensating variation of a wage change will be larger than the consumer surplus measure of the same change. So the Slutsky equation in (2.34) implies that for a factor price change, $EV \leq CS \leq CV$. Earlier we established that for consumer goods' price changes, $CV \leq CS \leq EV$. This apparently conflicting result is consistent with Figure 2.3, which shows how the variation and surplus measures of a factor price change can be found in relation to the demand functions for the factors' own-consumption. The CV of a wage increase equals $area\ A + B + C$, while the CS and EV measures are depicted by $area\ A+B$ and $area\ A$ respectively.

With these pieces of the puzzle, what can we now say about welfare measurement when a factor price change is involved? There is no problem with the compensating and equivalent variation measures. They can be precisely defined and, given compensated demand and supply curves, they could be uniquely calculated, even with multiple parameter changes (including consumer goods' prices, factor prices, and exogenous income changes). Difficulties arise, however, when all one has to work with are the Marshallian counterparts to the compensated demand and supply curves. Willig's bounds for multiple price changes are well-behaved, in part, because $CV \leq CS \leq EV$ for all goods' price changes. If factors' as well as goods' prices change, though, we do not have a unique ordering. CV may be larger or smaller than CS depending on whether the factors' or goods' price changes dominate. Likewise, the relationship between EV and CS for such a complex change will be indeterminate.

We will not encounter this conundrum often in environmental valuation, but it could occur if environmental policies have complex outcomes. In any event it is useful to bear in mind that the Willig results do not always set everything right. When the consequences of environmental events, projects or policies include factor price changes, the CV and EV measures will still be what we want, but the nature of the approximation from using a CS measure will be more complicated.[13]

2.6 Non-Linear Budget Constraints

The model that we have exploited so far is the standard neoclassical demand model. The properties that one desires and that we have assumed in preference functions are well known–increasing and quasi-concave in goods. Without these properties, the smooth Marshallian demand curves do not emerge from maximization. The critical nature of the linear budget constraint is less widely recognized. In fact welfare economics based on revealed behavior relies on the budget constraint being linear and separable in prices. This is the prices-as-parameters case. Linearity connects the direct and indirect utility functions and is the basis for duality results.[14]

We explore the implications of non-linear budget constraints for welfare measurement for price changes. The problems created by non-linear constraints extend to applications in which environmental quality changes, and so will be of particular interest to us in later chapters. Non-linear budget constraints arise for a variety of reasons. For example, quantity discounts may be available, or quantities may only be purchased jointly, such as a McDonald's 'happy meal' with soda, fries and a burger. Time costs in particular are likely to be non-linear because they depend on labor market institutions and inflexibilities.

To see the importance of the linear budget constraint, suppose that we replace the constraint

$$\mathbf{z} \cdot \mathbf{p} \leq y$$

[13]See Just, Hueth, and Schmitz (2004, chapter 7) for a more complete discussion.

[14]For duality in the absence of linearity, Epstein (1981) has a complete analysis of preferences but does not solve all of the empirical welfare measurement problems.

by the constraint

$$P(\mathbf{z},\gamma) \leq y \tag{2.35}$$

where $P(\mathbf{z},\gamma)$ is increasing and convex in $\mathbf{z}$ with known parameters γ. With a non-linear budget constraint, one loses the various hyperplane theorems that can be used to prove maximization because equation (2.35) is only a hyperplane when $P(\mathbf{z},\gamma)$ is linear and separable in the z's. Nevertheless, the optimization problem equivalent to equation (2.1),

$$\max_{\mathbf{z}} u(\mathbf{z}) \text{ subject to } P(\mathbf{z},\gamma) \leq y,\ \mathbf{z} \geq 0\},$$

is a well-defined problem.

For the sake of intuition, we approach the problem by writing the Lagrangian expression and ignoring temporarily the potential for corner solutions. From the Lagrangian expression

$$\max_{\mathbf{z},\lambda} u(\mathbf{z}) + \lambda(y - P(\mathbf{z},\gamma))$$

we have the first order conditions

$$\frac{\partial u(\mathbf{z})}{\partial z_i} = \lambda \frac{\partial P(\mathbf{z},\gamma)}{\partial z_i},\ i = 1, ..., n.$$

This set of first order conditions looks deceptively like the set one observes when prices are parameters. In both cases, the equilibrium conditions can be read as marginal value equals marginal cost:

$$\frac{\partial u(\mathbf{z})/\partial z_i}{\lambda} = \frac{\partial P(\mathbf{z},\gamma)}{\partial z_i}.$$

Further, there is at least an implicit solution

$$\mathbf{z} = \mathbf{f}(\gamma, y)$$

guaranteed by the quasi-convexity of the utility function and the convexity of the budget constraint. This is in effect a reduced form equation. It is equivalent to the Marshallian demand function in the sense that one has solved for the quantities. But it does not have the same convenient properties of the Marshallian function in the prices-as-parameters case.

Compensating and equivalent variation are still defined in terms of either the indirect utility function or the expenditure function. The indirect utility function is given by

$$v^*(\gamma, y) = \max_{\mathbf{z}}\{u(\mathbf{z})|\ P(\mathbf{z},\gamma) \leq y,\ \mathbf{z} \geq 0\}$$

so that CV can be defined as

$$v(\gamma^0, y^0) = v(\gamma^1, y^0 - CV).$$

One can also derive the marginal utility of income via the envelope theorem:

$$\lambda = \partial v^*(\gamma,y)/\partial y.$$

But because there is no parametric price in the budget constraint, Roy's identity no longer works. That is, there may be no parameter such that the derivative of $v^*(\gamma,y)$ equals $-\lambda z_i$.

The absence of prices-as-parameters also affects results related to the minimum expenditure function, given by

$$m^*(\gamma,\bar{u}) = \min_{\mathbf{z}}\{P(\mathbf{z},\gamma)|u(\mathbf{z}) \geq \bar{u}\}.$$

Again, CV can be easily defined as

$$CV = m(\gamma^0, u^0) - m(\gamma^1, u^0),$$

but making the connection between this expression and behavior (even Hicksian behavior) becomes difficult. There is a set of Hicksian demands,

$$\mathbf{z} = \mathbf{h}(\gamma,\bar{u}),$$

but their properties are not known and they do not derive from Shephard's Lemma which links demand so neatly with the expenditure function. Consequently integration such as in equation (2.8) is not available. Even the idea of the area under the Hicksian demand curve loses its meaning because there is no parametric price for the vertical axis. The basic problem with non-linear budgets arises whenever the marginal price depends on the quantity consumed. This means that demands cannot be solved as functions of marginal prices.

The purpose of this discussion of non-linear budget constraints is to raise awareness of the difficulties they pose, not to make the non-linearity seem insurmountable. There are several valuation applications where non-linearity is present but direct or indirect solutions are possible, as we will see in subsequent chapters.

2.7 Conclusions

The compensation principle is the starting point for modern applied welfare economics. It provides the grounds for defining equivalent and compensating variation. The strength of these measures is that they provide answers to well-defined questions. Whether the questions they answer are the right ones is another issue, but there are certainly many contexts in which they are useful for policy analysis. In theory, the compensation measures present some problems. While touted as unique measures (in the path independence sense), they do not produce unique results if one admits of different 'currencies' or numeraires used for measurement. Also, these measures are not observable, something that many would consider their greatest drawback. In practice, differences between the variation measures and observable measures are often small. When they are large, at least in the price change context, we can find ways to recover the variational measures or to bound them. A final complaint is that the use of compensating or equivalent variation measures for benefit-cost analysis typically ignores distribution effects. This is a consequence of using aggregate compensation measures. In concept nothing prevents the researcher from obtaining these measures for subgroups of the population.

The evaluation of programs and policies serves the public interest, and the compensation principle, when employed with good judgement, seems the least objectionable of methods of evaluation. Keeping in mind the drawbacks associated with the compensation principle, we set out to explain and explore the concepts of compensating and equivalent variation and approximations based on consumer surplus for environmental quality changes. The results from the price change case will occasionally carry over, but more often than not, we will need to begin with definitions and first principles to tease out useful answers.

Chapter 3

The Concept of Weak Complementarity

3.1 Introduction

The previous chapter presents the model for welfare measurement of price changes. This model relies on the result that a change in the price of a good is related to quantity demanded of that good by way of the envelope theorem. The same link naturally holds for factors (such as labor) sold at parametric prices. But many problems in the allocation of resources and the protection of the environment involve services that enter directly into a consumer's utility function or a firm's production function. For example, a household living in an industrial city will enjoy air quality determined not by their own consumption decisions but by the city's level and composition of transportation and manufacturing. A household may purchase the quantity of its drinking water, but the quality of the water will be determined by public water supply policies. In these cases, environmental quality is a direct determinant of utility, and government actions or exogenous events affect the level of the environmental good or service entering the individual's preference function. In such cases one cannot rely on the conceptual basis developed for price and income changes for measuring the welfare effects of changes in public goods. In this chapter we begin to develop the basic theory and extensions that support a more general set of welfare measures dealing with changes in the level of goods and services that enter preference functions exogenously.

The principle of welfare theory that supports welfare measurement developed for price and income changes applies to any argument of the indirect utility function or the expenditure function. Arguments of these functions are, by definition, exogenous elements of the individual's decision problem. In concept, one can define the change in income necessary to compensate for an exogenous change in air quality just as one can define the change necessary to compensate for an exogenous change in the price of gasoline. In contrast, it makes no sense to attempt to measure the compensating variation of a change in the quantity of a good consumed, unless that quantity is imposed on the individual. Asking how much exogenous income would be necessary to compensate for a change in the consumption of a freely chosen good is an ill-formed question to which no good answer can be given without some idea of the cause of the change.

The exogenous elements we address in the next several chapters can most usefully be thought of as the level of public goods or publicly determined quality levels of privately consumed goods. These public goods are exogenous to the individual in the sense that the level of fecal coliform at Santa Monica beaches or the amount of airborne particulate matter in downtown Baltimore or the existence of a hazardous waste site at Love Canal cannot be altered by the individual. However, and this is a key point, the individual may make decisions that alter his exposure to these exogenously determined public goods. The level of air quality or the distance and pathways to hazardous waste contamination at a particular residential location are both exogenous, but the individual can choose his residential location and thus choose his exposure. Public drinking water quality is determined by public actions and exogenous events, but a household can affect its exposure by altering the quantity consumed, by installing filtering devices, or by switching to bottled water. Exposure to fecal coliform can be avoided or lessened by choosing a cleaner recreation site.

Some kinds of public goods are so 'pure' as to make avoidance or mitigation almost impossible. One partakes of national defense regardless of what consumption decisions one makes. Altering one's level of national defense requires nothing less than changing one's country of residence. But these extreme cases will not be of interest to us. It is through their behavioral adaptations and adjustments that people reveal their preferences for improvements in public goods. By recognizing these behavioral adjustments, we will get the model for welfare measures right.

In this chapter we explore the most frequently relied upon restriction on preferences used in environmental valuation–that of weak complementarity. In theory, weak complementarity allows the value of a change in a public good to be measured in terms of a related private good, although complications arise when Marshallian rather than Hicksian measures must be used. Because weak complementarity forms the implicit or explicit basis of so much of non-market valuation, both mathematical and philosophical debates about its applicability can be found in recent literature. This chapter contains more than most readers wish to know about this restriction that holds such a central place in the welfare economics of environmental change.

3.2 The Basic Problem

Ultimately, we employ the individual's behavioral adaptations and adjustments to help measure welfare effects. Regardless of the types of adjustments a household might make, the optimization problem begins in the same way. As in the previous chapter, we start with the consumer's problem of allocating income among goods to maximize utility, only now utility is allowed to depend on the 'public good' or 'service', q, as well as the purchased goods, $\mathbf{z}$:

$$u = u(\mathbf{z}, q).$$

Although we begin the analysis with a single dimension of q, this can be generalized when useful. The utility function is assumed to be quasi-concave in $\mathbf{z}$ and q. In practice the services of this exogenous good can be undesirable or valuable, depending on how it is defined, but we will frame the problem so that q is desirable and increases in q increase utility. The household chooses the bundle $\mathbf{z}$ such that

$$v(\mathbf{p}, q, y) = \max_{\mathbf{z}}\{u(\mathbf{z}, q) | \mathbf{p} \cdot \mathbf{z} \leq y, \mathbf{z} \geq 0\}. \qquad (3.1)$$

There is no problem in *defining* the welfare measures for public goods. Suppose that the public good increases from q^0 to q^1. Then the compensating variation for this change satisfies[1]

$$v(\mathbf{p}, q^0, y) = v(\mathbf{p}, q^1, y - CV).$$

With the minimum expenditure function,

$$\min_{\mathbf{z}}\{\mathbf{p} \cdot \mathbf{z} | u(\mathbf{z}, q) \geq u, \mathbf{z} \geq 0\} = m(\mathbf{p}, q, u),$$

CV can also be defined as

$$CV = m(\mathbf{p}, q^0, u^0) - m(\mathbf{p}, q^1, u^0) \qquad (3.2)$$

where

$$u^0 = v(\mathbf{p}, q^0, y).$$

Equivalent variation is defined analogously, except that the reference utility is

$$u^1 = v(\mathbf{p}, q^1, y)$$

[1] We stick to the signing convention that compensating and equivalent variation will be positive for changes that increase the household's well-being so that $CV, EV \geq 0$ for increases in q when $u_q(\mathbf{z},q) \geq 0$.

rather than u^0. For this reference level of utility to matter, *i.e.* for CV and EV to differ, the changes in utility from the change in q must be large enough to alter the marginal utility of income.

We have used the terminology 'compensating and equivalent *variation*' to represent the welfare measures of interest, and throughout this book we will continue to do so irrespective of the parameter whose change is causing the welfare effect. Hicks (1939) made a distinction between compensating (equivalent) *variation* and compensating (equivalent) *surplus*, depending on whether price or quantity changes were at issue. At the risk of appearing ignorant of this distinction, we persist in using the term 'variation' for all changes. This is intended to emphasize the fact that the welfare measure, although sometimes difficult to obtain empirically, is conceptually the same for all parameter changes. It also avoids confusion, especially when the parameter of interest can be interpreted as a quantity when the story is told one way, but a quality characteristic when told in another.

Defining the variation measures is easy; measurement is something else. The compensating or equivalent variation measures could be calculated if either the indirect utility function or the expenditure function were available. But nothing about the implicit or explicit expressions for CV links behavior with welfare measurement. For the price change case, the expression for CV was quite analogous to equation (3.2). So why is the link absent for quality changes? In the price change case we relied on the envelope result. By Shephard's lemma, $\partial m/\partial p_i$ is a behavioral response. It is the demand for good z_i, albeit a Hicksian demand. Because the change, $m(p_i^0, \mathbf{p}_{-i}, u^0) - m(p_i^1, \mathbf{p}_{-i}, u^0)$, can be obtained by integrating over $\partial m/\partial p_i$ between p_i^0 and p_i^1, the same change can be obtained by integrating over the Hicksian demand for z_i. The only snag that arises is in holding u constant over the integration, as Hicksian demand functions are rarely observed. Fortunately, as we saw in the last chapter, there are ways around this obstacle.

For a change in q (the public good), CV can still be expressed as the integral over $\partial m/\partial q$ between the two levels of q:

$$CV(q^1, q^0) = -\int_{q^0}^{q^1} m_q(\mathbf{p}, q, u)dq = m(\mathbf{p}, q^0, u) - m(\mathbf{p}, q^1, u), \qquad (3.3)$$

but there is no theorem analogous to Shephard's lemma and therefore typically no direct link between $\partial m/\partial q$ and behavior. No general procedure exists for measuring compensation for changes in the public good.

Without more structure or complete information about preferences, we can go no further in measuring welfare for changes in public goods To this point, no specific information about q has been given–it is simply a good or service that enters individuals' preference functions exogenously. In the absence of a connection between behavior and welfare measures for changes in the public good, we need further restrictions on preferences that allow us to infer welfare for public goods from behavior. One such restriction can be found in Mäler's (1974) notion of weak complementarity, an idea implicit even in earlier applied work (see Stevens, 1966).

3.3 The Public Good as an Attribute

Recovering evidence about preferences for q from observations on behavior relies on observing individual decisions that are influenced by q. This means we need to know something more about preferences than simply that u is a function of q; we need a link between q and at least one good that is chosen freely by the individual. So let us start in that direction and see how restrictive this link needs to be.

3.3.1 Weak Complementarity

First, assume the least restrictive relationship: that changes in q influence the marginal utility of at least one good and hence affect the individual's choice of how much of the good to consume, where z_1 denotes that quantity. For now, ignore the distinction between Hicksian and Marshallian demand functions and assume that Hicksian behavior (i.e., behavior conditioned on constant utility rather than constant income) is observable. Suppose we *could* observe the choices the individual would make (with utility constant) given different levels of the public good. What could we learn from this? Because the CV measure of a change in z_1's *price* can be found by integrating behind the Hicksian demand function between the initial and final prices, let us consider a related measure as a possible candidate for the CV associated with a change in q. Integrate behind each of two Hicksian demand functions, conditioned on the two levels of the public good, and measure the net change in the entire area behind these curves. In other words, measure the *change* in the area behind the Hicksian demand curve caused by the change in q.

The Hicksian demand, which is the derivative of the minimum cost function with respect to its price, is now a function of q because of the maintained hypothesis that q affects the marginal utility of z_1:

$$z_1^h(\mathbf{p}, q, u^0) = \partial m(\mathbf{p}, q, u^0)/\partial p_1.$$

Calculating the change in the area under the Hicksian demand curve as it shifts with the change in q gives[2]

$$\begin{aligned}
&\int_{p_1^0}^{p_1^*} z_1^h(\mathbf{p}, q^1, u^0)dp_1 - \int_{p_1^0}^{p_1^*} z_1^h(\mathbf{p}, q^0, u^0)dp_1 \qquad (3.4)\\
=\ & [m(p_1^*, \mathbf{p}_{-1}, q^1, u^0) - m(p_1^0, \mathbf{p}_{-1}, q^1, u^0)]\\
& -[m(p_1^*, \mathbf{p}_{-1}, q^0, u^0) - m(p_1^0, \mathbf{p}_{-1}, q^0, u^0)]\\
=\ & [m(p_1^0, \mathbf{p}_{-1}, q^0, u^0) - m(p_1^0, \mathbf{p}_{-1}, q^1, u^0)]\\
& +[m(p_1^*, \mathbf{p}_{-1}, q^1, u^0) - m(p_1^*, \mathbf{p}_{-1}, q^0, u^0)].
\end{aligned}$$

This is the change in the area under a single demand curve with a change in q, imposing no restrictions on preferences that link q and z_1 in any specific way. Expression (3.4) comes within two terms of giving the correct compensating variation, as expressed in equation (3.2). The difference between the compensating variation and the change in the area under the Hicksian demand curve is

$$m(p_1^*, \mathbf{p}_{-1}, q^1, u^0) - m(p_1^*, \mathbf{p}_{-1}, q^0, u^0). \qquad (3.5)$$

This is the change in the expenditure function when the quantity demanded of z_1 is zero. Expression (3.4) is a lower bound on CV when q is desirable and $q^1 > q^0$, because (3.5) must be non-positive. Without further restrictions, we do not know how large an error (3.5) represents, and we can come no closer to the desired welfare measure.

Here is the role for weak complementarity. This concept was originally developed by Mäler (1974), who was instrumental in establishing the welfare basis of environmental valuation. Suppose that q does not affect utility when z_1 is zero. It may be that q is a quality dimension of z_1; the quality of z_1 will not matter to the individual if he does not consume z_1. As an example, suppose that q is the quality of the public drinking water supply and z_1 is the amount of water the individual demands from the public system. It is reasonable to assume that q is irrelevant to him if, because of price or other factors, he draws his water from a private well or purchases bottled water and consumes no publicly supplied water. If the relationship between q and z_1 is of this sort, then they are said to be weak complements.

Weak complementarity can be defined as a property of preferences, such that[3]

$$u(0, \mathbf{z}_{-1}, q^0) = u(0, \mathbf{z}_{-1}, q^1) \text{ } or \text{ } \partial u(0, \mathbf{z}_{-1}, q)/\partial q = 0. \qquad (3.6)$$

---

[2]As in Chapter 2, $\mathbf{p}_{-1}$ indicates the price vector excluding the 1st element and p_1^* is the price at which the Hicksian demand for z_1 is zero (*i.e.* the 'choke' price). The choke price will generally vary with q but we represent it simply as p_1^* to simplify notation.

[3]Weak complementarity can also be defined as a property of the indirect utility function: $v(p_1^*, \mathbf{p}_{-1}, q^1, y) = v(p_1^*, \mathbf{p}_{-1}, q^0, y)$.

We also require that the weak complement to q be a non-essential good. This means that the area behind the Hicksian demand function must be finite which in turn means that the compensation required by the individual for complete elimination of this good is finite. The existence of a finite choke price is a sufficient condition for non-essentiality. Given the existence of a Hicksian choke price, p_1^*, weak complementarity can also be written

$$m(p_1^*, \mathbf{p}_{-1}, q^0, u^0) = m(p_1^*, \mathbf{p}_{-1}, q^1, u^0) \tag{3.7}$$

because changes in q do not influence utility-constant, minimum expenditures when consumption of z_1 is driven to zero. When this condition holds, the two terms in equation (3.5) are equal and cancel, and the change in the area under the Hicksian demand function caused by a change in the public good equals the compensating variation for the public good:

$$\begin{aligned} & \int_{p_1^0}^{p_1^*} z_1^h(\mathbf{p}, q^1, u^0)dp_1 - \int_{p_1^0}^{p_1^*} z_1^h(\mathbf{p}, q^0, u^0)dp_1 \\ = \; & m(\mathbf{p}, q^0, u) - m(\mathbf{p}, q^1, u). \end{aligned} \tag{3.8}$$

This measure is illustrated in Figure 3.1 by the area $ABCD$.

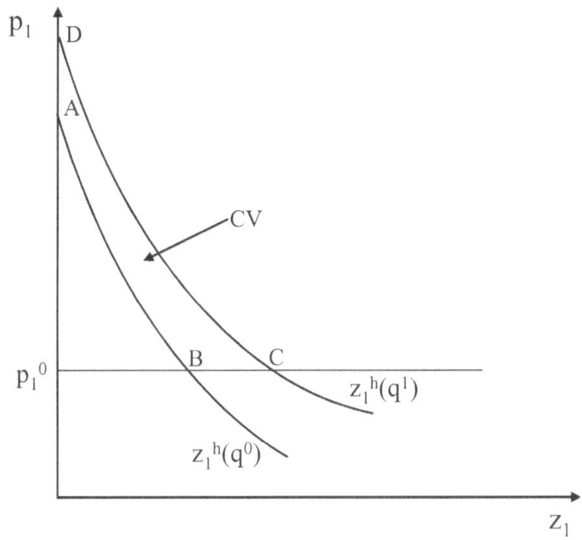

FIGURE 3.1. Weak Complementarity as the Area between Two Hicksian Demand Functions

Weak complementarity would appear to imply that the public good enhances only the marginal utility of the weak complement. For example, an improvement in water quality at a Los Angeles beach increases the marginal utility of a trip to the beach. The case may often hold but as we will see in the cross-product repackaging case later in this chapter, the preference structure need not be so restrictive. The assumption ties q to non-zero consumption of z_1, but as long as $z_1 > 0$, q can be linked to other goods in substantive ways.

Weak complementarity is intuitively appealing in the case of a single good. It also works when there is a *set* of goods that are weakly complementary with q (see Bockstael and Kling, 1988, for a treatment of this case). An example is provided by the PCB contamination of a New Bedford, Massachusetts harbor. The estuary has a number of beaches used by the households of New Bedford, and many households use several of the beaches. In the 1980's, a PCB contamination incident affected several beaches in the area. PCB's are a persistent and potentially carcinogenic chemical that were found in the sediments off these beaches. The reports of contamination led to changes in the use of these beaches by local residents. Both physical removal and deposition of new sediment were strategies considered to reduce concentrations of PCB's. Each of these potential strategies involved all beaches in a single clean-up activity. The value of the actual remedial effort chosen could be measured as the change in the areas under the demand curves for all contaminated beaches, as households responded to the cleanup activity. These demand functions are clearly inter-related and we must take that into account in the calculation.

The complication that arises in pursuing this valuation strategy is that the integrations of areas behind the set of demands for the affected beaches must be evaluated sequentially. The demand for trips to one beach is clearly a function of costs of access to the others, so that areas under relevant demands must be conditioned on a sequencing of price changes. Perhaps the easiest way of thinking about the valuation problem is to calculate the difference in the value of the existence of the set of beaches under the two possible quality levels (before and after the clean-up.) To value the set of beaches, holding quality constant at either of the quality levels, requires choosing a particular path of price changes for the set of beaches. That is, a path must be chosen from $\mathbf{p}_J^0$ to $\mathbf{p}_J^*$, where $\mathbf{p}_J$ is the vector of prices for the set of weak complements designated J. The welfare answer, when using Hicksian demands, will be independent of the order in which the beaches are evaluated because it will be independent of the path of price changes.

Supppose for illustration there are three goods in set J. Then one of the many equivalent ways of explicitly writing the change in these sequenced areas with a change in quality would be[4]

$$
\begin{aligned}
&[\int_{p_1^0}^{p_1^*} z_1^h(p_1, p_2^0, p_3^0, \mathbf{p}_{-J}, q^1, u^0)dp_1 + \int_{p_2^0}^{p_2^*} z_2^h(p_1^*, p_2, p_3^0, \mathbf{p}_{-J}, q^1, u^0)dp_2 \\
&+ \int_{p_3^0}^{p_3^*} z_3^h(p_1^*, p_2^*, p_3, \mathbf{p}_{-J}, q^1, u^0)dp_3] \qquad (3.9) \\
&-[\int_{p_1^0}^{p_1^*} z_1^h(p_1, p_2^0, p_3^0, \mathbf{p}_{-J}, q^0, u^0)dp_1 + \int_{p_2^0}^{p_2^*} z_2^h(p_1^*, p_2, p_3^0, \mathbf{p}_{-J}, q^0, u^0)dp_2 \\
&+ \int_{p_3^0}^{p_3^*} z_3^h(p_1^*, p_2^*, p_3, \mathbf{p}_{-J}, q^0, u^0)dp_3].
\end{aligned}
$$

Expression (3.9) reduces to the surprisingly simple measure of differences in expenditure functions:

$$
\begin{aligned}
&[m(p_1^0, p_2^0, p_3^0, \mathbf{p}_{-J}, q^0, u^0) - m(p_1^0, p_2^0, p_3^0, \mathbf{p}_{-J}, q^1, u^0)] \qquad (3.10) \\
&-[m(p_1^*, p_2^*, p_3^*, \mathbf{p}_{-J}, q^1, u^0) - m(p_1^*, p_2^*, p_3^*, \mathbf{p}_{-J}, q^0, u^0)].
\end{aligned}
$$

The multiple goods analogue is that the expenditure function is insensitive to changes in q when none of the goods in the *set* of weak complements is consumed, causing the last two terms in (3.10) to equal zero. Equation (3.10) is the compensating variation which can be measured by summing the three areas described in equation (3.9). The welfare measure obtained in this way represents the value to beach users only, and one might realistically expect other motivations for valuing the cleanup. But none of these motivations can be expected to lead to negative values for the clean-up, and as long as these values are non-negative, the weak complementarity measure from beach use at least serves as a lower bound on the value of the clean-up.

3.3.2 Can Weak Complementarity Be Tested?

Weak complementarity as originally conceived by Mäler and as discussed in this chapter is an idea about the kinds of values individuals place on resources. If resources are only of value when people use them, then weak complementarity holds. This is a story about motives that ensures that at least the Hicksian demand function for the related good(s) embodies all the necessary information about welfare. But how do we know if weak complementarity holds or fails to hold? Weak complementarity can fail for either of two reasons, both of which can be understood with stories about what people do or what their motives are.

---

[4] Any other price path that starts at p_1^0, p_2^0, p_3^0 and ends at p_1^*, p_2^*, p_3^* will do.

The first way in which weak complementarity may fail, and what researchers tend to think of most often, is that the public good may actually act as a symbol of something inherently valued by the individual–generating an 'existence value' irrespective of use.[5] If we represent this case as

$$U(\mathbf{z},q) = T(u(\mathbf{z},q),q), \tag{3.11}$$

where T is increasing in u and q, and if one of the private commodities, say z_1, is linked to the public good, then *some* of the economic value that can be attributed to the public good is provided in conjunction with a privately consumed good.[6] In this relationship, it may even be true that $u(0, \mathbf{z}_{-1}, q^1) = u(0, \mathbf{z}_{-1}, q^0)$. But the public good also provides some pure public value, in the sense that when z_1 is zero, utility still changes with a change in the public good:

$$dT(u(0,\mathbf{z}_{-1},q),q)/dq > 0 \tag{3.12}$$

In this case, even though observable behavior (*i.e.* demand for z_1) will be a function of the level of the public good, not all the value from changes in the public good can be measured as the area under z_1's Hicksian demand. For example, an angler might value reductions in nutrients in Florida Bay both because of improvements in the fishing, a use value, and enhancement of the functioning and resilience of the ecosystem, which could be valued irrespective of use. In the extreme, the preference function might be written as $T(u(\mathbf{z}), q)$ and nothing about q can be learned from observing behavior. An example of this might be the satisfaction an arm-chair traveler gets from the preservation of the Arctic National Wildlife Refuge.[7]

Weak complementarity can also fail to help us recover welfare measures even if the public good only matters to people when one or more private goods are consumed if those private goods are too numerous to take into account or too difficult to identify. The formal analysis of this is set forth in equation (3.9). It involves the standard preference function $u(\mathbf{z},q)$, but now it is no longer true that we can identify or enumerate the set of weak complements. As a result, the (sequenced) areas under the Hicksian demand curves for the set cannot be measured. As can be seen from equations (3.9) and (3.10), when the other private goods are ignored, the area under a single demand curve will underestimate the total value of a change in the public good as long as

---

[5] Krutilla (1967) was the first to consider existence value. Basic conceptual papers include Randall and Stoll (1983) and McConnell (1983) and reviews include Freeman (1993a) and Randall (1991). See Cummings and Harrison (1995) for a more critical view.

[6] See Herriges, Kling and Phaneuf (2004) for a more general analysis of this formulation, which owes its original specification to Michael Hanemann.

[7] No example is perfect. One might argue that such arm-chair travelers do make choices influenced by the ANWR, such as purchase of magazines or watching a TV documentary.

the public good is a *desirable* attribute of the other private goods or private actions.

Several papers have explored this issue from a strictly modeling point of view. To put these papers in perspective it is useful to remember that the task of obtaining welfare measures of quality changes diverges in an important way from the comparable task when prices change. In principle, one can integrate back from a Marshallian demand function to obtain a quasi-expenditure function (see Hanemann 1980 and Hausman 1981). The quasi-expenditure function is related to the full expenditure function as

$$\tilde{m}(\mathbf{p}, \phi(u)) = m(\mathbf{p}, u) \tag{3.13}$$

where $\tilde{m}$ is the quasi-expenditure function known only up to the unknown constant of integration, ϕ. The quasi-expenditure function can be solved from the following differential equation:

$$dm(p, u)/dp_i = z_i(p, m(p, u)). \tag{3.14}$$

As we saw in the last chapter, the constant of integration will be a function of the baseline utility, but it can not be a function of price. Therefore all the information necessary to evaluate how the expenditure function changes with a price change is retrievable from observable demand.

In the quality change case this is not true. Without further restrictions, the constant of integration implicit when integrating back from the Marshallian demand function will, in general, include q. Analogous to (3.13), the quasi-expenditure function could now have the form $\tilde{m}(\mathbf{p}, q, \theta(u, q))$ where θ is now the constant of integration. If so, then we can not learn everything we need to know about the changes in the expenditure function with changes in q from observable behavior. An infinite number of preference functions will be consistent with a given Marshallian demand function, because the constant of integration could in principle take the form of an infinite number of different functions of q. Further restrictions are necessary to learn anything about the welfare effects of changes in q.

A demand function consistent with weak complementarity would usually be considered one that integrated back to a quasi-expenditure function that exhibited weak complementarity, with the implicit assumption that this was the only piece of the preference function that contained q. Under these circumstances $m(\mathbf{p}, q, u) = \tilde{m}(\mathbf{p}, q, \theta(u))$. Larson (1991) argues that this need not be the case. He views weak complementarity as a further restriction on preferences that permits the recovery of welfare effects for changes in q, but he allows a more general class of preference functions to exhibit this property.

Specifically Larson suggests that by combining the differential equation in (3.14) with the weak complementarity condition

$$dm(p_1^*(q, u), q, u)/dq = 0, \tag{3.15}$$

any Marshallian demand function can be integrated back to a preference function that exhibits weak complementarity. To do so when weak complementarity is not a 'natural' outcome of the demand function specification, that is when demand does not integrate back to a *quasi*-expenditure function that itself implies weak complementarity, requires specifying a constant of integration that forces the condition. As an example, Larson considers the linear demand function

$$z_1 = \alpha + \beta p_1 + \gamma q + \delta y \tag{3.16}$$

where p_1 is really the price of z_1 normalized by the price of a numeraire and y is income similarly normalized. Demand function (3.16) integrates back to a quasi-expenditure function equal to[8]

$$\tilde{m} = \theta(q,u)e^{\delta p_1} - \frac{1}{\delta}[\alpha + \beta p_1 + \gamma q + \frac{\beta}{\delta}],$$

which by itself does not exhibit weak complementarity. However, one can impose weak complementarity by solving the differential equation implied by (3.15), and then obtaining a solution for θ which embodies the same parameters as are present in the demand function in a very specific way:

$$\theta(q,u) = \phi(u)e^{(\frac{\gamma\delta}{\beta})q}.$$

The resulting expenditure function

$$\tilde{m} = \phi(u)e^{\frac{\delta}{\beta}(\gamma q + \beta p_1)} - \frac{1}{\delta}[\alpha + \beta p_1 + \gamma q + \frac{\beta}{\delta}]$$

can be seen to be consistent with weak complementarity.

Herriges, Kling and Phaneuf (2004) look at the same result from another perspective. They examine another of Larson's examples–one in which there is a vector of public goods, $q_i, i = 1, ..., k$, where each element is a quality characteristic of a weakly complementary private good, z_i. Two preference functions based on the linear expenditure system are considered:

$$T(u(\mathbf{z},\mathbf{q}),\mathbf{q}) = \sum_{i=1}^{k} \Psi_i(q_i)\ln\{(z_i + \theta_i)/\theta_i\} + \ln(z_{k+1}) \tag{3.17}$$

and

$$T^*(u(\mathbf{z},\mathbf{q})) = \sum_{i=1}^{k} \Psi_i(q_i)\ln(z_i + \theta_i) + \ln(z_{k+1}). \tag{3.18}$$

[8] A slight notation change is involved here. To solve the differential equation, substitute m for y in equation (3.16).

The first function exhibits weak complementarity between z_i and q_i for any value of θ_i, while the second exhibits this property only when $\theta_i = 1$. Note that the T^* is obtained from T by adding the constant $\sum_{i=1}^{k} \Psi_i(q_i)\ln(\theta_i)$ so that the two preference functions lead to identical Marshallian demand functions and they are observationally equivalent. While the θ_i can be estimated, and $\theta_i = 1$ easily tested for, the authors recognize that this is a test of weak complementarity only if (3.18) represents underlying preferences. The exact same observable behavior could be generated by (3.17) which exhibits weak complementarity for any value of θ_i. Thus there is no way to test whether weak complementarity holds using observable data. In addition, the authors argue that treating the change in the area under the z_i demand function as a *lower bound* on the welfare effect is not even correct, by showing functional forms for which the additional unobserved 'piece' of preferences is negative.

Whether intended or not, both papers implicitly support the view that weak complementarity is not testable and must be a maintained hypothesis. We view weak complementarity as an assumption based on the understanding of the motives of individuals and the setting of the choices. For some resources and in some situations, the assumption of weak complementarity seems a plausible maintained hypothesis. For others, it is arguably a close lower bound, and for still others it may be a tenuous approximation at best. There will be circumstances in which researchers would like to call on this restriction, but have doubts about its applicability. From the Herriges *et al.* and Larson papers we know that significant differences can arise in welfare measures if one incorrectly assumes weak complementarity. It is difficult to test this restriction because behavior under the null hypothesis is not well defined. Weak complementarity implies that at a zero level of consumption of z_1 the individual is indifferent to changes in q. This is a troublesome condition, because at a zero level of consumption, there is no behavior with respect to z_1 with which to test a hypothesis. Also, by implication, weak complementarity is equivalent to the restriction that changes in q have no impact on *any* other goods, when $z_1 = 0$. This could be empirically tested for a finite set of identifiable goods, but the restriction requires that it be tested for *all* goods. Finally, there must be no appreciable existence value for q, but this is something that *by definition* cannot be tested from behavior.

Herriges, Kling, and Phaneuf provide an empirical example to show that it matters quite a lot which utility function is assumed. The model is one of recreational pheasant hunting in Iowa. The revealed value of a 20% increase in pheasant counts is three times as great when using the utility function in (3.18) as compared to that in (3.17). Also hypothesis tests on relevant estimated parameters from (3.18) are consistent with a violation of weak complementarity–if that form of preferences is the true one. So we are left with very different estimated welfare effects coming from two observationally equivalent Marshallian demand functions but different underlying preference structures, one of which exhibits weak complementarity while the other does not.

One might ask, if the second utility function is the right one, and weak complementarity does not hold, what motives would be consistent with such preferences? If weak complementarity fails, it is important to understand why. Failure of the weak complementarity assumption in this setting could occur because of substantial non-use values for the pheasants–for example hunters may appreciate the birds for their own sake. Or, perhaps there could be indirect use that would occur when hunters buy magazines to read about hunting pheasants. For this motivation to be sensitive to changes in q, the enjoyment from reading about pheasants would need to diminish if the stock of actual pheasants was somehow impaired.

To construct a model from these or other plausible motives, one would need to demonstrate the ways in which different motives are embodied in preference parameters and presumably provide more structure than is embodied in the authors' general functional forms. Herriges, Kling and Phaneuf give several possible explanations for what they call 'indirect use value'—the failure of weak complementarity because of the existence of other private goods or actions related to q. One is the idea that a quality change at one site might induce reductions in congestion at another site. This is true but will entail double counting if the congestion effect is appropriately modeled, and modeling a congestion effect is quite feasible. The second explanation given for indirect use value is altruism. The precise way in which altruism enters the utility function is not so obvious. Altruism has often been given as a rationale for existence value, which is not a use value at all, and is typically defined as a resource value present even when use is absent. There may well be good arguments about indirect use value, but all such values need a plausible intuitive explanation from which a model would follow.

Our point is that an intuitively plausible story about the type of behavior that would permit evidence of the failure of weak complementarity needs to be told. If the failure relates to the public good's complementarity with other private goods, then the model needs to reflect how this would show up in preference and demand functions. If the failure stems from non-use values, then at the very least the assumed underlying utility function should be consistent with this form of preferences, even though no observable behavior can reveal the size of the non-use value.

It is possible to define all sorts of mathematical representations of preferences, but mathematical formulas for utility functions are an economist's fiction. In the end the economic story is embedded in motives, not mathematics. For example, if weak complementarity is to be imposed through the constant of integration, as in the Larson paper, it is important to understand how information about the public good would be transmitted to other parts of the preference function in the absence of behavior, especially in the very restrictive way implied by his approach. In the absence of priors based on intuition or on knowledge of motives derived from other sources, there may be too little information to estimate a plausible preference function.

The conclusion to this story concerns practice. Given the specifications in equations (3.17).and (3.18), how should one proceed? As Herriges, Kling and Phaneuf have demonstrated through empirical analysis of these equations, quite different welfare measures can be derived from observationally equivalent models, depending on whether one assumes weak complementarity. This analysis has been substantially expanded by von Haefen (2004). The practice we would take would be to *assume* weak complementarity unless there is substantial support for the motives that would suggest otherwise, as well as a clear modeling strategy that links other use values with behavior. Indirect use value may be plausible in some cases, but trusting econometric results to inform us about indirect use value cedes undue leverage to the statistical process.

One alternative source of knowledge about motivations has been suggested by several authors and implemented by a few. Combining information on both behavior and stated preferences can provide a means of testing restrictions that we would normally have to treat as maintained hypotheses. Some authors have achieved this by specifying a commonly used demand function that integrates back to an expenditure function, but allows non-use value through the constant of integration. From this expenditure function, a willingness to pay function for q is derived which contains parameters associated with non-use value as well as those associated with use value. When the demand and willingness to pay functions are jointly estimated, the results allow testing hypotheses about weak complementarity and provide a means of estimating all the parameters necessary to calculate compensating variation whether or not weak complementarity holds. This approach has been taken by Eom and Larson (forthcoming).

3.4 Weak Complementarity and Marshallian Demands

In the price change case, where all the complications of the last section do not arise, economists infrequently take the trouble to integrate back to get exact CV or EV welfare measures. Instead they use the Marshallian counterpart of the Hicksian measure, because the errors from using the ordinary instead of compensated demands can often be expected to be quite small, especially relative to the errors that arise just in estimating the demand function. And, the Marshallian measure will certainly be bounded by CV and EV.

The theoretical defense for using the Marshallian measure analogous to Figure 3.1 to approximate the Hicksian measure for *quality* changes is weaker than in the price change case, and the conceptual path that leads from the Marshallian demand curves to changes in the expenditure function more serpentine (Bockstael and McConnell, 1993). Willig's (1976) results suggest that the area behind a Marshallian demand function above price is often a good approximation of the analogous area behind the Hicksian demand function. One might naively assume that, as a consequence, the *difference* between the areas behind the two Marshallian curves conditioned on different levels of quality would be a good approximation of the analogous difference using Hicksian curves.

It turns out that this is not so. To understand why, write out the Hicksian and Marshallian measures. The change in the area behind the Marshallian demand function is

$$\int_{p_1^0}^{\tilde{p}_1} z_1^m(\mathbf{p}, q^1, y)dp_1 - \int_{p_1^0}^{\tilde{p}_1} z_1^m(\mathbf{p}, q^0, y)dp_1, \tag{3.19}$$

where $\tilde{p}_1$ is the Marshallian choke price.[9] The change in the area behind the Hicksian demand function is

$$\int_{p_1^0}^{p_1^*} z_1^h(\mathbf{p}, q^1, u)dp_1 - \int_{p_1^0}^{p_1^*} z_1^h(\mathbf{p}, q^0, u)dp_1, \tag{3.20}$$

and this is CV, the exact welfare measure. Appealing to Willig's results, one might argue that the first term in equation (3.19) is a close approximation of the first term in equation (3.20), and the second term in equation (3.19) is a close approximation of the second term in equation (3.20). Therefore equation (3.19) must be a close approximation of (3.20).

There are at least two problems with this. First, Willig's results hold for the same finite price change applied to the Marshallian and Hicksian demands. The integration in equations (3.19) and (3.20) is from p_1^0 to the choke price, but the choke price will generally differ for the Marshallian and Hicksian demands. This in itself would not be fatal as we can rely on subsequent results by Randall and Stoll (1980) that show similar (although often not quite so tight) approximations for quantity as for price changes. We will have more to say about these results, but not before describing the additional and more difficult problem.

Willig's results assume that at the starting point of the integration, the values of the Hicksian and Marshallian functions are equal. That is

$$z_1^m(p_1^0, q^0, \mathbf{p}_{-1}^0, y^0) = z_1^h(p_1^0, q^0, \mathbf{p}_{-1}^0, u^0).$$

[9] With the Marshallian as well as the Hicksian function, the choke price depends on q and other arguments, but we suppress this dependence in the notation.

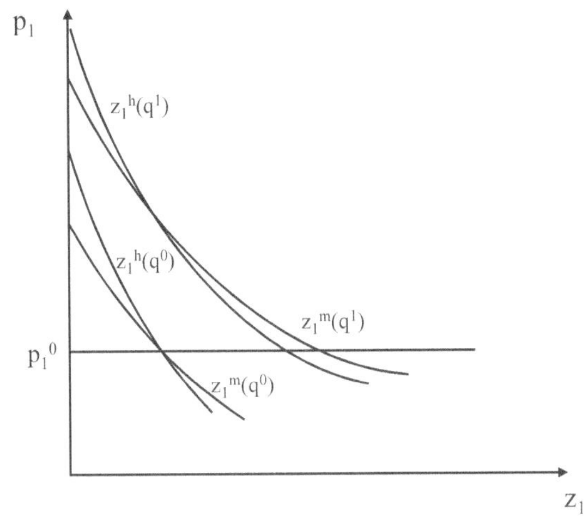

FIGURE 3.2. Shifts in Hicksian and Marshallian Demands Due to Quality Change

While this is true for the Hicksian and Marshallian functions conditioned on the initial level of quality, it will not be true of the two functions conditioned on the subsequent level of quality unless the income effect is zero. To understand what is happening, write a Slutsky-like equation that sets out the correspondence between the two functions at the initial level of prices and quality:

$$z_1^h(p_1^0, \mathbf{p}_{-1}^0, q^0, u^0) = z_1^m(p_1^0, \mathbf{p}_{-1}, q^0, m(p_1^0, \mathbf{p}_{-1}^0, q^0, u^0)).$$

A change in q implies that

$$\partial z_1^h/\partial q = \partial z_1^m/\partial q + \partial z_1^m/\partial y \cdot \partial m/\partial q. \tag{3.21}$$

An improvement in quality increases the Hicksian quantity demanded, but the adjustment in income implicit in the Hicksian response means that the Hicksian effect of a quality change will be smaller than the Marshallian effect when the good is normal. A graph illustrates this better than words. In Figure 3.2 we show that with a non-zero income effect the two quality-shifted demand functions cannot cross at p_1^0. At this point we know little about where they *do* cross, but one possible solution is portrayed in the figure.[10]

[10]In fact, it is possible with some preference structures for the Hicksian demand to shift backward at p_1^0. If weak complementarity holds, however, the areas between the current and shifted Hicksians will still be of the proper sign, as these Hicksians will cross at some price higher than p_1^0. We emphasize that the graph in 3.2 is only one possible configuration, but others are consistent with the story.

The standard Willig (1976) results (even as modified by Randall and Stoll, 1980) do not relate directly to changes in quality unless the shifted Hicksian and Marshallian demands cross at p^0, but this will happen only if there is no income effect. With a zero income effect the Marshallian and Hicksian curves are identical and the solution to the problem is trivial. For very small income effects, the Marshallian measure might be expected to approach the Hicksian one. But when sufficiently large income effects exist, not only have we no guarantee that the Marshallian measure will be a close approximation to the compensated one, we have no reason to believe that this measure will even be bounded by CV and EV. This is where a second Willig result proves useful.

3.4.1 The Willig Condition

When income effects are worth considering, additional restrictions on preferences must hold to ensure that the Marshallian measure is bounded by CV and EV and that the nature of the approximation is known. The necessary restriction, set out by Willig (1978), is

$$\partial \frac{v_q(\mathbf{p}, q, y)}{v_{p_1}(\mathbf{p}, q, y)} / \partial y = 0. \tag{3.22}$$

There are several equivalent statements of this condition. For example, equation (3.22) holds if and only if the marginal value of quality equals incremental consumer surplus:[11]

$$-\frac{\partial m}{\partial q} = \int_{p_1}^{\tilde{p}_1} \frac{\partial z_1^m(\mathbf{p}, q, y)}{\partial q} dp_1 = \frac{\partial \int_{p_1}^{\tilde{p}_1} z_1^m(\mathbf{p}, q, y) dp_1}{\partial q}. \tag{3.23}$$

Equation (3.23) provides a link between the Marshallian demand curve as a function of price, quality, and income and the expenditure function.

The Willig condition establishes an approximate relationship between the Marshallian and Hicksian measures. To make this connection, we employ the notion of Marshallian and Hicksian virtual prices. Begin with an exact measure of compensating variation when weak complementarity holds–equation (3.20), depicted in Figure 3.3A. Now we portray this same area by defining a Hicksian virtual price function for q that we call $\pi^h(\mathbf{p}, q, u)$.[12] For each value of q, π^h is the parametric 'price' that would cause the individual, facing the following

---

[11]Note that $\frac{\partial \int_{p_1}^{\tilde{p}_1} z_1^m(\mathbf{p},q,y) dp_1}{\partial q} = \int_{p_1}^{\tilde{p}_1} \frac{\partial z_1^m(\mathbf{p},q,y)}{\partial q} dp_1 + z_1^m(\tilde{p}_1, \mathbf{p}_{-1}, q, y)$ but the last term is zero at the choke price $\tilde{p}_1$.

[12]The idea of virtual price functions is developed in Neary and Roberts (1980) and used by Hanemann (1991) among others.

cost minimization problem, to freely choose that level of q (and to choose the same level of $\mathbf{z}$ as would be chosen when q is given and not priced):

$$\min_{\mathbf{z},q} \ \mathbf{p} \cdot \mathbf{z} + \pi^h q - \mu(u(\mathbf{z},q) - u).$$

The area under this Hicksian virtual price function is just the change in the expenditure function:

$$\int_{q^0}^{q^1} \pi^h(\mathbf{p}, q, u)dq = m(\mathbf{p}, q^0, u) - m(\mathbf{p}, q^1, u) \tag{3.24}$$

which is exactly equal to, in fact the definition of, our CV measure. As a consequence, the areas depicted in panels A and B of Figure 3.3 are just two ways of illustrating the same measure. Note that expression (3.24), evaluated at u^0, will always equal CV while expression (3.20) will equal CV only when weak complementarity holds.

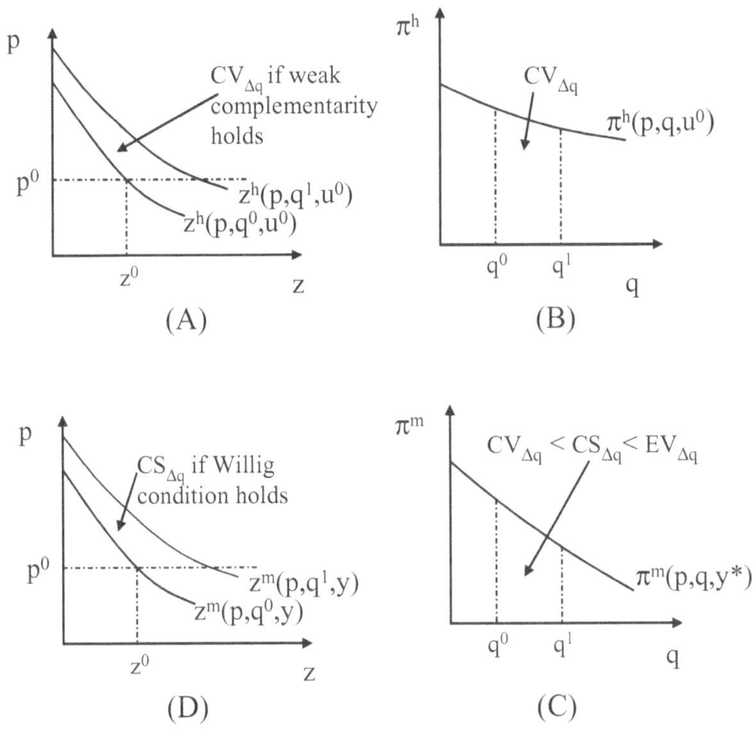

FIGURE 3.3. Connecting Marshallian Consumer Surplus with Compensating Variation for Quality Change

In virtual price-quality space, it is now much easier to describe the Randall-Stoll results. First, consider the Marshallian version of virtual price. The function $\pi^m(\mathbf{p}, q, y)$ is the price, for every value of q, that causes the individual facing the following imaginary maximization problem to choose that level of q (as well as the same level of $\mathbf{z}$):

$$\max_{\mathbf{z},q} \; u(\mathbf{z},q) + \lambda(y^* - \mathbf{p}\cdot\mathbf{z} - \pi^m q). \tag{3.25}$$

where y^* is treated as exogenous by the individual but must equal $y+\pi^m q$. This apparent sleight of hand serves to resolve the inconsistency that arises when we charge the consumer a virtual price but wish to ensure that the solution is such that the current levels of q (which in reality is actually free) and all other goods are still the levels that get 'chosen' in this new artificial situation. An individual solving (3.25) would set $u_q = \lambda\pi^m$. But u_q must equal $v_q(\mathbf{p}, q, y)$ from the original problem, and λ equals v_m. This means that π^m is really equal to v_q/v_m.

We can graph the area under the Marshallian virtual price function between the initial and final levels of q, as in Figure 3.3C. Randall and Stoll (1980) derive bounds that show that this area lies between compensating and equivalent variation. That is, the area in panel C is bounded by the area depicted in panel B conditioned on u^0 and the equivalent area conditioned on u^1. The Randall-Stoll error bounds are analogous to those developed by Willig (1976) for price changes, except that they depend on income flexibility rather than income elasticity, where the former is defined as $\xi = (\Delta\pi^m/\Delta y)(y/\pi^m)$. Labeling the area under the Marshallian function as CS, the errors in approximation will be smaller the smaller are $|CS|/y$ and ξ.

So we now know that the area in panel C is bounded by CV and EV and, under certain circumstances, may be a reasonable approximation of either. But the area in panel C is not observable because it depends on an abstraction–the virtual price function. What we have to work with are Marshallian demand functions. How does the area described in (3.19) and depicted in Figure 3.3D match up with panel C? The answer is that they are identical *if and only if* the Willig condition holds. Hence we see that what is feasible to observe–the change in the area under the Marshallian demand curve in Figure 3.3D–is bounded by the CV measure depicted in panel A and the analogous EV measure, *if* the Willig condition holds. Also the measure in panel D will have the same properties we claimed for panel C–it will be a good approximation to the measure in panel A–if income effects are relatively small and at least moderately close substitutes exist for the public good. (Naturally if income effects are approximately zero, Hicksian and Marshallian demand functions will converge and there will be no need to invoke this condition.)

Willig's 1978 paper developed this condition, implicitly in conjunction with the weak complementarity restriction, to deal with price indices of quality changes. When the Willig condition holds, price and quality changes can be translated into utility-equivalent pure price changes. Smith and Banzhaf (2004) expand on Willig and illustrate the implications of weak complementarity and the Willig condition diagrammatically, contributing to our understanding of these restrictions and their connection with price indices.

Palmquist (2005a) provides further insight into the Willig condition by relating it to path independence of line integrals. Palmquist points out that in Figure 3.3D we are implicitly changing both p_1 and q, so the problem is really a line integral problem. In fact the graph suggests the following path of price and quality changes: $[p_1^0, q^0] \to [\tilde{p}_1(q^0), q^0] \to [\tilde{p}_1(q^1), q^1] \to [p_1^0, q^1]$. But what is the integrand of the line integral? Analogous to how we often develop conventional consumer surplus measures, start by writing down the utility change that would result from a change in $[p_1, q]$. This is given by

$$\begin{aligned} \Delta u &= \int_L V_{p_1} dp_1 + V_q dq \\ &= \int_L \lambda \frac{V_{p_1}}{V_m} dp_1 + \lambda \frac{V_q}{V_m} dq \end{aligned} \tag{3.26}$$

where L denotes the line integral path and λ is the marginal utility of income. This expression has an interesting equivalent. By Roy's identity, $-\frac{V_{p_1}}{V_m} = z_1^m$, and the expression $\frac{V_q}{V_m}$ is simply the Marshallian virtual price function, π^m. If the marginal utility of income were invariant to changes in p_1 and q, then λ could be moved outside the line integral and the following would be a simple rescaling of (3.26):

$$\int_L (-z_1^m(\mathbf{p}, q, y) dp_1) + \pi^m(\mathbf{p}, q, y) dq. \tag{3.27}$$

Of course, there is no reason to believe that λ would be invariant to these changes. But analogous to the conventional price change case in Chapter 2, we can view the integrand in equation (3.27) as a consumer surplus measure. To see this, evaluate (3.27) over the path of price and quality changes described above:

$$\int_{p_1^0}^{\tilde{p}_1(q^0)} (-z_1^m(p_1, \mathbf{p}_{-1}^0, q^0, y) dp_1) + \int_{q^0}^{q^1} \pi^m(\tilde{p}_1(q), \mathbf{p}_{-1}^0, q, y) dq + \int_{\tilde{p}_1(q^1)}^{p_1^0} (-z_1^m(p_1, \mathbf{p}_{-1}^0, q^1, y) dp_1) \tag{3.28}$$

Weak complementarity ensures that the second term is zero and the remaining expression is the area depicted in panel D.

Now if this line integral is path independent, we could choose another path of change, and the alternative path would yield an answer equal to that produced by evaluating (3.28). Let's choose the simple path given by $[p_1^0, q^0] \to [p_1^0, q^1]$ that has the same initial and final values of parameters as the previous circuitous path but includes no change in p_1. Evaluating the line integral in (3.27) along this new path yields

$$\int_{q^0}^{q^1} \pi^m(p_1^0, \mathbf{p}_{-1}^0, q, y)dq, \tag{3.29}$$

the area in panel C. This suggests that if the line integral in (3.26) is path independent, then the areas in panels C and D are equal.

Path independence of the line integral in (3.26) requires that the following cross partials are equal:

$$\partial \frac{V_{p_1}}{V_m} / \partial q = \partial \frac{V_q}{V_m} / \partial p_1, \tag{3.30}$$

which implies

$$\partial z_1 / \partial q = \partial \pi^m / \partial p_1. \tag{3.31}$$

A little manipulation of the expression in (3.30) demonstrates that this is equivalent to the Willig condition in (3.22). With a bit more work we can write (3.31) as

$$\frac{y}{\pi^m} \frac{\partial \pi^m}{\partial y} = \frac{y}{z_1} \frac{\partial z_1}{\partial y},$$

which implies that the income flexibility of quality equals the income elasticity of the private good.

Beginning with two commonly used utility functions, Palmquist illustrates the implications of the Willig condition requirement. The repackaging model of Fisher and Shell (1971), for which the indirect utility can be written as $v = \ln(y) + \ln(q/p+1)$, exhibits weak complementarity and does not violate the Willig condition. For this case, the calculated area between the Marshallian demand functions, as pictured in Figure 3.3D, is bounded by CV and EV. However, for the Stone-Geary utility function, $\sum_{i=1}^{k} \Psi_i(q_i) \ln(z_i + \theta_i) + \ln(z_{k+1})$, weak complementarity holds for $\theta = 1$, but the Willig condition does not hold. Palmquist shows that when one calculates the equivalent to panel D, it falls outside the bounds of CV and EV calculated from the same utility function.

What are the implications of all this? To use the area between two Marshallian demands as an approximation of the CV measure of a change in the public good, certain restrictions on preferences are required: weak complementarity must hold and either income effects must be negligible or preferences must be consistent with the Willig condition. Weak complementarity is essential for even the conceptual Hicksian measure, described in equation (3.20), to have meaning. Because of its importance, there has been much discussion about whether one can empirically test for it. Testing for weak complementarity is not feasible, but the concept is intuitive, well defined, and easily understood. As such researchers themselves can usually determine when it is plausible. When they cannot, verification may be possible by direct questioning.

The Willig condition holds a different place in theory. It works, in the sense of showing us that changes in the area under Marshallian demand curves are bounded by CV and EV. We can provide no intuitive story about behavior that helps establish its plausibility. This, combined with its restrictiveness, makes it much less appealing than weak complementarity. In the price change case, the nicety of integrating back to the underlying preference structure is often neglected on the grounds that it really will not make much difference anyway. In the quality change case (especially when significant income effects are expected), working with the underlying expenditure function or indirect utility function rather than the uncompensated demands seems prudent.

3.5 Welfare without Weak Complementarity

Throughout this book we are concerned with situations where we hope to recover enough about preferences to permit the valuation of environmental services or access to natural resources. We have claimed that it is not generally possible to do this without imposing some structure on preferences–particularly with regard to how the public good is related to privately consumed goods. Much attention has been given here to the weak complementarity restriction, and we will continue to invoke this either explicitly or implicitly in the next several chapters. But weak complementarity is not the only possible restriction that can help us. In some cases, it is more intuitive to think of the public and private goods as being substitutes. In Chapters 8 and 9 we will consider how links between private and public goods as substitutes in either household or firm production can help us recover welfare measures for changes in the public goods.[13]

Ebert (1998) has asked the more general question: is it possible to avoid 'the arbitrariness' of any 'subjective restrictions' and still estimate the welfare effects of a change in a public good? CV is given by the change in expenditure functions conditioned on initial and final levels of q, but in the absence of further restrictions we will need information on the full preference ordering to obtain the relevant welfare measure.

---

[13]See also Smith, Evans, Banzhaf, and Poulos (2004) for a general treament of weak substitutability.

Ebert addresses the problem by exploiting the integrability conditions for complete demand systems with an extension. Specifically, he assumes that the researcher has both a full set of demand curves for private goods *and* inverse marginal willingness to pay curves for public goods. Suppose that we have the following problem:

$$\max_{\mathbf{z},\mathbf{q}}\{u(\mathbf{z},\mathbf{q})|\mathbf{z}\cdot\mathbf{p}+\mathbf{q}\cdot\boldsymbol{\pi}\leq y^*,\mathbf{z}\geq\mathbf{0},\mathbf{q}\geq\mathbf{0}\} \tag{3.32}$$

where temporarily we think of $\boldsymbol{\pi}$ as a vector of parametric prices for the $\mathbf{q}$ and y^* is income adjusted by $\mathbf{q}\cdot\boldsymbol{\pi}$. The full demand system for equation (3.32) is

$$\mathbf{z} = \mathbf{g}_z(\mathbf{p},\boldsymbol{\pi},y) \tag{3.33}$$

$$\mathbf{q} = \mathbf{g}_q(\mathbf{p},\boldsymbol{\pi},y) \tag{3.34}$$

where $\mathbf{g}_z$ and $\mathbf{g}_q$ are Marshallian demands from the problem in (3.32). Then the integrability conditions allow us to recover the minimum expenditure function and this permits welfare valuation of changes in prices or $\mathbf{q}$. The integrability conditions in this case include the symmetry and negative semi-definiteness of the Slutsky matrix for the *augmented* system.

Naturally we cannot actually invoke these conditions for (3.32) because in reality there are no prices for the vector $\mathbf{q}$. Instead, Ebert supposes that we have the following information:

$$\mathbf{z}=\mathbf{f}(\mathbf{p},\mathbf{q},y) \tag{3.35}$$

and the marginal willingness to pay functions

$$\mathbf{v}_q(\mathbf{p},\mathbf{q},y)/v_y(\mathbf{p},\mathbf{q},y)=\boldsymbol{\pi}^m(\mathbf{p},\mathbf{q},y). \tag{3.36}$$

The marginal value system is the set of marginal willingness to pay functions, one for each element of the quality vector, which we have called the virtual price functions in previous sections. When one has the information in equations (3.35) and (3.36), it is equivalent to having the complete demand system given by equations (3.33) and (3.34). In other words, knowledge of the demand curves and the marginal willingness to pay functions allows one to recover the utility function or the expenditure function.[14] Consequently, with the demand functions and marginal value functions, one can in theory completely recover preferences and hence perform any welfare measurement that can be asked of the preference function.

[14]The proof involves showing that the Slutsky matrices from the two different problems, equations (4.21) and (3.36), are the same as the augmented system in equation (3.32). See Ebert for details.

Ebert also demonstrates that the estimation of an incomplete system will provide full welfare measures. Suppose that the preference function is partitioned so that $u(\mathbf{z},\mathbf{q}) = u(\mathbf{z}^A,\mathbf{z}^B,\mathbf{q}^C,\mathbf{q}^{\mathbf{D}})$ where $\mathbf{z} = \mathbf{z}^A,\mathbf{z}^B$ and $\mathbf{q} = \mathbf{q}^C,\mathbf{q}^{\mathbf{D}}$. Then consider the demand functions $\mathbf{z}^A = \mathbf{f}(\mathbf{p}^A,\mathbf{p}^B,\mathbf{q}^C,\mathbf{q}^D,y)$ and the marginal value functions $\boldsymbol{\pi}^m_C(\mathbf{p}^A,\mathbf{p}^B,\mathbf{q}^C,\mathbf{q}^D,y)$. These demand functions and marginal value functions form an incomplete system for marketed goods $\mathbf{z}^A$ and public goods $\mathbf{q}^C$. The 'incompleteness' of the system is caused by the fact that the consumption of $\mathbf{z}^B$ and the levels of $\mathbf{q}^D$ are not observed. The estimation of these functions constitutes an incomplete system, but when the integrability conditions hold for this subsystem, full welfare measures can still be recovered.

The demonstration that a) the complete system of demands in equations (3.33) and (3.34) as well as the demand functions and marginal value functions in equations (3.35) and (3.36) provide equivalent information and b) one can completely recover preferences with a complete demand system and the marginal value functions is gratifying but should not be surprising. Likewise, if one possessed the equivalent functions for the incomplete demand system, one could recover the full welfare effects of a change in one of the $\mathbf{q}^C$. And all of this takes place without arbitrary restrictions such as weak complementarity on the preference function. We saw earlier in this chapter that if one has the virtual price functions in equation (3.36), no more information is needed for welfare measurement of changes in $\mathbf{q}$. We can calculate the area under the Marshallian virtual price functions for changes in the $q's$ and, using Randall and Stoll bounds, we can determine how close the approximation is to the exact CV or EV. For the single q case, the area marked in Figure 3.3B provides the correct CV measure and its approximation can be found in Figure 3.3C. The upshot is that when the researcher has the virtual price functions in equation (3.36), no more information is needed for valuation of changes in the $\mathbf{q}$. If one is lucky enough to have marginal value or virtual price functions, then no further research is necessary.

The key point, however, is that without restrictions such as weak complementarity, no behavioral function can give a complete welfare measure. This is because behavior can not reveal the marginal willingness to pay functions for the q's. Ebert alludes to acquiring this information through surveys, and Herriges, Kling and Phaneuf (2004) suggest using stated preference methods to recover these functions. Eom and Larson (forthcoming) demonstrate with an empirical study that a combination of behavioral and stated preference data may be useful, especially in sorting out use and non-use values. The use of revealed and stated preference data by Eom and Larson provides a template for decomposing total value into use and non-use value. The message is simply that despite Ebert's theoretical results, empirical results in the general case (i.e. in the absence of a restriction on preferences such as weak complementarity) would seem impossible if only behavioral data is relied upon.

3.6 Conclusions

Estimating benefits of public goods from observations on behavior requires careful construction of individual preference functions. The standard welfare economics based on the gains and losses of price changes provides some insights, but can also lead one astray. In this chapter we have explored an important restriction on preferences that will permit measurement of the welfare effects of changes in public goods. The most basic assumption is weak complementarity, which despite its rather pompous name, is a simple intuitive idea: individuals value changes in a resource only when they use it. Further, the Willig condition must hold if researchers wish to be assured that the difference in the area under Marshallian demand curves is bounded by CV and EV.

Throughout this chapter we have argued that weak complementarity is not a testable assumption but, in most cases where it is employed, it is quite plausible. As such, it is more akin to a research strategy than a parametric restriction. Not all researchers take this view. Ebert in particular argues that using weak complementarity as a maintained assumption makes results subjective, especially when the assumption is not testable. These two distinct positions cannot be resolved by tests or economic postulates, but will be based on personal predilections towards research. For our part, we believe that weak complementarity is often a good assumption to maintain. We also recognize that behavioral methods will be systematically wrong if, in fact, weak complementarity does not hold. But cases where weak complementarity would be a misguided assumption–for example a travel cost study of the demand for access to the Arctic National Wildlife Refuge–will often be obvious. The weak complementarity assumption about preferences must be buttressed by a sense of the importance of the resource and how it is used by individuals.

Chapter 4

Implementing Weak Complementarity

4.1 Introduction

The message from Chapter 3 is clear. To use revealed preference methods and only revealed preference methods to value changes in public goods, specifically changes in environmental quality, requires imposing some added restrictions on the individual's decision problem. The most commonly employed restriction is weak complementarity. In this chapter we deal with an array of conceptual and empirical problems that arise in making the weak complementarity model of environmental valuation operational in a conventional demand setting. The first section considers how one might go about specifying demand functions or systems that incorporate prices, income and quality characteristics. Subsequent sections treat conceptual issues that arise when the weak complement is really a household-produced good. When this is true, time enters the problem in a number of ways, complicating both specification and welfare measurement. Finally we consider how to make conceptual and empirical sense of welfare evaluation when individuals do not have perfect information about quality changes.

4.2 Specifying Demand as a Function of Quality

Eventually the researcher bent on empirical work must write down a demand function to be estimated. Economists have fewer priors and less experience in estimating demand functions with quality characteristics, compared with prices and income. The issues raised in the last chapter suggest this should be done with care, especially with regard to how the public and private goods interact.

Whether one begins with the demand function or a preference function, the implied relationship between q and one or more z's should make sense by connecting motives with behavior. This caution is relevant even if income effects are expected to be insignificant, because the weak complementarity condition still must hold for behavioral measures to make sense. If income effects are expected to be substantial, using the area determined by shifting an arbitrarily specified Marshallian demand function as a welfare measure is particularly dangerous. If the demand function is not consistent with an underlying preference relationship between q and $\mathbf{z}$ for which weak complementarity and the Willig condition hold, the area between the two Marshallian demands can not be assumed even to be bounded by CV and EV. We saw that one can force any analytically integrable demand function to be consistent with weak complementarity by adjusting the constant of integration, at the risk of generating a preference function that lacks plausibility.

For some types of preference structures consistent with plausible stories about the relationship between q and $\mathbf{z}$, it is possible for the Marshallian demand function to shift backwards at current prices with an increase in quality, even when that quality characteristic is desirable and adds to utility and even when weak complementarity and the Willig condition hold. In these cases, the Marshallian demands conditioned on different levels of the public good must cross at some price above current price, so that the area between the two curves ends up being a properly bounded and signed consumer surplus measure. Suppose one ignores the underlying preference structure and begins with an arbitrarily specified Marshallian demand. Then the finding that $\partial z_i/\partial q < 0$ could imply either a specification/measurement error or a recognition of this special form of preferences. Beginning with a demand function without appreciating the implications of motives can lead to ambiguity in interpreting estimation results.

The concern over quality changes originated in the development of price indices. Researchers were investigating approaches to correct price indices for changes in the quality of goods or services. In fact, this is what motivated Willig's 1978 paper as well as the earlier work by Fisher and Shell (1971). In estimating behavioral models, our goal is to measure welfare effects of quality changes rather than to calculate price indices. This would seem to make the specification of demand as a function of quality even more important and arguably of greatest importance for *systems* of demand functions. These considerations, while timely, are not new. Hanemann (1982, 1984) explored various approaches to modelling demand systems with quality components. Hanemann (1982) suggested three ways of making preferences depend on quality. One approach is to write utility as a function of one or more subfunctions, each of which is a function of the quantity of a quality-differentiated good or service and the level of its quality characteristics. A second approach involves writing the parameters of a conventional, utility-theoretic demand function as functions of the quality variables. The final approach is to specify demand functions and incorporate quality variables into these functions. In what follows we draw on Hanemann's work and on Willig (1978).

4.2.1 Translations of Utility Functions

Suppose we begin by considering utility as a function of subfunctions within which the private and public goods are related. If there is one good whose quality characteristic is of interest then the general form is

$$u(\mathbf{z}, q) = u(g_1(z_1, q), z_2, ..., z_n). \tag{4.1}$$

We will also consider a more general model, one that will be useful in the next chapter. A more general version is[1]

$$u(\mathbf{z}, \mathbf{q}) = u(g_1(z_1, q_1), g_2(z_2, q_2), ..., g_n(z_n, q_n)). \tag{4.2}$$

We can specify the relation between u and the g_i's in whatever way we wish as long as proper convexities are imposed. However, without more structure on the g_i subfunctions, there is no guarantee that weak complementarity holds. We explore whether weak complementarity is consistent with three common 'translations' of this general form.

The first translation is one in which each z_i and a function of the corresponding q_i enter the utility function as perfect substitutes, written as

$$u(\mathbf{z}, q) = u(z_1 + \psi_1(q_1), z_2 + \psi_2(q_2), ..., z_n + \psi_n(q_n)). \tag{4.3}$$

The implied Marshallian demands are of the form

$$z_i = f_i(\mathbf{p}, y + \boldsymbol{\psi} \cdot \mathbf{p}) - \psi_i(q_i) \tag{4.4}$$

where the function $f_i(\mathbf{p}, y)$ is defined as the Marshallian demand curve with all $\psi_i(q_i) = 0$. When only a single public good (quality characteristic) is of interest as in (4.1), the demand function for z_1 is given by

$$z_1 = f_1(p_{1,}, \mathbf{p}_{-1}, y + \psi_1(q_1)p_1) - \psi_1(q_1),$$

where $\mathbf{p}_{-1}$ denotes the vector of all prices other than p_1. This translation clearly does not meet the restriction of weak complementarity. The public good q_i can be substituted for z_i, but more to the point q_i continues to be

[1] Multiple public goods could be associated with each privately consumed good, but for notational simplicity we restrict q_i to be a scalar.

valued even if z_i is reduced to zero. The expenditure function for this utility function has the form

$$m(\mathbf{p}, q_1, u) = \tilde{m}(\mathbf{p}, u) - p_1\psi_1(q_1),$$

where $\tilde{m}(\mathbf{p}, u)$ is some subfunction which is not a function of q_1. The CV of a change in q_1 equals simply $p_1(\psi_1(q_1^1) - \psi_1(q_1^0))$. That is, one has only to estimate the function $\psi_1(q_1)$, which is really a technological relationship. It translates the environmental policy variable, q_1, into units comparable to the private good, z_1. If the units are the same, the welfare effect is just $p_1(q_1^1 - q_1^0)$. We investigate household production technologies that look identical to this restriction in Chapter 8 on averting behavior.

A second translation is the pure repackaging model of Fisher and Shell, the general form of which is

$$u(\mathbf{z}, \mathbf{q}) = u(z_1\psi_1(q_1), z_2\psi_2(q_2), ..., z_n\psi_n(q_n)). \tag{4.5}$$

In this utility function, z_i and q_i appear to be substitutes because the ith argument of the utility function increases by increasing z_i or q_i. However, weak complementarity holds because $\psi_i(q_i)$ is not an argument of utility when $z_i = 0$. The Marshallian demand curves consistent with repackaging have the form

$$z_i = f_i(p_1/\psi_1(q_1), ..., p_n/\psi_n(q_n), y)/\psi_i(q_i). \tag{4.6}$$

One can think of the ith component of the utility function, $z_i\psi_i(q_i)$, as a service level. The demand for this service is then a function of income and real prices adjusted to reflect quality. The indirect utility function is of the form:

$$v(p_1/\psi_1(q_1), ..., p_n/\psi_n(q_n), y).$$

One could alternatively start with the indirect utility function $v(p_1\zeta_1(q_1), ..., p_n\zeta_n(q_n), y)$ which is just the same function with $\zeta_i(q_i) = 1/\psi_i(q_i)$.

The implications of the pure repackaging model can be seen with a well-known example. Consider the single good analog to (4.6), corresponding to the pure repackaging translation of (4.1):

$$z_1 = f_1(p_1/\psi_1(q_1), \mathbf{p}_{-1}, y)/\psi_1(q_1).$$

Let z_1 be oranges and let $\psi_1(q_1) = q_1$ be the juice per orange. Then z_1q_1 is the quantity of orange juice in z_1 oranges and the price of orange juice is the price of oranges divided by the juice per orange. If an individual buys oranges for their juice content only, then a doubling of the juice content per orange (*i.e.* a doubling of q_1) is equivalent to halving the price of oranges. An increase in quality has two effects: it lowers the effective price of orange juice, increasing the demand for the good–oranges–and it makes oranges more efficient at producing juice, reducing the number of oranges needed to achieve a given level of utility. The two effects work in opposite directions.

It is not surprising, then, that the Hicksian demand for oranges can shift backwards–at the current price–with an increase in q_1. However, we know that the area between the two Hicksian demands conditioned on the two levels of quality will still be properly signed because of weak complementarity. When z_1 is a weak complement to q_1, it must be true that $\int_{p_1^0}^{p_1^*} z_1^h(\mathbf{p}, q_1^1, u)dp_1 - \int_{p_1^0}^{p_1^*} z_1^h(\mathbf{p}, q_1^0, u)dp_1$ is a measure of the CV of the quality change. If so, then at some $p_1 \geq p_1^0$, the two Hicksian demands must cross and the area behind the Hicksian curve at a higher level of quality minus the area behind the Hicksian conditioned on lower quality levels must be positive.

Even the Marshallian demand curve can shift *backwards* at the current price in this example. For this simple case

$$\partial z_1/\partial q_1 = -f/q_1^2 - f_1 p_1/q_1^3 = -(z_1/q_1)[1+\xi]$$

where ξ is the price elasticity of demand, which will be negative. If demand is inelastic $(1 + \xi > 0)$ then $\partial z_1/\partial q_1 < 0$. A shift backwards of the Marshallian demand at current price does not preclude the area between the two Marshallians from being positive any more than it does in the Hicksian case. As price rises, the elasticity of demand will tend to rise also, suggesting that at some point the two Marshallians will cross as well. The fact that the Willig condition holds for this preference function implies that the consumer surplus measure, $\int_{p_1^0}^{p_1^*} z_1^m(\mathbf{p}, q_1^1, y)dp_1 - \int_{p_1^0}^{p_1^*} z_1^m(\mathbf{p}, q_1^0, y)dp_1$, will be properly signed.

Pure repackaging may seem an extreme case, yet there are plausible examples in practical applications of importance to environmental economists. Consider the case of sportfishing, where z_1 is the number of days spent fishing and q_1 is the number of fish caught per day. When the consumer has preferences on the number of fish caught rather than the number of days, and the latter is the choice variable, the pure repackaging model holds.[2] In that case one could conceivably find a negative effect on the demand for fishing days from an increase in the number of fish caught per day at current prices. Other cases do not fit this story at all. For example, if the activity is swimming days and quality represents water quality as measured by clarity of the water, the quality may best be characterized as a complement to swimming. An increase in water clarity makes each unit of the good more enjoyable.

---

[2]This same set of circumstances could equally well be interpreted as one in which fishing trips and fish stocks are inputs into the production of sportfish caught. We will see in Chapter 8 that sometimes the same problem can be reframed in different ways, allowing the use of different restrictions to recover preferences.

A third translation is the cross product repackaging model, introduced in Willig (1978). Because its intuition is somewhat easier to grasp for the single good, we begin with the case in which q is a quality characteristic of one privately purchased good, z_1. The direct utility function for this form is

$$u = u(z_1, z_2 + z_1\phi_2(q), ..., z_n + z_1\phi_n(q)). \tag{4.7}$$

In Willig's words 'every unit of good 1 provides the services of $\phi_i(q)$ units of good i, in addition to fulfilling its own direct role in consumption.' Even this statement is more general than the most intuitive examples would support, since usually we would expect only a small subset (and probably only one) of the $\phi_i(q)$ functions to be other than zero. Consider another orange juice example, but this time the orange juice is prepackaged and calcium enriched. Orange juice, z_1, enters the utility function in its own right, and it also substitutes for calcium supplements in other forms. The extent of the substitution is determined by the amount of calcium, q, per unit of orange juice. Weak complementarity still holds because changes in the calcium content do not matter if the orange juice is not consumed.

One can imagine formulating some recreational demand models in the cross-product repackaging form. Suppose that the water at the local beach, used for swimming and fishing, is contaminated with a pollutant that does not hamper the recreational experience directly, but that has potential health risks. Perhaps it is PCB's or mercury in fish, which have no visual or olfactory signals. In this case q is a 'bad'. The individual still obtains the same amount of 'recreational' utility from trips to this beach, but the trips produce an increase in risk that diminishes another element of the utility function–some function of health.

The Marshallian demand function for z_1 associated with (4.7) is given by

$$z_1^m = f(p_1 - \sum_{j=2}^{n} \phi_j(q)p_j, \mathbf{p}_{-1}, y),$$

where any number of the $\phi_j(q)$ might be zero. The indirect utility function is

$$v(p_1 - \sum_{j=2}^{n} \phi_j(q)p_j, \mathbf{p}_{-1}, y). \tag{4.8}$$

The change in indirect utility from a change in q will be

$$\partial v(\mathbf{p}, q, y)/\partial q = -v_{p_1} \sum_{j=2}^{n} p_j \partial\phi_j/\partial q,$$

which is positive as long as $\partial\phi_j/\partial q \geq 0$ for all $j = 2, ..., n$ and at least one is strictly positive.

Increases in z_1 need not positively augment all other goods it affects. One can think of cases where 'augmentation' works in different directions for different $z_i's$. However, as long as $\partial \sum_{j \neq 1} \phi_j(q_1)p_j/\partial q > 0$, the demand for z_1 shifts out when quality increases, and does so at all levels of p_1. Weak complementarity holds for the cross-product repackaging model, because in (4.7) q has no effect on utility unless z_1 is non-zero. The Willig condition also holds, a result that can be shown by applying to (4.8) the form of the Willig condition developed by Palmquist (2005a) and presented in Chapter 3.

4.2.2 Utility Parameters as a Function of Quality

A second set of translations can help us adapt commonly used functional forms by introducing quality as a function of an existing parameter of the utility function. Writing the parameter as a function of quality characteristics does not ensure that any of the necessary properties–curvature, non-essentiality, or weak complementarity–hold, and so each case must be investigated separately.

As an example consider a form of the Stone-Geary utility function that received so much attention in the previous chapter:

$$u = \sum_{i=1}^{n} \alpha_i \ln(z_i - c_i).$$

In this form, the c_i are parameters of the utility function, originally interpreted as the subsistence level of the commodity in the sense that utility becomes undefined when consumption of the ith commodity is less than c_i. Because the utility function is additively separable, one can focus only on the parameters related to z_i. For non-essentiality, utility must be bounded as the quantity consumed of z_i goes to zero. This requires that c_i be less than zero, so that the term $\alpha_i \ln(-c_i)$ will be defined. Where should we introduce q_i? If the parameter c_i is a decreasing convex function of quality, we have a version of expression (4.3) and weak complementarity fails to hold. A better option is to follow the lead of Herriges, Kling and Phaneuf (2004) (also Phaneuf, Herriges and Kling, 2000) and make α_i depend on q_i. For the direct utility function to be quasi-concave in $\mathbf{q}$ and $\mathbf{z}$, one needs only that $\alpha_i(q_i)$ be increasing in q_i. The function actually estimated by Herriges, Kling and Phaneuf takes the following form:

$$u(\mathbf{z}, \mathbf{q}) = \sum_{i=1}^{n-1} \Psi_i(\mathbf{q}_i) \ln(z_i - c_i) + \ln(z_n)$$

where in their study, $\mathbf{q}_i$ is a vector of characteristics. The first $n-1$ of the z's are trips to pheasant hunting sites in Iowa while z_n is a composite commodity. The quality variables are formed as

$$\Psi(\mathbf{q}_i) = \exp(\boldsymbol{\delta} \cdot \mathbf{q}_i)$$

where $\boldsymbol{\delta}$ is a vector of parameters to be estimated.

As we saw in the last chapter, for weak complementarity to hold this specification requires in addition that $c_i = -1$. However, one can also achieve weak complementarity by incorporating elements of the pure repackaging model. Allow $\mathbf{q}$ to enter so that utility is given by $\sum_{i=1}^{n-1} \alpha_i \ln(b_i(\mathbf{q})z_i - c_i) + \beta \ln(z_n - c_n)$. In this form, weak complementarity holds for any value of c_i.[3]

A flexible model along these lines is estimated by von Haefen, Phaneuf and Parsons (2004). They posit the utility function[4]

$$u(\mathbf{z}, \mathbf{q}) = \sum_{i=1}^{n-1} \Psi(\mathbf{s}, d_i) \cdot (\psi(\mathbf{q}_i)z_i - c_i)^{\rho_i} + z_n^{\rho}$$

where the parameters to be estimated include c_i and ρ_i, and those embodied in the $\Psi(\mathbf{s}, d_i)$ and $\psi(\mathbf{q}_i)$ functions are

$$\begin{aligned} \psi(\mathbf{q}_i) &= \exp(\boldsymbol{\delta} \cdot \mathbf{q}_i) \\ \Psi(\mathbf{s}, d_i) &= \exp(\boldsymbol{\gamma} \cdot (\mathbf{s}, d_i)). \end{aligned}$$

For this model weak complementarity holds regardless of parameter values. It is, however, a repackaging specification, with the property that when Marshallian demands are price inelastic, they will be decreasing in quality arguments. Nevertheless, since the utility function is increasing in the quality attribute, welfare effects obtained directly from the associated expenditure function will be increasing in the quality improvement.

We have explored various approaches for introducing quality in preference functions and have emphasized starting with preference functions, because it is much easier to ensure that the desired properties hold. It is also worth observing that the typical model estimated will be a partial demand system, one that has a subset of prices and demands. In this case, the appropriate budget will be a sub-budget, a task difficult in concept and still more in measurement.

4.3 Weak Complementarity and Household Production

The demand for recreational trips is by far the most frequent context in which weak complementarity is invoked. Although many of our examples have been recreational ones, we have not yet explicitly resolved the dilemma that this

---

[3]As long as $c_i < 0$, required for utility to be defined when the commodity is zero.

[4]von Haefen (2004) also estimates a limiting form of this model where the ith component of utility is $\Psi_i \cdot \ln(\psi(\mathbf{q}_i)z_i - c_i)$, which is the model that holds when $\lim_{\rho_i \to 0}(\psi(\mathbf{q}_i)z_i - c_i)^{\rho_i}/\rho_i$.

creates: recreational trips are not marketed goods purchased at market equilibrium prices. In this section, we address cases in which the weak complement is 'produced' by the household.

The household production approach to commodities began with the work of Gorman (1953), Becker (1965), and Lancaster (1966). Its use in understanding household decisions comes from the observation that households frequently buy goods that are not valued in their own right but, when combined with other goods, produce valued services. In other settings, households buy household cleaning goods such as sponges, brooms, and vacuum cleaners not for the direct enjoyment they provide but because they produce cleanliness, which does provide utility. In this example, cleanliness is the 'commodity' (output) that the household values, while the sponges and other purchases are designated as 'goods' (inputs) which yield utility only when they produce cleanliness. More generally, households buy goods and combine them with their own time to produce commodities such as domestic services, recreational services, and health status.

Before investigating the model, we reiterate a previous concern and caveat: despite the designation of 'household' production, the distinction between the individual and household is a difficult one for which there is, to date, no adequate treatment in the valuation literature. In the original paper on household production, Becker treated the household as the decision making unit, suggesting that intra-household allocations of consumption and production activities would be made 'optimally' (p 512). Although much progress has been made in solving this problem in consumption and labor supply, (see Bourguignon and Chiappori, 1992, Browning et al. 2004), little of this beyond Smith and Van Houtven (2004) has made its way to valuation models. Hence we continue to use the terms individual and household interchangeably, recognizing that potentially important differences are being suppressed.

Although there are many settings in which a household produced good could be a weak complement to an environmental good (especially in developing country settings), most of the examples and modeling considerations we address in this section will be specific to recreation demand modeling. Recreational trips are often logically viewed as weak complements to environmental quality, a connection that is practically important because many environmental improvements are believed to impact households through this pathway. Early in the history of the Clean Water Act, Freeman (1982) and others pointed out that the benefits of the Act were likely to accrue largely through recreation. Other environmental improvements, such as habitat protection, will also benefit individuals through some recreational pathway, at least in part. Even air quality, so often linked with health considerations, can affect individuals via recreation when air quality improvements imply improvements in visibility. Weak complementarity is a plausible restriction (and certainly a convenient one) when it seems reasonable that individuals are not concerned about the environmental quality at recreational locations they do not visit. When weak complementarity is a reasonable assumption, demand functions for recreational trips to a site can be used to provide the information we need to value quality changes at the site. Where it is not, it is often a lower bound (in absolute value terms) of the value of a change.

While the household production concept is straightforward, applications that require modeling the demand for the commodity may not be so, as the critique by Pollak and Wachter (1975) made clear. Joint production and economies of scale create problems that can make these empirical applications difficult. A common case of joint production arises when a household uses time as an input in production and also values the time directly. The result is that the marginal cost of the commodity depends on the level of that commodity or other commodities. The difficulties are both conceptual and econometric, and they stem from the nature of the cost of producing commodities. The conceptual problem is that under certain cost conditions, the Marshallian demands for commodities are not unique. The econometric problem is created by non-constant marginal costs, which means that a given household has both a supply curve and a marginal value curve, and the Marshallian demand is determined by their intersection. The concept of an ordinary Marshallian demand function fails when marginal cost is not constant.

4.3.1 Household Production and Constant Marginal Costs

To see the empirical implications of the household production function, we begin with one of the simplest of models which, when applied to recreational demand, is really the old travel cost model thinly disguised.[5] For purposes of exposition, we focus on trips to a single site, a limitation that will be relaxed in the next chapter, and we ignore time as an input and return to this issue shortly. Individuals maximize utility which is a function of trips (z_1) taken to a site, environmental quality at the site (q), and a composite commodity (z_2). Trips produced by combining purchased inputs (such as gasoline, lodging, etc.) denoted $\mathbf{x}$. The individual's optimization problem is

$$\max_{z_2,\mathbf{x}} u(z_1, z_2, q) \text{ subject to } y = \mathbf{r} \cdot \mathbf{x} + z_2 \text{ and } h(z_1, \mathbf{x}) = 0 \tag{4.9}$$

[5]The travel cost model, originally conceived by Hotelling, treats the cost of travel to a site as the appropriate 'price' of the trip. The model was further developed by Clawson (1959) and many others. Hotelling (1947) originally wrote out the travel cost model as a model of trips per capita from different distance zones to a recreational site as a function of travel costs. Cooper and Loomis (1993) provide an updated version of this zonal travel cost model.

where y is exogenous income, $\mathbf{r}$ is a vector of prices corresponding to the input vector $\mathbf{x}$, the price of the composite commodity is normalized to 1, and $h(z_1, \mathbf{x})$ represents the household production technology. The variable q continues to denote the level of the environmental good for which z_1 is a weak complement. In this formulation, q is a quality characteristic of the household produced good, not an input into production. For example, z_1 might measure the number of deep sea sportfishing trips taken while q is the expected fish catch for such a trip.[6]

The household production function implies a cost function that is the solution to

$$c(z_1, \mathbf{r}) = \min_{\mathbf{x}}\{\mathbf{r} \cdot \mathbf{x} | h(z_1, \mathbf{x}) = 0\} \tag{4.10}$$

so that the utility maximization problem can be alternatively stated with a budget constraint that incorporates this cost function:

$$\max_{\mathbf{z}} \{u(z_1, z_2, q) | z_2 + c(z_1, \mathbf{r}) \leq y\}. \tag{4.11}$$

If $c(z_1, \mathbf{r})$ is linear in z_1, equation (4.11) gives the individual demand curve that evolved from the original travel cost model proposed by Hotelling and Clawson.[7] The problem becomes one in which the marginal cost per trip equals the average cost per trip $(\bar{c}(\mathbf{r}))$, since both are constant over trips:

$$\max_{\mathbf{z}} \{u(z_1, z_2, q) | z_2 + z_1\, \bar{c}(\mathbf{r}) \leq y\}. \tag{4.12}$$

This simple and classic formulation is completely equivalent to earlier formulations in this chapter because $\bar{c}(\mathbf{r})$ is a function only of parameters and so can serve as a parametric price for trips.

Although this simple model is consistent with previous models, it is useful to develop the theory in this notation so that subsequent modifications can easily be illustrated. The expenditure minimization problem associated with the utility maximization of (4.12) is

$$\min_{\mathbf{z}}\ \bar{c}(\mathbf{r})\, z_1 + z_2 \text{ subject to } u = u(z_1, z_2, q), \tag{4.13}$$

which leads to a compensated demand of the form $z_1^h(\bar{c}(\mathbf{r}), q, u)$. Weak complementarity between q and z_1 implies that the compensating variation of a

[6]For some problems, the distinction between a quality characteristic of a household produced commodity and an input into the household production process is a semantic one, as we will see in Chapter 8.

[7]The original zonal travel cost model was based on the number of trips per capita from different origin zones located at increasing distances from the site. Our notation assumes that data on individuals are available, as is typical of modern applications.

change in q equals

$$CV = \int_{\bar{c}^0}^{\bar{c}^*} z_1^h(\bar{c}, q^1, u)d\bar{c} - \int_{\bar{c}^0}^{\bar{c}^*} z_1^h(\bar{c}, q^0, u)d\bar{c} \tag{4.14}$$

where $\bar{c} = \bar{c}(r)$, $\bar{c}^0$ stands for the observed level of constant marginal costs, and $\bar{c}^*$ is the constant marginal cost that causes compensated demand for z_1 to be zero.[8] As before, the area between the two compensated demands equals compensating variation because $z_1^h(\bar{c}(\mathbf{r}), q^1, u) = \partial m / \partial \bar{c}$ from Shephard's lemma and $m(\bar{c}^*(q^1), q^1, u) = m(\bar{c}^*(q^0), q^0, u)$ due to the weak complementarity assumption. The Marshallian demand for trips can, in theory, be estimated as a function of the trip 'price' ($\bar{c}$), environmental quality, income, and other relevant variables. The area between the Marshallian demands conditioned on two levels of q,

$$\int_{\bar{c}^0}^{\bar{c}^*} z_1^m(\bar{c}, q^1, y)d\bar{c} - \int_{\bar{c}^0}^{\bar{c}^*} z_1^m(\bar{c}, q^0, y)d\bar{c}, \tag{4.15}$$

can then be used as an approximation of the compensated variation, if the underlying preference structure is consistent with the Willig condition or if income effects are negligible.[9]

Before moving on to more complicated specifications of the household production model, we return briefly to the motivation implicit in Hotelling's original story. Hotelling proposed the travel cost model as a means of valuing the site itself rather than valuing a public good as a quality dimension of a recreational experience. Recreational demand models continue to be used in this way, as well as being used to value changes in environmental amenities. The procedure for valuing a site is subsumed in the procedure for valuing a quality dimension of the site and offers no additional challenges. The value of the site is simply the entire area behind the Hicksian demand function for trips to the site: $\int_{\bar{c}^0}^{\bar{c}^*} z_1^h(\bar{c}, q^0, y)d\bar{c}$. Raising the cost of site access to its choke price is equivalent to eliminating the site, as long as the site only has value to the recreationist if he uses it.

4.3.2 Incorporating Time Costs

While the demand model implied by the utility maximization problem in equation (4.12) matches that posed by Hotelling, it is not the model used today in recreation demand nor is it a model that would normally be useful in a more

[8] As in other cases, the choke price typically will depend on the arguments of the demand curve: $\bar{c}^* = \bar{c}^*(q)$.

[9] Alternatively CV and EV can be calculated directly from knowledge of the expenditure function associated with this Marshallian demand.

general household production setting. This is because it ignores the central role of time, a key input of the household production model of Becker. This is certainly true in recreation as well as other environmental applications of household production, most notably in developing countries. We will return to some of these developing country applications of the household production model in Chapters 8 and 9.

What makes these applications special is that producing a household good takes a significant amount of time. In the conventional utility maximization decisions of the household, money is considered the only scarce resource and the only constraint on utility maximization is a monetary budget constraint. Now a constraint that limits the amount of time that can be spent on all production and leisure activities explicitly recognizes that time is also a scarce resource to the household. When individuals face a parametric wage in the labor market and can choose their hours of work, the two constraints can be written as one–usually in terms of money–and the problem reduces to one that looks quite conventional. However, these simplifying labor market assumptions may not hold for everyone, even in a developed country setting, and will be even less likely in developing countries where well-functioning labor markets are often absent.

The time dimension of household production has been treated most extensively in recreational demand modeling. For many recreational experiences, the time required to access a site is at least as great a deterrent to its use as the money costs of access. Researchers recognized that ignoring time had serious consequences for valuation (e.g. Cesario and Knetsch, 1970). To illustrate this, consider the problem in (4.9) modified to include time. The following restatement assumes that labor is paid a parametric wage rate and the amount of labor supplied is freely chosen. $\bar{T}$ is total available time to the individual, t_w is the time spent working at parametric wage, w, and $\bar{y}$ is exogenous (non-wage) income. At this point we assume each trip has a constant money cost, $\bar{c}$, and a constant time cost of access, t. Further, we assume that the time spent actually recreating, on-site time, is zero. We return to this issue below. The individual's problem is

$$\begin{aligned}&\max_{\mathbf{z}} u(z_1, z_2, q) \text{ subject to}\\&\bar{y} + wt_w \geq \bar{c}z_1 + z_2\\&\text{and}\\&\bar{T} \geq t_w + tz_1.\end{aligned} \tag{4.16}$$

Because the individual is free to allocate his available time between work and leisure in this example, the two constraints can be combined into one:

$$\max_{\mathbf{z}} u(z_1, z_2, q) + \lambda(\bar{y} + w\bar{T} - (wt + \bar{c})z_1 - z_2), \tag{4.17}$$

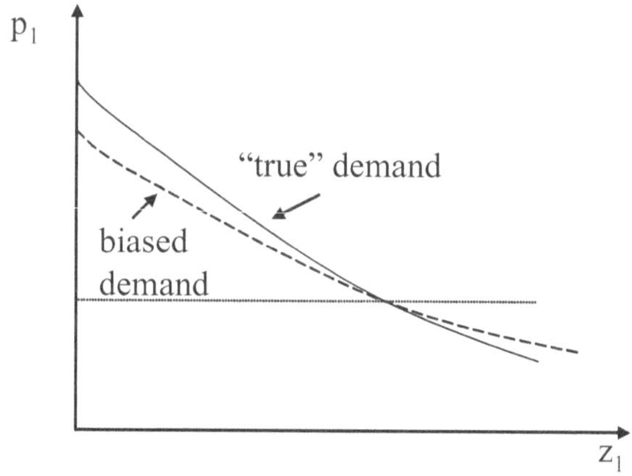

FIGURE 4.1. Direction of Bias when Time Costs are Omitted but Correlated with Money Costs

implying that the opportunity cost of time spent in leisure equals the wage rate. The 'price' of a trip is now the money cost of the trip plus the opportunity cost of time, and the income endowment includes both exogenous income and Becker's full income ($w\bar{T}$) so that $z_1^m = f(wt + \bar{c}, \bar{y} + w\bar{T})$. If time and money costs are correlated, and they usually are in recreational demand settings,[10] omission of the time cost term in the econometric estimation of this demand function would produce an estimated cost coefficient whose absolute value is biased upward, resulting in an error such as that portrayed in Figure 4.1. This bias leads to an underestimate of the consumer surplus for access. The more significant are time costs in the total cost of a trip, the more bias will be introduced. Since the coefficient on costs figures prominently in consumer surplus calculations irrespective of functional form, including it and measuring it accurately is critical to consistent estimation of welfare effects.

The importance of valuing time is apparent. The means of doing so are not. In the problem in equation (4.16), we assumed that the individual faced a constant wage that could be earned for any freely chosen amount of work time. Even when this is an accurate picture of labor market alternatives, defining and eliciting from respondents the 'true' effective wage is not straightforward. For one thing, income taxes and other forces will drive a wedge between wages and the opportunity cost of time. For another, wage rates are difficult to obtain and rarely, if ever, gathered from surveys. At best, respondents report total

[10]Time and money costs will be correlated if individuals with high (low) money costs of access also have large (small) travel times. This feature is characteristic of most recreational demand data because both types of costs are generally functions of distance to the site.

household incomes and often only within broad ranges. These total incomes may be the result of one or more workers' labor and may include non-wage income. McConnell and Strand (1981) argued that, for all of these reasons, an individual recreationist's opportunity cost of time will tend to be less than the reported total household income divided by typical hours worked per year. How much less is an empirical issue and likely to vary across samples, so they suggest estimating this fraction as part of the estimation process. An alternative means of attributing an effective wage rate to an individual was proposed by Smith, Desvousges and McGivney (1983). In their study, an initial hedonic analysis of wage rates using a separate data source allowed the authors to impute wage rates to recreationists on the basis of their location and personal characteristics.

Neither of these approaches addresses a more fundamental problem. A non-trivial share of recreationists in most recreational activities of interest (e.g. swimming, sport fishing, etc.) are not employed at income earning jobs, either because they are retirees, students, homemakers or simply among the ranks of the involuntarily unemployed. Among those that are employed, only some have flexible work alternatives as portrayed in (4.16). In practice, labor market constraints can be quite complex. Individuals often are committed to a fixed work week and fixed vacation allotments from their primary jobs. For some, overtime work is possible at a higher wage rate; for others additional part time work can be gotten at a lower wage rate. Others choose to be unemployed because their value of time is higher than the wage of available jobs. These possibilities suggest discontinuous labor market constraints that may lead to either corner or interior solutions (Bockstael, Strand, and Hanemann, 1987).

Figure 4.2 portrays a variety of labor market situations and helps to illustrate why we often can learn little about the value of time from the effective wage. In this graph income is on the vertical axis and leisure on the horizontal axis. Individuals facing labor market constraints such as those portrayed in Figure 4.2 have the opportunity of earning a wage represented by the slope of the budget constraint between points A and B for a fixed work week of t_F hours. That is, the individual can choose not to work and be at point A or choose to work t_F hours at a constant implied wage rate, but no option in between is available. Employment alternatives available after the t_F hours pay a lower wage indicated by the flatter slope of the line between points B and D, but permit free choice of work time.

For individuals at points such as C in the graph, the time and money constraints can be collapsed into one to form a full income constraint as in (4.17), which accords with Becker's original household production model. The demand for z_1 is a function of full income $(\bar{y} + w\bar{T})$ and full price $(\bar{c} + wt)$, and the marginal value of time is the wage rate. Yet if the individual is found at points A or B, his labor market experience will reveal nothing useful about his value of time. The two constraints do not collapse, leaving the problem as stated

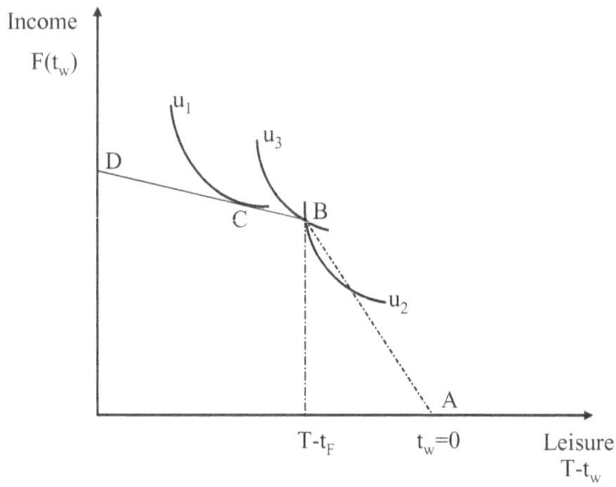

FIGURE 4.2. A Complex Labor Market Constraint with Interior and Corner Solutions

in equation (4.16), and demand is a function of the parameters $(\bar{y}, \bar{T}, t, \bar{c})$, entering separately.[11] This is not to say that individuals at corner solutions in the labor market do not have a monetary value for their time. Rather, their value of time is not equal to the wage rate. To see this, note that individuals whose indifference curves are labeled u_2 would choose to work t_w hours, where $0 < t_w \leq t_F$, but are constrained by institutional inflexibilities to choose all or nothing. They value time at more than the wage rate they receive. Individuals with indifference curves such as u_3 are found at exactly the same solution, but value time at something less than the wage rate they receive but more than the available wage rate if they took an additional part-time job.

To make these more general models operational requires eliciting information about labor market circumstances from respondents, a task which has proved difficult in practice. Beginning with the same theoretical model, Feather and Shaw (1999, 2000) utilize data from a survey that asks respondents a series of contingent behavior questions. The responses to these questions identify individuals at interior and corner solutions and help differentiate u_2 and u_3 circumstances, and when combined with information on wage rates for active

[11]Recognition that time, as well as money, is a scarce resource leads to the conclusion that compensating and equivalent variation could be measured in either a time or money metric. While the latter is generally more useful, the existence of an alternative provides some insights. Alternative projects or policies can be ranked differently depending on which metric is used to calculate benefits if individuals who value one alternative have a distinctly different monetary value of time than those who prefer the other (Bockstael and Strand 1985). Individuals with less available money (time) will have a higher marginal utility for money (time) and will therefore bid less money (time) for a given utility gain.

workers, allow the estimation of a sample selection model. Under the assumption that labor market decisions are longer run than recreation decisions, marginal values marginal values from the labor market can be taken as constant for recreational decisions. Feather and Shaw (1999) use predicted marginal values of time derived from labor markets to estimate recreation demand models.

For some recreational demand problems the model pictured in Figure 4.2 may seem quite reasonable, but for many (especially short trips to local recreational sites) the notion that the recreationist trades time spent visiting recreation sites with work time may seem 'farfetched' (Cesario, 1976). Smith, Desvousges and McGivney argue that time constraints are especially complex as the 'timing of the time' is important. The question is whether free time is available when the demand for certain activities occurs. For example, individuals with fixed work weeks may not be constrained if the recreational activity of interest occurs on evenings or weekends or is seasonal and vacation time is adequate. Smith *et al.* represent time allocation through two types of time constraints, one of which relates only to recreational activities.

4.3.3 Time in Incomplete and Partial Demand Systems

The issue of time is most relevant in recreation demand systems. The systems specified in the empirical literature are rarely more than a demand model for a handful of sites and, therefore, should not be confused with a complete system of demands where expenditures arising from the set of demands equal income. In fact, researchers usually estimate partial or incomplete demand systems as discussed in Chapter 2. Shaw and Feather (1999) raise the very important question of how the demand should be specified when time allocation is a central consideration. Here we review the intersection of issues in estimation of models that involve the allocation of time and incomplete or partial demand functions. These models may involve corner solutions in the labor market but we will avoid complicating the problem with non-linear budget constraints.

Collapsible Time and Income Constraints

The model of the previous section is adapted to include an additional good so that a sub-utility function for recreational activity can be considered. Specifically, z_1 and z_2 are now demands for recreational trips to two different sites and z_3 is the composite commodity, so that

$$\max_{\mathbf{z}} u(\psi(z_1, z_2, q), z_3) \text{ subject to} \tag{4.18}$$

$$\bar{T} = t_w + t_1 z_1 + t_2 z_2 + t_3 z_3 \tag{4.19}$$

$$\bar{y} + w t_w \geq c_1 z_1 + c_2 z_2 + p z_3. \tag{4.20}$$

In this model, the recreation goods z_1 and z_2 form a weakly separable group. Further all goods including the composite commodity z_3 are assumed to have both time costs and money costs. Because the individual is free to allocate time between work and leisure in this example, the two constraints can be combined into one and the choice problem becomes

$$\max_{\mathbf{z}} u(\psi(z_1, z_2, q), z_3) + \lambda(\bar{y} + w\bar{T} - (wt_1 + c_1)z_1 - (wt_2 + c_2)z_2 - (wt_3 + p_3)z_3). \tag{4.21}$$

The complete demand system for this problem is just the analogue of the complete system in Chapter 2 with the time constraint included. The complete system would be

$$z_i = f_i(\bar{p}_1, \bar{p}_2, \bar{p}_3, q, y),\ i = 1, 2, 3 \tag{4.22}$$

where $\bar{p}_i = wt_i + c_i$, $i = 1, 2$; $\bar{p}_3 = wt_3 + p_3$; and $y = \bar{y} + w\bar{T}$. Estimating the complete system–that is, estimating all three demands–is equivalent to obtaining the expenditure or the indirect utility function, and allows one to recover welfare estimates that can be formulated from the preference function in equation (4.18). Of course, this requires that one estimate a complete system of demands for all goods and services, including demand for recreational sites. This imposes practically impossible data demands.

An *incomplete* system might be estimated with observations only on the recreation subgroup with all of the arguments from the complete system. The full specification would be

$$z_i = f_i(\bar{p}_1, \bar{p}_2, \bar{p}_3, q, y),\ \ i = 1, 2.$$

While easier to estimate than the full system, this specification is never observed in recreation demand models. The closest specifications would be those that assume that the price $\bar{p}_3$ is constant across observations. This model takes the form

$$z_i = \hat{f}_i(\bar{p}_1, \bar{p}_2, q, y),\ \ i = 1, 2.$$

Recreation prices are assumed to vary across individuals, and are accounted for. Other prices do not vary and can be subsumed into the parameters of the demand function $\hat{f}_i$. Although the correct measure for income is 'full' income, $y = \bar{y} + w\bar{T}$, one typically finds in its place some measure of nominal income.

Alternatively, the specification of a recreational subfunction within the preference function, as in equation (4.18), leads naturally to the estimation of a *partial* demand system. Consider a two stage budgeting problem in which the recreation subgroup is the second stage. Suppose that y_R is the amount of full income to be optimally allocated to the second stage. Then the partial demand functions will be

$$z_i = g_i(\bar{p}_1, \bar{p}_2, q, y_R),\ \ i = 1, 2. \tag{4.23}$$

These demand functions are consistent with the maximization problem

$$\max_{\mathbf{z}} \psi(z_1, z_2, q) \text{ subject to } y_R \geq (wt_1 + c_1)z_1 + (wt_2 + c_2)z_2.$$

Prices are the full income prices, the same as in the incomplete demand functions, but income is not. The sub-budget y_R is the optimal level of expenditures (in terms of full costs–both money and time converted to money by the wage) allocated to the recreational subgroup, given all of the prices and income faced by the household. This two stage allocation works as long as the second stage choices are feasible–i.e., they don't use more time than is needed in the first stage for income. The solution for the z_i needs to be checked against the time constraint in the complete problem.

One might be tempted to argue that the quantities chosen in the partial system, equation (4.23), consisting of time that might otherwise be used for work, would not provide the income consistent with optimal choices. This, however, cannot be true because the quantities that satisfy the demand curves in equation (4.23) must also satisfy the full maximization problem of (4.21). Hanemann and Morey (1992) show that the welfare effects from the demand curves in equation (4.23) when y_R is held constant are bounded by welfare effects from the complete system.

Time and Income Not Fungible

Now consider these partial and incomplete systems when the time and budget constraints cannot be collapsed. These systems have been examined in detail by Larson and Shaikh (2001, 2004). The two constraints would be

$$T^* = t_1 z_1 + t_2 z_2 + t_3 z_3 \quad (4.24)$$

$$y^* \geq c_1 z_1 + c_2 z_2 + p z_3 \quad (4.25)$$

where now the time constraint T^* is net of work time and income y^* includes exogenous as well as wage income. The complete system would be

$$z_i = f_i(c_1, c_2, p_3, q, t_1, t_2, t_3, y^*, T^*),\ i = 1, 2, 3. \quad (4.26)$$

There are several differences between these demand functions and the demand functions with collapsible time and budget constraints in equation (4.22). First, the time and cost arguments are separate; i.e. c_k and t_k enter separately rather than as the single argument $c_k + wt_k$, the sum of transportation and time costs. Second, income is now income received rather than full income. Third, the time budget is included in the demand function.

Like other complete demand systems, the set of equations in (4.26) is never estimated for recreational demand models. The incomplete version of this model is

$$z_i = f_i(c_1, c_2, p_3, q, t_1, t_2, t_3, y^*, T^*),\ i = 1, 2,$$

where z_3 is simply omitted from estimation. Suppose that the arguments associated with the composite commodity are constant across individuals. Their effects will be incorporated into the demand function parameters, letting us write

$$z_i = \hat{f}_i(c_1, c_2, q, t_1, t_2, y^*, T^*),\ i = 1, 2.$$

This demand function is similar to a number that have been estimated, for example, by Bockstael, Strand, and Hanemann. A model close to this has been estimated by Shaikh and Larson (2002) using the almost ideal demand system.

Now consider partial demand functions with the time and income constraints in equations (4.24) and (4.25). The partial demand functions will be based on the maximization problem

$$\max_{\mathbf{z}} \psi(z_1, z_2, q) \text{ subject to } y^*_{RN} \geq c_1 z_1 + c_2 z_2 \text{ and } T^*_{RN} = t_1 z_1 + t_2 z_2$$

where y^*_{RN} is the optimal amount of expenditure allocated to the recreational sector and T^*_{RN} is the optimal amount of time devoted to the recreational subgroup when time and income are not fungible. This maximization yields the demand curves

$$z_i = g_i(p_1, p_2, q, t_1, t_2, y^*_{RN}, T^*_{RN}), i = 1, 2.$$

This specification, although not typically estimated, appears to be the most plausible for settings where labor-leisure choices are not made smoothly at the wage rate. This model can be interpreted as a conditional demand function, the condition being that demands depend implicitly on quantities in other groups through y^*_{RN} and T^*_{RN}.

4.3.4 On-Site Time and Non-linear Budget Constraints

As if there were not already enough difficulties in properly modeling time costs, there is one more feature of the problem worth considering, one the astute reader would notice was absent from our previous discussion. Time is often so closely aligned with the commodity being 'produced' by the household that the commodity itself is often measured in units of time.

Time spent in on-site recreation (not just time spent accessing the recreational site) is an example. Time on-site is a scarce resource that has an opportunity cost and also a measure of the consumption of the recreational activity. Some researchers have argued that on-site time is not a cost, just a utility generating activity. This view has misleading implications. For example, when access costs are small and on-site time is not considered a part of the time constraint, enthusiastic golfers would be expected to choose a huge consumption of golf activities. Most do not do so because golf takes an appreciable amount of on-site time, which must be reallocated from innumerable other demands. Others have ignored the problem by arguing that on-site time is exogenous and that it is constant over all recreationists. Where the first and not the second assumption seems reasonable, researchers have often estimated different demand functions for recreational experiences of different lengths. Whether one or more demand functions are estimated, researchers have rarely given sufficient attention to the specification of the problem.

To help focus on the on-site time issue, we assume interior solutions in the labor market so that time can be valued at some function of the wage rate, although the results hold up as well when labor market corner solutions are possible. The problem in (4.16) is adapted to reflect the fact that time on-site is appreciable and enters both the utility function and the time constraint. Two dimensions of the recreational experience will now be treated explicitly: the number of trips taken to a site and the amount of time spent on-site for each trip. We will assume for convenience that the amount of on-site time per trip does not change across trips.

The on-site time problem could be modeled in various ways. The most general model allows both trips and on-site time to be choice variables, entering directly into the utility function. Presumably both are weak complements of q. Label the number of trips, x, and time on-site, $\tilde{t}$, where, once chosen, it is approximately the same for every trip. Both money costs and travel time per trip are assumed exogenous and constant at levels c and t, respectively. Assuming an interior solution in the labor market, time is valued at the wage rate, w. We also allow for money costs per unit of time on-site equal to $\tilde{c}$. The maximization becomes

$$\max_{x,\tilde{t},z_2} u(x,\tilde{t},z_2,q) + \lambda(y - z_2 - (c + wt)x - (\tilde{c} + w)\tilde{t}x), \tag{4.27}$$

where y is full income in the Becker sense. It is clear that utility must depend on on-site time. Without this, on-site time would be set to zero. Now replace the parametric expression $c + wt$ by p_x and $(\tilde{c} + w)$ by $\tilde{p}$, so that

$$\max_{x,\tilde{t},z_2} u(x,\tilde{t},z_2,q) + \lambda(y - z_2 - p_x x - \tilde{p}\tilde{t}x). \tag{4.28}$$

If we impose on this general model the requirement that on-site time be truly exogenous, then the problem is quite straightforward. There are certainly cases in which this could be true. If the recreational activity involves participating in an organized nature walk or attending the theater, then the on-site time is determined exogenously to the recreationist. The budget constraint is linear in the two remaining choice variables x and z_2, and the parametric price of a unit of x becomes $p_x + \tilde{p}\tilde{t}$. The proper demand specification takes the form:

$$x = f(p + \tilde{p}\tilde{t}, \tilde{t}, q, y). \tag{4.29}$$

As McConnell (1992) argues, even though $\tilde{t}$ is exogenous, it still enters (4.29) both through the price of the trip and as a utility producing dimension of the trip. Omission of $\tilde{t}$ in the latter capacity will tend to bias the coefficient on the trip 'price' downward, suggesting a steeper demand for trips than is actually true. As long as (4.29) is specified correctly, estimation of the welfare effect of a change in q can be achieved in a straightforward way using the demand for trips. If expression (4.29) were actually the Hicksian rather than the Marshallian demand, then CV would equal the area between the two Hicksian demands, one conditioned on the initial and one on the final level of q. As always, we need to add the caveat that the result holds only approximately when using Marshallian demands (unless there are zero income effects) and then only if the Willig condition holds.

It is surprising to learn that the answer does not change appreciably when $\tilde{t}$ is endogenous, even though this is no longer a standard utility maximization (or expenditure minimization) problem. When on-site time is endogenous, the budget constraint is nonlinear in the three decision variables, z_2, x, and $\tilde{t}$. As a result, a well-defined Marshallian function does not exist for on-site time, $\tilde{t}$, because $-v_{\tilde{p}}/v_y = x\tilde{t}$. Likewise, Shephard's Lemma does not yield the usual results because $\partial m/\partial \tilde{p} = x\tilde{t}$. However, the usual envelope theorem results still hold for x: $-v_{p_x}/v_y = x^m$ and $\partial m/\partial p_x = x^h$. The Marshallian function can be estimated as a function of $(p_x, \tilde{p}, q, y)$ rather than as in (4.29) and the Hicksian as a function of $(p_x, \tilde{p}, q, u)$. Can this help us obtain welfare measures?

Once again we temporarily side-step the difficulties inherent in using Marshallian rather than Hicksian demands and show that, if we had the latter, resolution is easy. CV is defined as

$$CV = m(p_x, \tilde{p}, q^0, u^0) - m(p_x, \tilde{p}, q^1, u^0)$$

which can be shown to equal the change in the area under the Hicksian demand for trips:

$$\Delta area = \int_{p_x^0}^{p_x^*} x^h(p_x, \tilde{p}, q^1, u^0)dp_x - \int_{p_x^0}^{p_x^*} x^h(p_x, \tilde{p}, q^0, u^0)dp_x. \tag{4.30}$$

Expression (4.30) can be rewritten as

$$m(p_x^*, \tilde{p}, q^1, u^0) - m(p_x^0, \tilde{p}, q^1, u^0) - m(p_x^*, \tilde{p}, q^0, u^0) + m(p_x^0, \tilde{p}, q^0, u^0),$$

which equals CV if $m(p_x^*, \tilde{p}, q^1, u^0) = m(p_x^*, \tilde{p}, q^0, u^0)$. This will be true because q is a weak complement of both x and $\tilde{t}$, and when x equals 0, $\tilde{t}$ must also equal 0. No time can be spent on site when no trips are taken to the site.

Pollak and Wachter (1975) were the first to point out that non-linear budget constraints are likely to arise frequently in the context of household production.

For one thing, jointness in production, such as the case above, will generate non-linearity. As they argue more generally 'jointness is pervasive in situations involving the allocation of time', and the allocation of time is central to the household production story.

Pollak and Wachter also note that non-linearity in the budget constraint can arise from a nonlinear cost function for a commodity whose production technology exhibits non-constant returns to scale. In fact, an alternative way of telling the on-site time story might have been to define a household produced commodity, a 'recreational experience', that was produced from the inputs: trips and on-site time in the form of $z_1 = g(x, \tilde{t})$. If the production function for recreational experiences were simply the product of trips and on-site time, then the cost function would be linear in z_1 and there would be no problem. However, this is an unrealistic 'technology', as it implies that a recreational experience made up of one trip lasting 40 hours is exactly equivalent to ten trips each lasting four hours. In this unrealistic case, the cost minimization solution would always involve taking one and only one trip. In general, the problem would seem to imply a nonlinear cost function for z_1, unless on-site time is exogenous.

Once again, we assume a Hicksian world, so as to avoid one layer of complication, and specify the expenditure minimization problem in terms of the nonlinear cost function for z_1 as

$$m(\mathbf{r}, q, u) = \min_{z_1, z_2} z_2 + c(z_1, \mathbf{r}) + \mu(u^0 - u(z_1, z_2, q)).$$

Given information on preferences and the form of the cost function, one can derive a function that expresses z_1 as a function of $\mathbf{r}$ and u, but this is not a conventional Hicksian demand function. One can not directly connect this function back to the expenditure function using Shephard's lemma. There is no parameter that can be used to differentiate the expenditure function to produce a behavioral demand function for z_1. This does not mean that the compensating variation of a change in the public good is not well defined. Compensating variation is defined using the expenditure function as

$$CV = m(\mathbf{r}, q^0, u^0) - m(\mathbf{r}, q^1, u^0). \tag{4.31}$$

The absence of an envelope result for the household-produced commodity eliminates the usual way in which we connect observable behavior with underlying preferences.

If the household production technology embodies an essential input, there may be a practical way of obtaining a welfare measure in the nonlinear budget constraint case. Just, Hueth and Schmitz develop the notion of essential inputs and their usefulness for welfare measurement in the context of the firm. Setting out the problem from the perspective of inputs helps establish welfare results in the household production case as well (Bockstael and McConnell, 1983). The input, x_1, is considered essential in the production of z_1 if no z_1 can be produced when x_1 is not employed. Focusing our attention on the demand for an essential input is helpful, as the demands for inputs are often well-defined even if the demand for the commodity is not.

The result is easy to prove as long as we have an envelope result. That is, we need the Hicksian demand for the essential input to be derivable as $\partial m/\partial r_1 = x_1^h$, where r_1 is a parameter or function only of parameters. Now, evaluating the expression for the area between the Hicksian demands for x_1, conditioned on two levels of the public good, q, gives us

$$\int_{r_1^0}^{r_1^*} x_1^h(r_1, \mathbf{r}_{-1}, q^1, u^0)dr_1 - \int_{r_1^0}^{r_1^*} x_1^h(r_1, \mathbf{r}_{-1}, q^0, u^0)dr_1 \tag{4.32}$$

where r_1^* is the level of r_1 at which the Hicksian demand for x_1 is zero and $\mathbf{r}_{-1}$ reflects all other arguments. Given the envelope theorem result, equation (4.32) equals

$$\begin{aligned}&\{m(r_1^*, \mathbf{r}_{-1}, q^1, u^0) - m(r_1^*, \mathbf{r}_{-1}, q^1, u^0)\} - \\ &\{m(r_1^*, \mathbf{r}_{-1}, q^0, u^0) - m(r_1^0, \mathbf{r}_{-1}, q^0, u^0)\}.\end{aligned} \tag{4.33}$$

Expression (4.31) equals (4.33) if $m(r_1^*, \mathbf{r}_{-1}, q^1, u^0) - m(r_1^0, \mathbf{r}_{-1}, q^0, u^0) = 0$. The set of conditions that are sufficient for this to be true are: a) x_1 is an essential input into z_1 so that at r_1^* the household will not produce z_1, and b) z_1 is q's weak complement so that when z_1 is not produced, changes in q do not matter to the household.

The on-site time problem of (4.28) is a special and particularly interesting case of this result. The price of x (trips to site) is not strictly speaking parametric–it equals $p_x + \tilde{p}\tilde{t}$, where $\tilde{t}$ is endogenous, although the envelope results still hold for x. The dual to the problem in (4.28) is written as:

$$m(p_x, \tilde{p}, q, u) = \min_{x, \tilde{t}, z_2} z_2 + p_x x + \tilde{p}\tilde{t}x + \mu(u^0 - u(x, \tilde{t}, q, z_2))$$

so that we have the necessary envelope result that $\partial m/\partial p_x = x^h$–the derivative of the expenditure function with respect to travel cost equals the level of trips. Substituting p_x for r_1 in expressions (4.32) and (4.33) and interpreting x_1 as the demand for trips, the results described above follow. The fact that we can use the demand for trips to reveal welfare effects of changes in q even in the presence of on-site time, should not bother us, since on-site time must be zero when trips to the site are zero.

As usual this rather neat result holds for Hicksian demands. For the equivalent Marshallian measure to be bounded by CV and EV, the Willig condition must hold. The version of the Willig condition most useful in this case is

(3.23), where the incremental consumer surplus is now defined in terms of that measured using the essential input demand. Expression (3.23) is now rewritten as

$$-\frac{\partial m(p_x,\tilde{p},q,u)}{\partial q} = \frac{\partial \int_{p_x^0}^{p_x^*} x_1^m(p_x,\tilde{p},q,y)dp_x}{\partial q}.$$

This result shows that the idea of an essential input can be used in certain circumstances. Further, we find that in taking care to model time on site we can formulate models that are not difficult to estimate and account for the considerable cost of time spent enjoying the recreational activity.

4.4 Information and Behavioral Change

Revealed preference methods depend on people adjusting to changing circumstances. The appeal of such methods is that they depend on actual behavior. The drawback is that when individuals cannot perceive change, they cannot be expected to alter their behavior. PCB contamination of fish is invisible and if public agencies do not publicize the contamination, then no change in behavior will take place. The PCB case is an apt one, because even after ingestion, consumers have no way of knowing levels of fish contamination without information from expert sources. Revealed preference methods would seem to suggest that if no behavioral change takes place, no gains or losses have occurred. Yet in the fish consumption case, consumers are worse off because their health risks have increased. Should we abandon revealed preference methods when information is imperfect?

In an application that involved similar unobservable changes in health risks, Foster and Just (1989) provided a novel and insightful notion of welfare loss. In the early 1980's high levels of heptachlor, a carcinogen, were found in the fluid milk supply in Hawaii. Pineapple leaves, treated with the highly toxic pesticide, were found to be the source, as these leaves were used as feed for dairy herds. Without knowledge of the carcinogen, consumers did not change their consumption of milk, until ultimately the public health service issued warnings. At that time, dramatic shifts in demand took place.

Naive application of revealed preference methods would suggest that there were no losses to consumers until contamination was made public. Foster and Just argued that not only were there damages before the contamination was publicized, but the damages were all the greater because of the absence of information. In developing their welfare measures they drew on the weak complementarity restriction: the heptachlor contamination only affected households if they consumed fluid milk produced in the islands. Here we paraphrase their arguments, simplifying their story and recasting it in our own notation and retaining their essential idea.

Suppose the household consumes a good, z_1, such as milk, which has at least one quality characteristic that is unobservable. Households have expectations on the level of this unobserved quality, which we call q, and the initial expectations are correct.[12] Consumption decisions depend on the believed level of q and are the solution to the problem

$$v(p_1, q, y) = \max_{z_1, z_2} u(z_1, z_2, q) - \lambda(y - p_1 z_1 - z_2)$$

where z_2 is a numeraire. The dual to this problem is the usual expenditure function, $m(p_1, q, u)$.

Now suppose there is a change in q. If that change is known to consumers, the appropriate welfare measure is the compensating variation, given by

$$CV = m(p_1, q^0, u) - m(p_1, q^1, u). \tag{4.34}$$

This measure will equal the change in the area behind the Hicksian demand function for z_1 because of weak complementarity. However, what if no information about the change reaches consumers? There are two consequences. The first is an expected one–we will not be able to use the behavioral change to calibrate the welfare loss. The second is less expected but conceptually more important. When the consumers are ignorant of the contamination, their losses are actually greater because they have not been able to adjust their exposure to the increased risk.

Foster and Just define the welfare effect of the unrecognized quality change or lack of information as

$$CV_I = m(p_1, q^0, u) - \tilde{m}(p_1, q^1, u\,; z_1^0) \tag{4.35}$$

where z_1^0 is the level of milk consumption that is freely chosen when the household believes quality to be equal to q^0 and $\tilde{m}(p_1, q^1, u\,; z_1^0)$ is a restricted expenditure function equal to the income necessary to achieve the original level of utility after the quality change, when the level of consumption of z_1 is not allowed to change.[13] So this new welfare measure is the difference between

---

[12]Foster and Just consider the case in which the quality characteristic occurs with some variation, so that each consumption occurrence is a draw from a probability distribution that can be described by one or more parameters. The contamination incident alters the true value of the parameters of the distribution. The household possesses expectations on these parameters that were correct before the contamination but remained unchanged until the contamination incident was made public. The authors also consider more complex adjustments until the final stage of resolution is obtained. We simplify the problem to make the key conceptual point.

[13]Foster and Just use a different signing convention, and as a result all their welfare measures will be reversed in sign.

the expenditure function at initial conditions and the restricted expenditure function. The motivation for using this measure is explained by the authors:

> if consumers purchase a product they believe to be safe (because adverse information is withheld) and find out after consumption that it will cause cancer, then they experience essentially the same adverse effects as if they were forced to consume the product with perfect information....Consumers thus incur a cost of ignorance beyond the loss they would experience if free to adjust. (Foster and Just, p 272)

Having established a convincing conceptual measure, how would we obtain it empirically? Short of asking stated preference questions, there is no substitute for observing *some* behavior that reflects a response to changes in q. In the heptachlor case, households ultimately learned about the contamination so that data on their behavioral response could be collected. Even with this information, an additional problem remains–the restricted expenditure function in (4.35) cannot be derived from a demand function. Yet this restricted expenditure function is necessary if one is to calculate the welfare loss during the period of 'ignorance'.

Foster and Just's solution is to begin with a form for the indirect utility and expenditure functions (such as those explored in the first section of this chapter), derive the implied Marshallian demand, and estimate parameters using behavioral data from pre- and post-knowledge periods as a function of the *believed* levels of quality. Having recovered the expenditure function, they then solve for the price of milk, p_1^1, that would cause the individual to freely choose z_1^0 in the face of the new level of quality, q^1. The expenditure function, $m(p_1^1, q^1, u^0)$, must equal the restricted expenditure function, $\tilde{m}(p_1^0, q^1, u^0; z_1^0)$, adjusted by the difference in the amount of money spent on z_1. Either of the two expenditure functions, $m(p_1^1, q^1, u^0)$ or $\tilde{m}(p_1^0, q^1, u^0; z_1^0)$, yields the same consumption choices. Both are consistent with z_1^0 being chosen and if $u = u^0$ and $z_1 = z_1^0$, then the level of z_2 must also be the same in both minimization problems. However, the *values* of the two expenditure functions are different because in the former z_1^0 is purchased at price p_1^1, while in the later it is purchased at price p_1^0. Therefore the value of the actual expenditure function evaluated at p_1^1, q^1 must be adjusted by the difference in expenditures on z_1 in order to be equated to the restricted expenditure function. The restricted expenditure function can be written in terms of the actual expenditure function as

$$\tilde{m}(p_1^0, q^1, u^0; z_1^0) = m(p_1^1, q^1, u^0) - (p_1^1 - p_1^0)\, z_1^0.$$

In the case of contamination, $p_1^1 < p_1^0$ and the restricted expenditure function will naturally be larger than the unrestricted expenditure function. The loss due to contamination *during the period of ignorance*, equation (4.35), can now be expressed as

$$CV_I = m(p_1^0, q^0, u^0) - m(p_1^1, q^1, u^0) + (p_1^1 - p_1^0)\, z_1^0.$$

The loss due solely to lack of information (the 'cost of ignorance') is given by the difference between these losses and the usual CV measure in (4.34) and equals

$$m(p_1^0, q^1, u^0) - m(p_1^1, q^1, u^0) + (p_1^1 - p_1^0)\, z_1^0.$$

In the heptachlor case, monthly losses were estimated to be three to four times as great during the period in which consumers were ignorant of the contamination as compared to the period in which they were informed and could adjust consumption.

The details of this application can be challenged, but the concept behind it is a powerful one. Behavioral response to changes in environmental circumstances provides information to the researcher about how much the individual values the change. It also provides the individual with a means of partially mitigating losses or taking advantage of gains. Losses from degradation will be greater and gains from improvements will be smaller when individuals are prevented from adjusting their behavior in the face of change. Ignorance of changing circumstances can be thought of as the same as restrictions on these behavioral adjustments.

4.5 Quality Changes and Induced Price Effects

So far we have evaluated the welfare consequences of a change in the public good in its role as a quality characteristic of a privately consumed good, assuming that changes in quality have no effect on the price of the weak complement. For those who work mainly with recreation demand models, this implicit restriction probably went unnoticed. To economists who deal exclusively with market goods, ignoring the price consequences of a change in the quality of a good must seem bizarre. Two types of circumstances allow us to assume away price changes. The first is when price is administered and not set in the market, public drinking water being a good example. The second is when the privately consumed good is actually produced by the household. Recreation is the most prominent example of this in the environmental valuation literature. The price of a recreational experience is the cost the household must incur to access the recreational site. A change in the environmental quality at a site will generally have no bearing on the cost of access and therefore will not induce a price change for the privately consumed good. However, there are cases in which changes in the public good–which serves as a quality characteristic of a privately consumed good–do induce changes in one or more prices, and it is these cases that we explore in this section.

Enough confusion surrounds one aspect of welfare measurement that we feel obligated to make a special point of it before moving on to induced price changes. We summarize this point as follows: welfare measurement need only be made in the 'market' or 'space' in which a parameter change takes place. This statement requires explanation. In Chapter 2 we found that the entire welfare effect of a price change could be found as the area behind the demand function between the initial and final prices. This change in price may induce shifts in other demands that are substitutes or complements, but we know from the envelope theorem that welfare measurement need not consider changes in areas behind these other shifting demands. If a change in p_1 is the only parameter change that takes place, then the entire welfare measure can be found as an area associated with the demand for z_1 because $\partial m/\partial p_1 = z_1^h$ and the full welfare measure can be gotten by integrating over z_1^h between the price levels. Similarly, an improvement in an environmental good, q, will shift the demand for a weakly complementary good, z_1, because q acts like a quality dimension of z_1. The welfare effect of this change in q can be measured as the change in the area behind the (Hicksian) demand for z_1 and above its price.[14] The shift in the demand for z_1 might lead to a shift in the demand for some other good, but even if this second shift takes place, no additional welfare measure need be calculated. The area depicted in Figure 3.1 for good z_1 captures the entire welfare effect, as long as q is the only parameter that changes and as long as weak complementarity holds.

To make this more concrete, suppose the quality of the drinking water from a public water system is degraded through a pollution event. If informed, the household's demand for water from this source is likely to shift backwards. At the same time its demand for a substitute such as bottled water is likely to shift outwards. Nonetheless, the welfare effect of the pollution event can be measured in its entirety as the area in Figure 3.1. From the envelope theorem we know that this is a measure of the entire effect, given that the individual only cares about the quality of the public water system if he uses it. There is no additional welfare effect that needs to be measured in substitute good 'space', even though the demand for the substitute good has shifted. This does not deny that *equivalent* measures can sometimes be found in other ways as we saw in the last section. Nevertheless no *additional* measure need be added once we obtain the measure associated with the demand for the weak complement of q.

[14]This result can be generalized to a set of weakly complementary goods, as mentioned earlier, and the arguments in this section will still hold.

4.5.1 Induced Price Changes

Sometimes the environmental quality change can cause so substantial and widespread a shift in demands as to *induce* one or more prices to change. When this happens, other exogenous arguments in the expenditure function (or equivalently, the indirect utility function) change and welfare measurement becomes more complicated. The price of the weakly complementary good (z_1) might change if z_1 is a marketed good and/or prices of substitute (or complement) goods might be induced to change as a result of the initial shift in z_1.

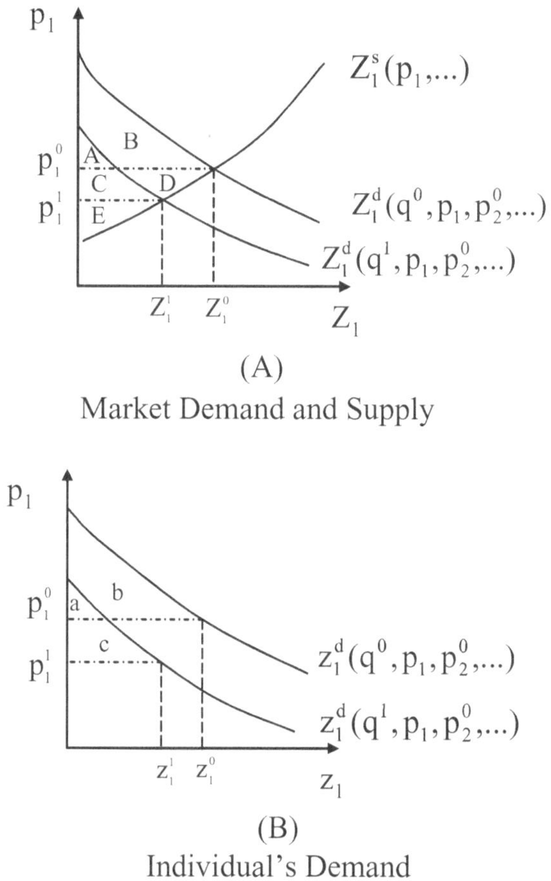

FIGURE 4.3. A Quality Change and an Induced Price Change

As an example of the price adjustment, suppose we are interested in evaluating the welfare consequences of an event that contaminates a food product. In

Foster and Just's paper, heptachlor was found to be contaminating Hawaiian milk products. Since this affected all households in Hawaii, the ultimate result was a shift in the *market* demand for milk and a resulting decline in market price. Here is a case in which the environmental 'bad' (heptachlor contamination) becomes a quality dimension of a privately consumed good, milk. A welfare effect arises because of the change in q, as usual, but the induced price change induces additional welfare effects for both consumers and producers.

This complicates welfare measurement somewhat, but it adds no new conceptual challenges. It also highlights the need to deal with markets as well as individuals. The applications that we have dealt with so far have involved collecting data on household behavior. Now, not only will it be convenient to use aggregate market data, it will be necessary to do so in order to estimate potential price changes.

The leap to aggregate data is simple so long as no non-market interdependencies among agents exist. In the absence of interdependencies, the area between two price levels and behind the aggregate market demand function for a good is simply the sum of the corresponding areas behind each consumer's individual demand for the good between the same prices. Therefore, if the latter is an acceptable approximation to the more desirable compensating variation measure of that price change for the individual, the former will be an acceptable approximation of the sum of all consumers' welfare effects, as long as the estimated aggregate demand function is consistent with aggregation over individual demands. The same is true for the producer side.

Consider what happens when there is a change in q, and q is weakly complementary to a marketed good, z_1. Aggregate data would allow us to estimate the market demand, denoted Z_1 and illustrated in Figure 4.3A. The point (p_1^0, Z_1^0) marks the initial market equilibrium, and the shift in demand brought about by a deterioration in q leads to an induced decline in p such that the new market equilibrium is given by (p_1^1, Z_1^1).

To ensure the correct result for consumers, we define the compensating variation measure using Hicksian demands. First interpret the demand function in Figure 4.3B as a graph of the individual's Hicksian demand before and after the quality change, with the induced price change represented as well. If the CV measure is the change in the effective area behind the Hicksian demand function and above price, then it must equal the area $(a+c)-(a+b)=-b+c$ and is represented by

$$\begin{aligned} & \int_{p_1^1}^{p_1^*(q1)} z_1(p_1, q^1, \mathbf{p}_{-1})dp_1 - \int_{p_1^0}^{p_1^*(q^0)} z_1(p_1, q^0, \mathbf{p}_{-1})dp_1 \qquad (4.36) \\ = \; & m(p_1^*(q^1), q^1, \mathbf{p}_{-1}, u^0) - m(p_1^1, q^1, \mathbf{p}_{-1}, u^0) \\ & -\{m(p_1^*(q^0), q^0, \mathbf{p}_{-1}, u^0) - m(p_1^0, q^0, \mathbf{p}_{-1}, u^0)\}. \end{aligned}$$

When weak complementarity holds, this expression equals

$$m(p_1^0, q^0, \mathbf{p}_{-1}, u^0) - m(p_1^1, q^1, \mathbf{p}_{-1}, u^0).$$

This difference in expenditure functions is by definition the compensating variation of the joint price and quality change. As usual, we observe and estimate Marshallian demand functions. If we accept the substitution of Marshallian for Hicksian demands, given the caveats expressed in the last chapter, the aggregation step yields the aggregate consumer surplus estimate equal to areas $-B+C$ in Figure 4.3A. Adding this to the producer surplus associated with the price change $(-C-D)$ yields $-(B+D)$ as an approximation of the welfare effect of the quality change and subsequent induced price change for both groups.

Induced price changes can occur in other markets besides that for the weakly complementary good. In our drinking water example above, we might find that if the pollution event is extensive enough and the shift in the demand for bottled water significant, then the increase in aggregate demand for bottled water may lead to an increase in its price. In this case only the price of a substitute good is changed, not the price of the weakly complementary good. Nonetheless, this constitutes an additional exogenous parameter change to individuals–both consumers and producers of bottled water–and leads to additional welfare consequences. It would be represented in the expression for welfare changes by additional prices that differ in the initial and terminal expenditure functions.

In a more general case, a change in the quality, q, induces a shift in the demand for z_1 (the weakly complementary good) as well as a shift in a substitute good, z_2. The fact that z_2 might shift with a change in q does not negate the weak complementary status of q. Weak complementarity only means that when z_1 is not consumed, q does not matter. In this example, both goods are marketed goods and the shifts in demand are sufficiently substantive to induce price changes for both goods. The actual path of quality and price changes may be quite complex, since the demand for z_1 is conditioned on the quality of z_1 and the price of z_2, while the demand for z_2 is conditioned on both the price and quality of z_1. We know these parameter changes must be sequenced to obtain the proper welfare measure. We also know that if we had Hicksian demand functions to deal with, the order in which we chose to evaluate the parameter changes would make no difference. In this case, with quality changes as well as price changes, the Marshallian approximation to the Hicksian measures will involve more justification. By reasons of data and inference, we are generally restricted to using Marshallian rather than Hicksian demands. In addition, this more complex setting imposes on the researcher the added burden of accurately identifying and estimating *market* supply and demand curves.

4.6 Conclusions

This chapter concludes our analysis of models of weak complementarity. We have explored a few of the many issues that arise in making the weak complementarity strategy for welfare measurement operational. We begin with the idea that implementation requires the researcher to write down a demand function that depends on a public good. Introducing the public good, which in the weak complementarity story is generally a quality characteristic of a privately consumed good, into the preference structure in a sensible way requires thought. *Ad hoc* specifications of demand functions may not be consistent with weak complementarity or may have other undesirable characteristics and pose particular difficulties in the context of systems of demands.

Further specification problems arise if the weak complement is a household produced good. Many applications of interest to us, such as recreational demand models, are best thought of as household production models and as such involve the additional challenge of accounting for time. Yet a full exploitation of the household production function approach provides additional means of obtaining welfare measures, as well as better insights into the modeling of time in recreational demand studies.

Chapter 5

Measuring Welfare in Discrete Choice Models

5.1 Introduction

In their daily lives, consumers choose discretely–what model of car to buy, which beach to visit, whether to use public transportation, and so on. On the surface, the neoclassical model of preferences and demand appears ill-suited to analyzing such discrete choices or to providing a framework for welfare analysis. Nevertheless, there is a well developed and useful literature on econometrics and welfare measurement in this choice setting. The heart of this literature is the McFadden (1974) random utility model, which found its initial applications in transportation. Hanemann (1978) was the first to develop and apply this approach to valuing environmental and natural resources.

The random utility model, as a basis for discrete choice modeling and welfare measurement, departs from the standard neoclassical model in two ways. First, it models an individual's behavior on a *choice occasion*–that is, it models a single choice among a finite set of mutually exclusive alternatives. This contrasts with the neoclassical model that characterizes consumption decisions as a budget allocation process within a period of time, for example a year. Second, the random utility model incorporates a stochastic term reflecting the researcher's ignorance right from the start, rather than adding it in an *ad hoc* way to the demand function after the entire constrained utility maximization process has been rationalized. Both departures from the neoclassical model give the discrete analysis an element of realism.

Modeling the choice occasion rather than modeling choices across a season also simplifies the conceptualization of welfare measures. Specifying utility in terms of random and deterministic influences makes the modeling ideally suited for the econometric analysis of choices and it provides the probability distributions for the estimates of the pertinent welfare measures.

The genius of the random utility model is the integration of randomness and behavior. Individuals know their preferences and act accordingly. Researchers make mistakes in the specification of models and in observing and measuring variables with error. Researchers work in a world without full knowledge so they can only formulate hypotheses about the probability of behavior.

The discrete choice model handles choices among alternatives successfully without excessive econometric problems and yields welfare estimates with comparative ease. This is not surprising as the discrete choice model is simpler in structure than the models of more complex optimization decisions that give rise to ordinary demand functions. The distinction in conventional demand analysis between first order conditions and demand functions essentially disappears from the discrete choice model.

This simplification, however, comes at the cost. In the context of the utility theoretic version of the discrete choice model, often called the random utility model, one can derive the model of behavior, get logically consistent measures of welfare, and find empirical means of implementation. It does not provide a complete model; the connection between a single discrete choice and the frequency of such choices per period of time is not addressed. When policies or events make the alternatives in a discrete choice more or less attractive, we might expect that some people will want to alter the frequency of choices. Persistent pursuit of a solution has led to a model that is more general in some ways than the discrete choice model, the generalized corner solution model of Phaneuf, Herriges, and Kling (2000). This model construction is a substantial improvement over the random utility model but lacks its ease of implementation and simple interpretation.

Throughout the chapter we use examples from recreation because almost all efforts to use the random utility model to estimate welfare measures are found here. This is not surprising. The random utility model solved a perplexing problem for researchers who were attempting to value changes in environmental quality using the logic of Chapter 3. While it might make perfect sense to view an environmental amenity as a quality characteristic of a recreational experience or trip, such as in the case of water quality and beach use, it is generally difficult to *observe* variation in the demand for trips in the face of differing quality levels. Obtaining data embodying enough variation in quality at a single recreational site, together with behavioral response to this differing quality, is impossible in a cross-section analysis because all recreationists face the same quality characteristics.[1] One type of behavior observable in cross-section analysis held the promise of revealing something about how people valued changes in quality–that of the choice among recreational sites. By studying individuals' choices among available recreational sites with different access costs and different quality characteristics, researchers hoped to reveal how these individuals traded money for improved quality. Hanemann (1978) saw the random utility model as an ideal vehicle for modeling this type of choice.

The literature is replete with examples of random utility models applied to recreational choices such as beach choice, sportfishing mode and location, national park visitation, etc. Although other applications of random utility models can be found in the environmental literature, they tend to be used to explain choices but not to estimate welfare effects. Examples include producers' choice of technology adoption (including farmers' choice among best management practices), commercial fishermen's choice among fisheries and farmers' choice of crops, fuel choice of rural households in developing countries, participation in preservation/conservation programs, and of course transportation mode choice to name just a few.

All of these discrete behaviors have environmental implications, but few have been used to reveal benefits and costs of environmental improvements or degradation. There are some exceptions. Persson (2002) looks at rural household choice among sanitation facilities and values changes in their effective prices. Curtis and Hicks (2000) measure the welfare loss to Hawaiian longliners from a sea turtle protection policy that circumscribes their choice of fishing locations. Hegan, Hauer, and Luckert (2003) model choice among fuelwood collection sites by Zimbabwe villagers, where the estimated welfare effects of site elimination are used to test hypotheses about rent dissipation for an open access resource. Morey, Sharma, and Karlstrom (2003) value introductions of additional health providers to malaria victims in Nepal. There are, no doubt, other non-recreational environmentally-related applications in which random utility models are used for welfare analysis, but they are few relative to the recreation literature.

In the first sections of this chapter we will explore the random utility model and how it is used to obtain welfare measures but will avoid econometric issues to the extent possible. Several works deal solely with discrete choice econometrics, including the comprehensive and accessible book by Train (2003). Subsequent sections of this chapter will address the various solutions to the missing piece of the problem–the frequency of choice occasions. This chapter also reviews the hedonic travel cost model, which is occasionally proposed as

---

[1] Collecting panel data over a long enough period to detect quality changes is often impractical because of recall problems. Besides it would in general be impossible to separate out other factors affecting the demand for trips that would also be varying over time.

an alternative to the random utility model. The hedonic travel cost approach attempts to model the choice among heterogeneous and spatially diverse alternatives, but does so by drawing on hedonic concepts. We explore these arguments and show how this model attempts to characterize choices and estimate welfare effects.

5.2 The Basic Discrete Choice Model

Random utility models begin with an opportunity for choice–the choice occasion. On this occasion, an individual chooses among a set of known alternatives. In the simple version of the model, this choice is modeled as a single event. A choice occasion might be linked to the need to buy a new refrigerator, for example, in which case the individual would choose among alternative refrigerator makes and models. One of the earliest applications of discrete choice models arose in transportation, where the individual chose among modes of commuter travel (e.g. bus vs car vs subway) based on costs and on attributes such as proximity and waiting times. (See for example, Domencich and McFadden, 1975, and Ben-Akiva and Lerman, 1985.) Our examples will most often cast the problem as one of choice of site on an outdoor recreation trip.

To illustrate the nature of the general problem, begin by specifying the indirect utility that arises from a particular choice when the individual chooses among J alternatives. The alternatives are mutually exclusive; only one can be consumed given the decision period. For example, on a given choice occasion, an individual chooses one and only one beach to visit or transportation mode for his commute to work. To derive the indirect utility, suppose direct utility is

$$u(z_1, ..., z_J, \mathbf{q}_1, ..., \mathbf{q}_J, z_{J+1}, \varepsilon). \tag{5.1}$$

Utility depends on the quantity consumed of each of the J alternatives, $z_1, ...z_J$, where only one $z_j > 0$ and all others equal 0. Utility also depend on the attributes or quality characteristics, $\mathbf{q}_1, ...\mathbf{q}_J$, of the alternatives; a composite commodity (with a price of 1), z_{J+1}; and ε, a random vector of tastes, known to the individual but not to the researcher. For most examples, the non-zero z equals 1 unit of that z, although in principle it could equal any number as long as the quantity of the z is not a choice variable of the individual. The consumer's only choices are which of the J alternatives and how much of z_{J+1} are consumed.

The random utility model most naturally characterizes a choice among quality-differentiated goods. Quality is treated as exogenous in the sense that an individual cannot pick an alternative and then change the set of characteristics embodied in the alternative. However, by choosing a particular alternative, the individual is effectively choosing 'exposure' to a specific set of quality characteristics. So in the end, the individual implicitly chooses quality by choosing an alternative, even though the vector of quality characteristics embodied in each alternative is exogenous to the individual.

If the goods were marketed, different alternatives, constituted as different bundles of attributes, would likely have prices reflecting their differing quality levels. As we will see in Chapter 6, prices tend to be bid up for marketed goods with higher quality. This is insured by profit maximizing suppliers and/or competition among buyers for a fixed quantity of a marketed good, as in the housing market case. In the case of non-marketed goods that have the characteristics of a public good, units of the good are not allocated to the mutually exclusive use of individuals. One individual's consumption of the good does not preclude the use of another's, at least until congestion is reached. Particularly in the case of outdoor recreation, it is often nature rather than the market that defines the characteristics of the alternatives. Consequently bundles of different qualities may be arranged randomly. In addition, 'prices' are not market clearing prices but marginal costs from a household production process. As a result, sometimes a low cost bundle will have high quality–a situation that would be arbitraged out of existence if a market were operating. More implications of the absence of a systematic relationship between price and quantity will be evident as we proceed.

The budget constraint for the utility maximization problem in equation (5.1) is

$$y = \mathbf{p} \cdot \mathbf{z} + z_{J+1}, \tag{5.2}$$

but only one of the $z_j^{\prime}s$ will be consumed. From now on we will assume that the fixed amount of that chosen z is one unit (something that could be ensured in any case by redefining the units of measurement). So, a single choice occasion implies a single unit of one of the mutually exclusive alternatives will be chosen. The remaining z's, except for the numeraire, are zero.

Suppose the individual chooses alternative j. The indirect utility conditioned on this choice is

$$v^*(y - p_j, \mathbf{q}_j, \varepsilon_j) = u(0, .., 0, 1, 0..0, \mathbf{q}_1, ..., \mathbf{q}_J, y - p_j, \boldsymbol{\varepsilon}) \tag{5.3}$$

where the '1' is in the jth place because one unit of z_j is purchased, $\mathbf{q}_j$ is the relevant vector of characteristics embodied in alternative j, and $y - p_j$ remains to be spent on the numeraire good and so is substituted for z_{J+1}. We will call this a *conditional* indirect utility function because it is conditioned on the individual having chosen alternative j and because it is a function only of variables that are parameters to the individual. Writing the conditional indirect utility function as a function of $\mathbf{q}_j$ and no other $\mathbf{q}$'s would imply that the other vectors, $\mathbf{q}_k$, $k \neq j$, do not affect utility once alternative j is chosen. This is consistent with a weak complementarity relationship between the characteristics and their associated alternatives: $\mathbf{q}_j$ only affects utility if alternative j is chosen.

The development of the model from the preference function in equation (5.1) and the budget constraint in equation (5.2) is standard fare in the random utility model, but involves a significant departure from typical models. The role of income in the conventional utility maximization model is meaningful because the income allocation process applies to a period of time–for example a year. But the random utility model as described in equation (5.3) applies to a choice occasion, making the role of income no longer so obvious. As we discuss later in the chapter, the role of income as a constraint in a model of choice occasions is ambiguous. Consequently, income effects, when they can be measured, are difficult to interpret.

In conventional utility maximization problems such as those presented in Chapters 2 through 4, the individual solves the problem by adjusting choices so that marginal benefits equal marginal costs. In a choice among discrete alternatives, the utility maximization problem is, at least conceptually, vastly simplified. The individual is viewed as evaluating what his utility would be (on the choice occasion) were he to choose each of the available alternatives. The solution is a simple comparison across the alternatives to determine which is associated with the highest utility. Of course, if there are a large number of alternatives or if it is difficult to learn about the attributes of the alternatives, then the individual's cognitive problem may not be simple, but in concept it is far more straightforward than utility maximization in continuous dimensions. From the modeler's perspective, the simple comparison is probabilistic because the researcher does not know all the factors that the individual takes into account in the comparison. From the start, the researcher models the choice in a stochastic framework.

Under the assumption of uniform sampling of the population,[2] the probability that a given individual is observed by the researcher to choose some alternative k is

$$\Pr(k) = \Pr(v^*(y - p_k, \mathbf{q}_k, \varepsilon_k) \geq v^*(y - p_j, \mathbf{q}_j, \varepsilon_j)) \ \forall j \neq k. \tag{5.4}$$

Further assumptions on the distribution of the errors and the functional form of the conditional indirect utility functions are required to make the probability statement operational. Random utility model applications from the time of McFadden have specified the error as additive and distributed as a type I extreme value. By treating the error term as additive, we divide the model into

[2]The estimation of random utility models, like the estimation of other behavioral models, depends greatly on sampling. For a full treatment of sampling in random utility models, see Manski and McFadden (1981). For a concrete introduction, see Haab and McConnell (2002).

deterministic and random components for the convenience of the estimation:

$$v^*(y-p_j,\mathbf{q}_j,\varepsilon_j)=v(y-p_j,\mathbf{q}_j)+\varepsilon_j.$$

The density function for the type I extreme value error is

$$f(\varepsilon)=\theta e^{-\theta\varepsilon}e^{-e^{-\theta\varepsilon}} \tag{5.5}$$

where $-\Gamma'(1)/\theta$ is the mean and $\pi^2/6\theta^2$ the variance.[3] The size of the parameter θ determines how much the researcher knows of the deterministic component relative to the random component. A lower θ implies a higher variance and a larger mean of the random term. This parameter is typically normalized to one except in nested models (discussed later in the chapter) where it is possible to estimate different values of the parameter across nests.

This distribution was initially employed because of its tractability. Unlike the normal distribution, it has a closed form expression for the cumulative density. Spectacular gains in electronic computing efficiency and greatly improved algorithms have reduced the advantages of simple specifications, and many more complicated forms are now computationally feasible.

A model with additive error and a type I extreme value distribution is called a logit. With these assumptions, one can show that the probability of choosing alternative k is[4]

$$\Pr(k)=\frac{\exp[v(y-p_k,\mathbf{q}_k)]}{\sum_j\exp[v(y-p_j,\mathbf{q}_j)]}. \tag{5.6}$$

This logit probability is the basic behavioral relationship for the random utility model. It implies that *differences* in utilities motivate choices. To show this, rewrite (5.6) equivalently as

$$\Pr(k)=\frac{1}{\sum_j\exp[v(y-p_j,\mathbf{q}_j)-v(y-p_k,\mathbf{q}_k)]}.$$

Only differences in the utility index matter, not absolute levels.

An important consequence of this property of the random utility model is that when the utility function is linear in arguments (as is the most common specification), the characteristics that do not vary across alternatives do not influence choices.[5] Suppose that the deterministic part of the preference function

[3] $\Gamma'(1)=-0.57721$, the Euler constant. For a good discussion of the extreme value distribution, see Ben-Akiva and Lerman (1985).

[4] See Ben-Akiva and Lerman (1985) or Haab and McConnell (2002) for a derivation of this probability.

[5] This implies that characteristics of the individual cannot be included in a straightforward way in the conditional logit. However, it is always possible to introduce such characteristics

is given by $v(y - p_j, \mathbf{q}_j) = \boldsymbol{\alpha} \cdot \mathbf{q}_j + \beta(y - p_j)$. Then the choice will be independent of income because $v(y - p_j, \mathbf{q}_j) - v(y - p_k, \mathbf{q}_k) = \boldsymbol{\alpha} \cdot (\mathbf{q}_j - \mathbf{q}_k) - \beta(p_j - p_k)$. This is a common feature in random utility models and one that in many cases may be desirable. Later, we will revisit this issue and consider the problems that arise when we allow for an income effect.

5.3 Welfare in the Random Utility Model

When individuals face a choice among discrete alternatives, their behavior is a simple comparison among values of conditional indirect utility functions. This means that the researcher estimates parameters of the conditional indirect utility function directly instead of estimating parameters of a demand function solved from first order conditions derived from constrained utility maximization. In consequence, welfare measurement in the discrete choice model is, in one respect, more straightforward than for conventional demand models.

In another respect, it is more complicated though. An appealing feature of the random utility model is that the random component is an integral part of the story. But that makes for greater complexity in defining welfare effects. To define a compensating variation measure, we must first address the special way in which uncertainty enters into the random utility model in characterizing which of the discrete alternatives is chosen under any given set of circumstances.

Let's return to the general case in which we place no restrictions on the functional form for v^* nor on the distribution of ε. Following the analysis of McFadden (1999), the implicit definition for the compensating variation of a change in the quality characteristics of one or more alternatives is

$$\max_{j \in J} v^*(y - p_j, \mathbf{q}_j^0, \varepsilon_j) = \max_{j \in J} v^*(y - CV - p_j, \mathbf{q}_j^1, \varepsilon_j). \tag{5.7}$$

CV is the amount of money taken from income that will equate the utility of the preferred choice after the change in q with the utility of the preferred choice before the change. At first this expression seems incorrect. The right hand side of the expression requires that the maximization procedure be applied after the quality change and after the compensation is paid or taken away. This will not necessarily result in the discrete choice that the individual would have chosen under the new quality regime in the absence of the compensation. But that is exactly the point. In the conventional case, we calculate the CV of a price

---

if one allows the arguments to enter in cross-products with alternative characteristics. For example, one might hypothesize that individuals with different levels of education react differently to some quality characteristic and allow for this by including the product of quality and education as an argument.

change, for example, by first changing price and then adjusting income (through some compensation paid or received) to move the individual back to the original utility level. But in moving him back, we are allowing him to choose his new location freely. We are calculating that amount of money given or taken away that causes him freely to choose to be at the original utility level. He will not typically choose the same amount of the good after the compensation is made as he would after the price change but before the compensation and we do not constrain him to do so in making our CV measurement. So the expression in (5.7) is completely consistent with the conventional notion of compensating valuation.

In notational terms, define the event Λ_{ba} to be the choice of b before and a after the change and after compensation is paid, where CV_{ba} is the compensating variation associated with the change. Because CV is part of the maximized expression, the probability of the event Λ_{ba} occurring is the probability that b solves the left hand side of equation (5.7) *and* a solves the right hand side, where the solution is conditioned on CV. Given that the expression in (5.7) is probabilistic from the researcher's perspective, we will ultimately be looking for the expected value of the compensating variation. In other words, we are looking for

$$E(CV) = \sum_{j=1}^{k} \sum_{i=1}^{k} \Pr(\Lambda_{ij}) \cdot E(CV_{ij}|\Lambda_{ij}). \tag{5.8}$$

This is a simple definition. Expected CV is the expected value of CV conditioned on an event, weighted by the probability that that event occurs, and summed over all possible events. In the general case, where errors are not additive (and, therefore, the random part of preferences can affect the utility of income), the computation of the probability and the conditional expectation equation (5.8) can be difficult. In fact, this is the problem that the literature on non-linear income effects must solve. We will see that three commonly made restrictions on the preference function greatly simplify the calculation.

To help develop some intuition about the nature of the stochastic element in the problem, it is instructive to think about a case that on the surface would appear to simplify the calculation a good deal–when we have a complete set of observations on the behavior of individuals before and after a change. Suppose a regulation is put into place that forces sewage treatment plants to adopt new technology. Sewage overflow events are reduced in frequency, resulting in reduced levels of fecal coliform in the waters off a subset of beaches–those that happen to be near such plants. If we do a *retrospective* analysis of this regulation and have information on the behavior of the full population of beach users before and after the reduction, then are compensating variation estimates of the policy simplified because the researcher no longer has uncertainty about the alternatives chosen? The answer is 'only partially'. If both before and after the regulation, the preferred beach for an individual is one of those not affected by the new sewage treatment technology, then the individual will experience no change in utility and CV will be zero. In this case, welfare measurement is simple because there is no welfare effect. However, suppose that an individual chooses alternative b before and a after the change. Then the researcher knows the choice in the initial situation, but not the value of the indirect utility function because v_b^* is still a function of ε_b. What's more, even though she knows the choice made in situation a, the researcher does not know the choice that would be made if income were adjusted by the compensation necessary to return the individual to the initial utility function. The compensating variation of the change is implicitly defined by

$$v^*(y - p_b, \mathbf{q}_b^0, \varepsilon_b) = \max_{j \in J} v^*(y - CV - p_j, \mathbf{q}_j^1, \varepsilon_j). \tag{5.9}$$

As always, compensating variation is defined as the post-change adjustment in income necessary to equate utility after the change and utility before the change. Note that if one could solve this expression explicitly for CV it would be obvious that CV depends on determinants of utility in the before and after cases, including the random elements.

Let's return to (5.7) and attempt to evaluate the general expression for CV by adding more structure to the problem. Because the expression is stochastic to the researcher, we are effectively looking for the CV that equates the expected maximum utility before the change with expected maximum utility after the change when compensation is paid or received. Expected maximum utility,

$$\tilde{V}(\mathbf{p}, \mathbf{q}, y) = E[\max_{j \in J}\{v^*(y - p_j, \mathbf{q}_j, \varepsilon_j)\}], \tag{5.10}$$

does not in general have a closed form solution. But if we adopt the structure imposed in the last section, such that v^* is additive in the $\varepsilon's$

$$v^*(y - p_j, \mathbf{q}_j, \varepsilon_j) = v(y - p_j, \mathbf{q}_j) + \varepsilon_j, \tag{5.11}$$

and the $\varepsilon's$ are distributed independently and identically type I extreme value, then convenient results follow. For the type-I extreme value distribution of equation (5.5), $E[\max_{j \in J}(K_j + \varepsilon_j)]$ equals the log-sum expression, $\ln(\sum_{j=1}^{J} \exp(K_j)) + \bar{C}$, where $\bar{C}$ is an unrecoverable constant.[6] Therefore the expected maximum utility function in (5.10) is given by the log-sum formula:

[6] Mathematically, the distributional assumption implies that $\bar{C}$ is Euler's constant. However, the absolute level of utility can never be measured, so that (5.12) is true only up to an unknown constant in any event.

$$\tilde{V}(\mathbf{p}, \mathbf{q}, y) = \ln\left(\sum_{j=1}^{J} \exp(v(y - p_j, \mathbf{q}_j))\right) + \bar{C}. \tag{5.12}$$

This is an indirect utility function in that it depends only on parameters exogenous to the individual. In addition it reflects the fact that the researcher does not know with certainty which discrete alternative will be chosen. If, furthermore, the conditional utility function is linear and separable in income so that $v(y - p_j, \mathbf{q}_j) = \boldsymbol{\alpha} \cdot \mathbf{q}_j + \beta(y - p_j)$, then the expected maximum utility function is given by

$$\tilde{V} = \ln\left(\sum_{j} \exp(\boldsymbol{\alpha} \cdot \mathbf{q}_j + \beta(y - p_j))\right) + \bar{C}. \tag{5.13}$$

Now, suppose we wish to measure the compensating variation associated with a quality change for alternative j from $\mathbf{q}_j^0$ to $\mathbf{q}_j^1$.[7] Make the three assumptions above: the conditional utility function is (i) linear in income, (ii) additive in a stochastic element, and (iii) the random error is distributed type I extreme value. The first two assumptions allow us to write v^* as

$$v^*(y - p_j, \mathbf{q}_j, \varepsilon_j) = \beta(y - p_j) + \boldsymbol{\alpha} \cdot \mathbf{q}_j + \varepsilon_j. \tag{5.14}$$

Substituting this expression into (5.7), gives us

$$\max_{j \in J}[\beta(y - p_j) + \boldsymbol{\alpha} \cdot \mathbf{q}_j^0 + \varepsilon_j] = \max_{j \in J}[\beta(y - p_j - CV) + \boldsymbol{\alpha} \cdot \mathbf{q}_j^1 + \varepsilon_j],$$

which implies

$$\max_{j \in J}[-\beta p_j + \boldsymbol{\alpha} \cdot \mathbf{q}_j^0 + \varepsilon_j] = -\beta CV + \max_{j \in J}[-\beta p_j + \boldsymbol{\alpha} \cdot \mathbf{q}_j^1 + \varepsilon_j] \tag{5.15}$$

because the expressions βCV and βy remain constant over the maximization. Note that given the linear structure, we can actually solve for the compensating variation explicitly, but it is still a function of random variables:

$$CV = \{\max_{j \in J}[-\beta p_j + \boldsymbol{\alpha} \cdot \mathbf{q}_j^1 + \varepsilon_j] - \max_{j \in J}[-\beta p_j + \boldsymbol{\alpha} \cdot \mathbf{q}_j^0 + \varepsilon_j]\}/\beta. \tag{5.16}$$

Since the ultimate choice of the individual is unknown to the researcher, the expected value of CV must be computed. To do so, evaluate the expectation

---

[7]Much of the welfare measurement for this model is worked out in Hanemann's well-known working paper (1982). This paper has been reprinted and can be found in Hanemann (1999b).

of the max v^* terms in the above expression. When the error is additive and type I extreme value, this is easily done using the log-sum expression in (5.13). The $\bar{C}$'s cancel and the solution is:

$$E(CV) = [\ln\left(\sum_{j=1}^{J} \exp(\boldsymbol{\alpha} \cdot \mathbf{q}_j^1 - \beta p_j)\right) - \ln\left(\sum_{j=1}^{J} \exp(\boldsymbol{\alpha} \cdot \mathbf{q}_j^0 - \beta p_j)\right)]/\beta. \tag{5.17}$$

Explicit solution of CV, accomplished in the steps implied by (5.15) and (5.16), is possible only when utility is linear in income and independent of the error term. If not, then the classic log-sum measure in (5.17) for the expected value of CV is no longer correct.[8]

Welfare calculations with this preference function are especially easy because the marginal utility of income is assumed to be constant. In the random utility model, this implies only that the marginal utility of income (β) remains constant over the vector of alternatives that could be chosen and over the ranges of the attributes and prices being considered and only within the context of the single choice occasion. When this is not reasonable and the conditional utility function is not linear and separable in income, then the general formulation in equation (5.8) is appropriate. We consider this property in more detail later in the chapter.

5.3.1 More Welfare Calculations with the Linear Model

When the conditional utility function is linear and separable in income, and has an additively separable error term that is type I extreme value, welfare measurement is fairly straightforward. The expected value of CV associated with any change in the vector of attributes of one or more alternatives is given by expression (5.17). This form reflects the fact that the researcher does not know which alternatives were chosen before and after the policy, project or event being evaluated. The welfare calculation takes into account the utility levels that would accrue with each possible choice and the associated likelihood of each choice.

For some problems, expression (5.17) simplifies even further. For example, suppose that a clean-up activity causes quality dimension m to improve by the exact same amount, Δq_m, for all alternatives. Then the compensating variation

---

[8] For example, McFadden (1999) shows that if the conditional utility function is given by $\sqrt{y - p_j} + \boldsymbol{\alpha} \cdot \mathbf{q}_j + \varepsilon_j$, then the welfare measures that are calculated using the expected maximum utility expression are biased compared with the true expected CV.

for this change is defined by

$$\begin{aligned} E(CV) &= \{\ln\left(\sum_j \exp(\boldsymbol{\alpha}\cdot\mathbf{q}_j + \alpha_m\Delta q_m - \beta p_j)\right) \\ &\quad -\ln\left(\sum_j \exp(\boldsymbol{\alpha}\cdot\mathbf{q}_j - \beta p_j)\right)\}/\beta \\ &= \{\ln\left(\left[\sum_j \exp(\boldsymbol{\alpha}\cdot\mathbf{q}_j - \beta p_j)\right]\exp(\alpha_m\Delta q_m)\right) \\ &\quad -\ln\left(\sum_j \exp(\boldsymbol{\alpha}\cdot\mathbf{q}_j - \beta p_j)\right)\}/\beta \\ &= \alpha_m\Delta q_m/\beta. \end{aligned}$$

This welfare measure is just the change in utility from a change in the mth quality divided by the marginal utility of income–simply the monetized value of the change. Since the quality change occurs at all alternatives, the probabilities of choosing different alternatives are not relevant.

There is another type of welfare question that arises frequently in environmental and natural resource applications and that can be treated with ease in this framework. One often wants to calculate the loss from eliminating an alternative (for example, the loss from closure of a recreational site). In a demand model we represent the elimination of a site by driving the price of access to infinity. In the discrete choice framework we need only eliminate the site from the set of alternatives. In the context of a simple random utility model, expected CV is then defined by

$$E(CV) = \frac{1}{\beta}[\ln\left(\sum_{j\epsilon J_{-k}} \exp(\boldsymbol{\alpha}\cdot\mathbf{q}_j - \beta p_j)\right) - \ln\left(\sum_{j\epsilon J} \exp(\boldsymbol{\alpha}\cdot\mathbf{q}_j - \beta p_j)\right)]$$

where J_{-k} is the set of alternatives excluding the eliminated alternative, k. Note that the summation in the left hand side term is only over the J-1 alternatives remaining after elimination of the kth site. The above can be expressed alternatively as

$$E(CV) = \frac{1}{\beta}\ln\left[\frac{\sum_{j\epsilon J_{-k}} \exp(\boldsymbol{\alpha}\cdot\mathbf{q}_j - \beta p_j)}{\sum_{j\epsilon J} \exp(\boldsymbol{\alpha}\cdot\mathbf{q}_j - \beta p_j)}\right] = \frac{\ln[(1-\Pr(k))]}{\beta}$$

where $\Pr(k)$ is the probability of choosing alternative k in the initial circumstance when all alternatives are available. Since $(1 - \Pr(k))$ is less than 1, CV will be negative. The magnitude of the probability of choosing alternative k when all alternatives are available is critical in determining the welfare loss should this alternative be eliminated. As that probability approaches zero, the welfare loss also approaches zero. As the probability approaches one, the welfare loss grows without limit. This apparently extreme result stems from the nature of the random utility model, although it is similar in spirit to Hanemann's 1991 result that the CV of a quantity change in a public good approaches infinity as the elasticity of substitution between the public good and other goods approaches zero. In the framework of discrete choice, one of the alternatives must be selected on any given choice occasion, so the relevant substitution is among available alternatives on that occasion. Hence, if the site that is chosen with (close to) certainty is also the site to be eliminated, the welfare measure becomes undefined. Not choosing an alternative is not an option in this model, though we will later encounter a model that incorporates as a choice the possibility of foregoing all of the J concrete alternatives.

5.3.2 Welfare Measurement with Imperfect Information

For behavior to reveal something about preferences over environmental goods, it must be predicated on accurate information about those goods. As we discussed in the last chapter, if individuals have imperfect information about quality–and in particular if they do not perceive or are not informed of changes in that quality–then their actions will not reveal how they value those changes. Imperfect information complicates welfare analysis but it also adds a new dimension of policy that can be evaluated–the provision or withholding of information.

Many environmental characteristics of resources are difficult to perceive without scientific instruments. Pollutants such as PCB's or carbon monoxide are odorless, tasteless and invisible. While some air and water pollutants produce visible or olfactory clues, others do not. Other environmentally related quality characteristics, such as fish catch at recreational sites, are truly stochastic and their variance may be so large as to make formation of accurate expectations difficult. When perceptions differ from reality, two types of problems arise. First, because behavior will be based on perceived and not actual levels of q, the researcher will need to use perceptions of q in estimating the decision model and recovering parameters of the preference function. But perceptions are typically far more difficult to come by than objective data. Second, welfare consequences will depend on the actual levels of q but at the sites chosen under imperfect information, making the welfare measures in (5.17) erroneous.

Leggett (2002) developed a solution to the problem of imperfect information for the type-I additive error with constant marginal utility. He considered cases in which quality outcomes are drawn from stochastic distributions, much like the Foster and Just (1989) argument, and that individuals may not have perfect perceptions of those distributions when making discrete choices among sites. Here we simplify notation by assuming a scalar indicator of quality, q. When making his discrete choice, the individual expects quality to be $\tilde{q}_j$ at site j but upon arrival finds quality to be equal to q_j. From Leggett's results we learn that the expected maximum utility function in such a case is given by:

$$\tilde{V} = \ln\left(\sum_{j=1}^{J} \tilde{v}_j\right) + \sum_{j=1}^{J} \tilde{\pi}_j (v_j - \tilde{v}_j) + \bar{C} \tag{5.18}$$

where

$$\begin{aligned} \tilde{u}_j &= \tilde{v}(y - p_j, \tilde{q}_j) + \varepsilon_j, \\ u_j &= v(y - p_j, q_j) + \varepsilon_j, \\ \text{and } \tilde{\pi}_j &= \exp(\tilde{v}_j) / \sum_{k=1}^{J} \exp(\tilde{v}_j). \end{aligned}$$

In the last expression, $\tilde{\pi}_j$ is the probability the individual will choose alternative j, based on his *perceptions* of quality at all sites.

With expression (5.18) in hand, we can now evaluate the expected compensating variation associated with a change in *actual* quality levels under different assumptions about perceptions. For example, if perceptions are never accurate–either before or after a change, then a change in the actual quality at one or more sites has expected CV equal to

$$\begin{aligned} E(\widetilde{CV}) &= \frac{1}{\beta}\{\ln \sum_{j=1}^{J} \exp(\tilde{v}_j^1) - \ln \sum_{j=1}^{J} \exp(\tilde{v}_j^0) \\ &\quad + \sum_{j=1}^{J} \tilde{\pi}_j^1 [v_j^1 - \tilde{v}_j^1] - \sum_{j=1}^{J} \tilde{\pi}_j^0 [v_j^0 - \tilde{v}_j^0]\}, \end{aligned} \tag{5.19}$$

where the superscripts indicate initial (0) or subsequent (1) levels of quality. If perceived and actual quality converge in either or both periods, the expression simplifies.

Intuitively, the first line of (5.19) equals the usual expression for expected CV but conditioned on perceived quality levels. The second line provides two correction factors that pertain to behavioral adjustments induced by the difference between measured and perceived qualities. The first adjustment term is the weighted sum of the differences between actual and perceived quality in the subsequent period over all sites, where the weight is the probability of choosing the site. The second is the same weighted sum of differences but calculated using initial period values.

Suppose individuals' perceptions are initially correct but a change occurs that goes unperceived and so individuals continue to act as they did before the change. In this case, initial perceptions will be accurate, $v_j^0 = \tilde{v}_j^0$, but subsequent ones will not be. In fact, $v_j^0 = \tilde{v}_j^1$ because individuals are unaware of the actual change that has taken place. Likewise, $\tilde{\pi}_j^1 = \pi_j^0$ because no change in the probabilities of selection will occur. The resulting expected compensating variation of the change is

$$E(\widetilde{CV}_{\Delta q}) = \frac{1}{\beta}\{\sum_{j=1}^{J} \tilde{\pi}_j^1 [v_j^1 - \tilde{v}_j^1]\} = \frac{1}{\beta}\{\sum_{j=1}^{J} \pi_j^0 [v_j^1 - v_j^0]\}. \tag{5.20}$$

If, subsequently, information is publicly provided about the quality change and no further change takes place, the value of this information (VOI) on a given choice occasion is

$$VOI = \frac{1}{\beta}\{\ln[\sum_{j=1}^{J} \exp(v_j^1)] - \ln[\sum_{j=1}^{J} \exp(\tilde{v}_j^1)] - \sum_{j=1}^{J} \tilde{\pi}_j^1 [v_j^1 - \tilde{v}_j^1]\} \tag{5.21}$$

$$= \frac{1}{\beta}\{\ln[\sum_{j=1}^{J} \exp(v_j^1)] - \ln[\sum_{j=1}^{J} \exp(v_j^0)] - \sum_{j=1}^{J} \pi_j^0 [v_j^1 - v_j^0]\}. \tag{5.22}$$

This expression can be thought of as the value of information because it shows the welfare gain that emerges simply from obtaining full information about a past quality change. As Leggett shows, the sum of the two effects (equations 5.20 and 5.21) equals the welfare effect under perfect information, and is equal to expression (5.17). This result has important implications. If the change is an improvement, then $E(\widetilde{CV}_{\Delta q})$ will be positive. For the VOI measure (5.21) also to be positive, it must be true that $\frac{1}{\beta}\{\ln[\sum_{j=1}^{J} \exp(v_j^1)] - \ln[\sum_{j=1}^{J} \exp(v_j^0)]\}$ can not be less than $\frac{1}{\beta} \sum_{j=1}^{J} \pi_j^0 [v_j^1 - v_j^0]$. But this must be true because the first term equals the CV under perfect information and the second term equals CV

when the choice is constrained to the one made initially. From Le Chatelier's Principle, the constrained choice can not yield greater utility than the unconstrained choice. This implies that the full gains from the improvement will not be enjoyed until accurate information allows individuals to adjust their behavior optimally. Reasoning similarly, if the change is a degradation then $E(CV_{\Delta q})$ must be negative, but $E(\widetilde{CV}_{\Delta q})$ must be more negative, making the value of information positive as we would expect.[9] Any delay in disseminating information when perceptions are incorrect is costly.

5.4 Generalizing Discrete Choice Models

The choice probability implied by the simple logit model written in equation (5.6) exhibits the well-known characteristic of independence of irrelevant alternatives (IIA). This means that the odds of choosing alternative j relative to alternative k are independent of other alternatives:

$$\frac{\Pr(j)}{\Pr(k)} = \frac{\exp[v(y - p_j, \mathbf{q}_j)]}{\exp[v(y - p_k, \mathbf{q}_k)]}.$$

It is easy to think of examples in which this property fails. In the classic example, an individual faces two alternative commuting modes: private car and bus. With the introduction of a new alternative–a second bus route–then one would expect that the probability of choosing the first bus route would decline more than would the probability of driving. With the addition of the new alternative that is more similar to one of the existing alternatives, one would expect a disproportionate shift in demand for the two pre-existing alternatives. But the nature of the logit formulation forces the odds of two alternatives being chosen to be independent of anything having to do with the other alternatives, including the introduction or elimination of other alternatives. Introducing the second bus route within the simple logit formulation forces an equal proportional decline in the probabilities of choosing private car and the first bus route. As Train (2003) points out, if it were possible to take into account in the specification all dimensions of the alternatives that mattered to individuals, then the IIA property would be a natural outcome of the model rather than a restrictive property of it. But we can never do so.

---

[9] In fact, we can show that the value of information is never negative. Expression 5.21 can be written $\frac{1}{\beta}\{\ln[\sum_{j=1}^{J}\pi_j^0\exp(v_j^1 - v_j^0)] - \ln[\exp\sum_{j=1}^{J}\pi_j^0[v_j^1 - v_j^0]]\}$ where $\pi_j^t = \Pr(j)$ and $t = 0$ for before and $t = 1$ for after. This expression will always be non-negative because by Jensen's inequality $\sum_{j=1}^{J}\pi_j^0\exp(v_j^1 - v_j^0) \geq \exp\sum_{j=1}^{J}\pi_j^0[v_j^1 - v_j^0]$.

The IIA property of the logit is closely related to restrictions in substitution among alternatives (see Train, 1999). When we use equation (5.6) to calculate the effect of a marginal change in a characteristic of a site k on the probability of choosing site j, an interesting and very restrictive property emerges. Writing the impact as a *percent* change in the probability of choosing alternative j with respect to a marginal change in characteristic m of alternative k, one finds that

$$\frac{\partial[\ln(\Pr(j)/\Pr(k))]}{\partial q_{mk}} = -\partial v(y - p_k, \mathbf{q}_k)/\partial q_{mk}. \tag{5.23}$$

Note that subscript j does not appear on the right hand side of (5.23). The same expression will result on the right hand side, irrespective of the alternative in the numerator on the left. As a result, the same proportionate change will occur for every alternative (except k) with a marginal change in a characteristic of k. If the change represents an improvement in alternative k, then the probability of choosing k will rise and all other choice probabilities will fall by the same proportion. The way to relax this restriction is to adopt a more flexible preference function. The logit model can be relaxed by adopting a more general random component of preferences from the extreme value family, although one could achieve the same goal with the multivariate normal.

The property of independence of irrelevant alternatives is both a weakness and a strength of the logit model. The weakness is particularly apparent as we explore real world examples with differential patterns of substitutability among alternatives. However, if the real world problem is one in which the IIA restriction does not seem to be a gross violation of the truth this property can be useful. Suppose the model is estimated using existing alternatives faced by individuals and incorporating carefully measured generic characteristics of these alternatives. The results of such an estimated model can then be used to infer demand for a proposed but not yet realized alternative with characteristics that are known or could be designed by public action, such as a new public transportation system (see Domencich and McFadden, 1975).

5.4.1 Nested Models: Relaxing the IIA Property

The nested logit model offers a means of increasing flexibility and reducing the limitations posed by the independence of irrelevant alternatives property. The nested model divides choices into stages much like a tree diagram with branches. The usual terminology, however, is that one decision stage is "nested" within another. Figure 5.1 depicts a typical two level nested structure in which the individual chooses between A1 and A2 at one level based on the possible alternatives available at the second level. The individual chooses among the 'B' alternatives, conditional on having chosen A1; and chooses among the 'C' alternatives, conditional on having chosen A2. Alternatives within a single

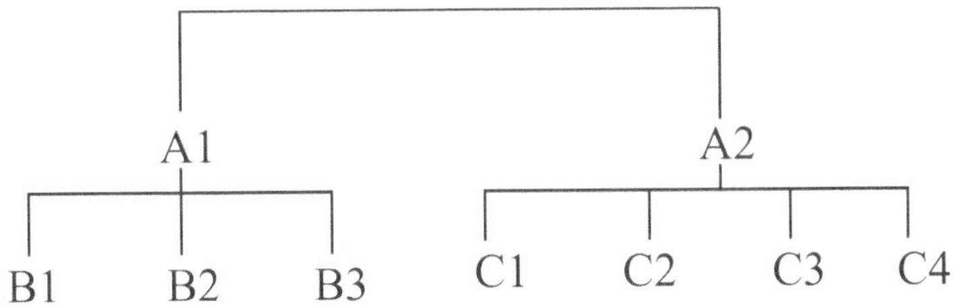

FIGURE 5.1. Example of Nested Decision Tree

branching structure or nest (such as the alternatives B1 and B2) should embody greater similarity than alternatives across branches or nests (such as alternatives B1 and C1). Most important, the IIA property now characterizes choices within nests but not across them, so that the IIA property must hold within sets (A1, A2); (B1, B2, B3); and (C1, C2, C3, C4), but not across these sets. The approach is implemented by assuming a more general distribution than the type I extreme value distribution of equation (5.5).

Figure 5.1 resembles the nesting structure that was present in the first random utility model estimated for recreation.[10] In that model, beach users were viewed as making a binary choice between saltwater (A1) and freshwater (A2) beaches at one decision stage (often called the upper level of choice) and choosing the actual beach to visit within the chosen saltwater or freshwater 'nest' at another decision stage (often called the lower choice level). Intuitively, the nature of the decision process being mimicked is the following. The individual is viewed as choosing the best alternative (the best beach) within each nest (i.e. salt or fresh water) and then comparing the utility obtained from visiting the preferred saltwater beach with that from the preferred freshwater beach in order to decide on the preferred nest (or beach type). By dividing the choices in this way, the independence of irrelevant alternatives assumption need only hold across salt water beaches and across fresh water beaches but not across all beaches. This makes sense since adding a new fresh water beach would not be expected to reduce the probability of choosing *all* beaches proportionally, although it might reduce the likelihood of choosing all salt water beaches by a similar proportion and all fresh water beaches by another proportion.

The nested random utility model is based on conditional indirect utility functions that look like those in (5.4) with the addition of an extra subscript. In the following, n is a nest (e.g. $n = 1$ implies the set of salt water beaches) and j_n is an alternative within the nth nest (e.g. salt water beach j_n). Here we restrict q at any site to be a scalar rather than a vector, so as to avoid extra

[10]The first random utility model in the recreational literature was estimated in Hanemann's 1978 dissertation. The same data set was revisited in Bockstael, Hanemann, and Kling (1987).

notational complexity. In general, the probability of choosing alternative j_n in nest n is given by[11]

$$\begin{aligned} \Pr(n, j_n) &= \Pr(v^*(y - p_{nj_n}, q_{nj_n}, \varepsilon_{nj_n}) \\ &\geq v^*(y - p_{mk_m}, q_{mk_m}, \varepsilon_{mk_m})) \; \forall m \neq n, \forall k_m \neq j_n. \end{aligned}$$

In order to accommodate the greater flexibility in substitution pattern implied by this type of story, McFadden used the generalized extreme value distribution (GEV). The class of models that rely on the GEV distribution has the same advantage as the simple logit choice probabilities in that they generally have closed form solutions. But the GEV is consistent with jointly distributed errors so that different patterns of correlation across alternatives are possible. The nested model stems from one type of GEV model and allows for correlation in ε's within nests but not across them.

To derive the nested model, assume that the random elements of preferences have a generalized extreme value distribution where the cumulative density function is

$$F(\varepsilon) = \exp\left[-\sum_{n=1}^{N}\left[\sum_{j_n=1}^{J_n}\exp\left(-\frac{\varepsilon_{nj_n}}{\theta_n}\right)\right]^{\theta_n}\right]. \tag{5.24}$$

The θ_n's are parameters of the distribution and inversely proportional to the variance. N is the number of nests and J_n is the number of alternatives within the nth nest. In the two-nest structure of the saltwater/freshwater beach example, expression (5.24) would be written as

$$F(\varepsilon) = \exp\left[-\left[\sum_{j_1=1}^{J_1}\exp\left(-\frac{\varepsilon_{1j_1}}{\theta_1}\right)\right]^{\theta_1} - \left[\sum_{j_2=1}^{J_2}\exp\left(-\frac{\varepsilon_{2j_2}}{\theta_2}\right)\right]^{\theta_2}\right].$$

When this distribution prevails, the probability of choosing alternative (n, k_n) in the two nest case is given by

$$\Pr(n, k_n) = \frac{\exp(\frac{v_{nk_n}}{\theta_n})\left[\sum_{j_n=1}^{J_n}\exp(\frac{v_{nj_n}}{\theta_n})\right]^{\theta_n - 1}}{\exp\left[\left[\sum_{j_1=1}^{J_1}\exp\left(\frac{v_{1j_1}}{\theta_1}\right)\right]^{\theta_1} + \left[\sum_{j_2=1}^{J_2}\exp\left(\frac{v_{2j_2}}{\theta_2}\right)\right]^{\theta_2}\right]} \tag{5.25}$$

where v_{ij_i} is the systematic portion of the indirect utility function conditional on choice of (i, j_i). Using this probability statement, one can see that IIA no

[11]For a full study of the nested logit model and various alternatives, see Morey (1999) or Train (2003). Here we try to be intuitive, not comprehensive.

longer holds across nests or branches of the decision tree. For example, the odds of choosing the first saltwater beach over the first freshwater beach are given by

$$\frac{\Pr(1,1)}{\Pr(2,1)} = \frac{\exp(\frac{v_{11}}{\theta_1})\left[\sum_{j_1=1}^{J_1}\exp(\frac{v_{1j_1}}{\theta_1})\right]^{\theta_1-1}}{\exp(\frac{v_{21}}{\theta_2})\left[\sum_{j_2=1}^{J_2}\exp(\frac{v_{2j_2}}{\theta_2})\right]^{\theta_2-1}}.$$

The odds of choosing these two alternatives do not remain constant as another alternative is added to the choice set. The new alternative will either be added into nest 1 or nest 2, changing only the numerator or the denominator, but not both. In this two nest problem, the probability ratio for alternatives from each nest depends on the utilities in both nests.

But what about the odds for two alternatives within a given nest, say alternatives 1 and 2 in the first nest? The odds are

$$\begin{aligned}\frac{\Pr(1,1)}{\Pr(1,2)} &= \frac{\exp(\frac{v_{11}}{\theta_1})\left[\sum_{j_1=1}^{J_1}\exp(\frac{v_{1j_1}}{\theta_1})\right]^{\theta_1-1}}{\exp(\frac{v_{12}}{\theta_1})\left[\sum_{j_1=1}^{J_1}\exp(\frac{v_{1j_1}}{\theta_1})\right]^{\theta_1-1}} \\ &= \frac{\exp(\frac{v_{11}}{\theta_1})}{\exp(\frac{v_{12}}{\theta_1})}.\end{aligned}$$

The odds of choosing these two alternatives are independent of other alternatives in the choice set. Hence, IIA still holds within nests for the nested logit model. Consequently a model with nests restricts the IIA property to within nest choices but does not eliminate it.

The parameter θ_n is a measure of the degree of independence among alternatives within nest n. If $\theta_n = 1$ then the errors within nest n are independent. If all $\theta's$ equal 1,then expression (5.25) becomes a simple logit model. While $1-\theta_n$ is not precisely a correlation coefficient, the correlation between any pairs of alternatives within nest n increases as $\theta_n \to 0$.

It is possible to decompose the choice probabilities in (5.25). By Bayes theorem, the choice probability $\Pr(n, j_n)$ can be shown to equal the product of two simple logits:

$$\Pr(n, j_n) = \Pr(j_n|n) \cdot \Pr(n).$$

To write this out more explicitly, v_{nj_n} must be partitioned into elements that describe alternatives at the lower choice level and those that describe alternatives at the upper choice level. Assuming a linear utility function, v_{nj_n} could be written as $\beta\mathbf{L}_{nj_n} + \gamma\mathbf{D}_n + \varepsilon_{nj_n}$, where the variables in $\mathbf{D}$ vary only between nests but not within them, but those in L vary over all alternatives. The conditional choice probability (conditional on choice of nest n) is

$$\Pr(k_n|n) = \frac{\exp(\beta\mathbf{L}_{nk_n}/\theta_n)}{\sum_{j_n=1}^{J_n} \exp(\beta\mathbf{L}_{nj_n}/\theta_n)} \tag{5.26}$$

and the marginal choice probability is[12]

$$\Pr(n) = \frac{\exp(\gamma\mathbf{D}_n + \theta_n I_n)}{\sum_{g=1}^{N} \exp(\gamma\mathbf{D}_g + \theta_g I_g)}, \tag{5.27}$$

where I_n is called the 'inclusive value' of nest n and is defined as

$$I_n = \ln \sum_{j_n=1}^{J_n} \exp(\beta\mathbf{L}_{nj_n}/\theta_n). \tag{5.28}$$

This is a terrific result. The complex expression in (5.25) is equal to the product of the two simple logit specifications (equations (5.26) and (5.27)), with the inclusive value term (5.28) serving as link between the two pieces. The inclusive value, which takes the form of a log-sum, is calculated using the results of the lower choice level estimation and used as an argument in the upper level choice problem.

Welfare Measures in Nested Models

The change in the distributional assumption for the random components of preferences implies a change in the form of the unconditional indirect utility function. This is because the parameters change the way in which the researcher must take account of uncertainty about preferences. When random preferences have the distribution in equation (5.24), the unconditional indirect utility function becomes

$$\tilde{V} = \ln\left(\sum_{n=1}^{N}\left[\sum_{j_n=1}^{J_n} \exp\left(\frac{v_{nj_n}}{\theta_n}\right)\right]^{\theta_n}\right) + \bar{C}. \tag{5.29}$$

[12] Although θ is sometimes constrained to be equal across nests, the more general case is notated above. Expression (5.27) must be estimated with implicit dummy variables associated with the different $\theta_n I_n$ and only $n-1$ different θ's can be identified.

where $\bar{C}$ is once again an unrecoverable constant.

With approximate constant marginal utility of income across alternatives, the welfare measurement is again computationally simple. Suppose the conditional indirect utility function is given by

$$v_{nj_n} = \alpha q_{nj_n} + \gamma b_n + \beta(y - p_{nj_n}) + \varepsilon_{nj_n}, \tag{5.30}$$

where $\mathbf{L}_{nj_n}$ in (5.26) has been decomposed into q_{nj_n} and p_{nj_n} as both quality characteristics and price would normally vary across all alternatives. The nest-specific attribute vector, $\mathbf{D}_n$, is represented here by one variable, b_n, which varies across nests but does not vary across alternatives within a nest. For example, McConnell, Strand, and Blake-Hedges (1995) in a study of marine recreational fishing, define the lower level choice as a choice among fishing sites and specify it as dependent on such variables as catch rates and costs. The upper level is a choice among modes of fishing. One of the b_n variables equals 1 for the private boat alternative if the individual owns a boat. Now,

$$\tilde{V} = \ln\left(\sum_{n=1}^{N} \exp(\gamma b_n)\left[\sum_{j_n=1}^{J_n} \exp\left(\frac{\alpha q_{nj_n} + \beta(y - p_{nj_n})}{\theta_n}\right)\right]^{\theta_n}\right) + \bar{C},$$

which appears complex but is easy to deal with because the marginal utility of income is treated as constant. The compensating variation measure of a change in q at one or more sites is given by

$$\begin{aligned} E(CV) \;=\; & [\ln\left(\sum_{n=1}^{N}\left[\sum_{j_n=1}^{J_n} \exp(\frac{\alpha q^1_{nj_n} - \beta p_{nj_n} + \gamma b_n)}{\theta_n})\right]^{\theta_n}\right) - \\ & \ln\left(\sum_{n=1}^{N}\left[\sum_{j_n=1}^{J_n} \exp(\frac{\alpha q^0_{nj_n} - \beta p_{nj_n} + \gamma b_n}{\theta_n})\right]^{\theta_n}\right)]/\beta \end{aligned} \tag{5.31}$$

where q^0 indicates initial levels of the attribute and q^1 subsequent levels after the policy, project or event being evaluated.

In moving from the conditional logit to the nested logit, we assumed a more general distribution for the unobserved portion of preferences, implying a richer model of behavior. This is not quite the same as 'tacking on' a different error term. The random component of a random utility model actually stands for a part of the individual's preferences, but a part that is unknown by the researcher. Therefore, the parameters of the distribution of this random component are dictated by the nature of preferences.

5.4.2 Mixed Logit Models: A Further Generalization

In addition to the nested model, a number of other more general models have been developed. For example, one can suppress the IIA property of random utility models entirely by supposing that the random terms are jointly normal. This approach has been examined by Chen and Cosslett (1998) and by Chen, Lupi and Hoehn (1999). But the generalization that has received the most attention of late is the mixed logit model or random parameter logit developed by McFadden and Train (2000). For a thorough analysis of this model, see Train (2003).

The random parameters version of the mixed logit model allows one or more parameters in the conditional indirect utility function to be stochastic. We begin with a simple (non-nested) random utility model where conditional indirect utility is a function of price and a scalar quality measure; and we introduce subscripts on characteristics for the individual (denoted i) to be sure that we are cognizant of the possible variation in these variables across the sample as well as across alternatives:

$$v^*(y_i - p_{ij}, q_{ij}, \varepsilon_{ij}) = \alpha q_{ij} + \beta(y_i - p_{ij}) + \varepsilon_{ij}.$$

The mixed logit allows one or more of the parameters in this model to vary across individuals. Illustrating using the coefficient on quality, the model is now:

$$v^*(y_i - p_{ij}, q_{ij}, \varepsilon_{ij}) = \tilde{\alpha}_i q_{ij} + \beta(y_i - p_{ij}) + \varepsilon_{ij} \tag{5.32}$$

where $\tilde{\alpha}_i$ varies over individuals in some unknown way, but β is assumed constant.[13] This is consistent with a common argument that the greatest source of variation in preferences relates to preferences for quality characteristics. The way in which $\tilde{\alpha}_i$ varies over individuals is unknown, implying that $\tilde{\alpha}_i$ is random to the researcher. Express $\tilde{\alpha}_i$ as $\alpha + \varphi_i$ where φ_i is drawn from the distribution $N(0, \sigma)$. Rewriting φ_i as $\varphi_i = \sigma\mu_i$, where μ_i is a random draw from a standard normal distribution, allows us to write the conditional indirect utility function in terms of the three parameters to be estimated, α, β, and σ:

$$v^*(y_i - p_{ij}, q_{ij}, \varepsilon_j) = \alpha q_{ij} + \beta(y_i - p_{ij}) + \varepsilon_j + \sigma\mu_i q_{ij}.$$

Because the parameter, $\tilde{\alpha}_i$, is random, but assumed to be distributed independently of the extreme value error term, the probability that individual i chooses

---

[13] In theory it is feasible to make all parameters random, but a random coefficient on the travel cost parameter (i.e. the marginal utility of income) tends to induce especially large dispersion in welfare measures. We will have more to say about this, as well as examples of other ways to introduce variation in the marginal utility of income, later in this chapter.

alternative k becomes

$$\Pr(i \text{ chooses } k) = \int_{-\infty}^{\infty} \frac{\exp(\alpha q_{ik} + \beta(y_i - p_{ik}) + \sigma\mu q_{ik})}{\sum_j \exp(\alpha q_{ij} + \beta(y_i - p_{ij}) + \sigma\mu q_{ij})} f(\mu) d\mu \qquad (5.33)$$

where $f(\cdot)$ is the unit normal density. The model no longer exhibits IIA because the ratio of the probabilities of choosing two alternatives no longer reduces to a simple function of the ratio of the conditional indirect utility functions of only those alternatives. Further, the substitution pattern varies, depending on the alternatives.

Train (1999) illustrates this model with an application from a study of river fishing in Montana. Utility for a site depends on seven characteristics (and a correction for aggregation over sites, which we will ignore here). For three of the characteristics, the direction of the impact on utility is known *a priori*: increases in travel cost have a negative effect on indirect utility while increases in fish stocks and an aesthetics rating have a positive effect. Hence the distributions of these parameters can be restricted as non-positive for travel cost and non-negative for fish stocks and aesthetics. The other four measured characteristics are attributes that might appeal to some anglers but not to others, for example the availability of campsites. Changes in these four characteristics can have positive or negative effects on utility and thus the distributions of their random coefficients are not restricted.

Restrictions on the distributions of the parameters must be determined by *a priori* knowledge. For cases where the direction of the impact is known, the parameters are distributed lognormally; for others the distributional assumption is that of normality. In the case of unrestricted parameters, the results reveal the proportion of the sample that places a positive value on the attribute. For example, Train (1999) finds that about 50 percent of anglers preferred sites with campgrounds, while the remainder did not.

Breffle and Morey (2000) provide an application of the mixed logit, but they also introduce heterogeneity in preferences in another way as well. In their model, Maine salmon fishermen choose among several Atlantic salmon fishing sites along the Maine and Canadian coastlines. Response to expected catch rates is hypothesized to vary over fishermen in their model as it is in Train's, but in the Breffle and Morey model the heterogeneity is captured by interacting catch rates with known angler characteristics. In this way, the authors explain the heterogeneity systematically rather than allowing it to be entirely random. Breffle and Morey do use the random parameters feature of the mixed logit to model the effect on site choice of whether the site is in Canada or the U.S., however. Preferences for Canadian over U.S. sites are presumed to vary randomly over anglers because the authors have no hypotheses about the factors that would affect the willingness of fishermen to cross over into Canada to fish.

The mixed logit model is more intuitive when interpreted as an error components model. As we saw in the last section, patterns of correlation among the ε's can be introduced into the conventional random utility model by nesting the decisions. Alternatives within the same nest are allowed to be more similar than alternatives in different nests. The mixed logit can be used to accomplish the same thing and more. Start with the conditional indirect utility function represented in (5.11), but write it as $v_{ij}^* = \boldsymbol{\beta} \cdot \mathbf{X}_{ij} + \varepsilon_{ij}$, where j indexes the alternatives and i the individual making the choice. In the standard logit, the ε_{ij} are assumed to be distributed independently and identically extreme value, but in the mixed logit ε_{ij} could be replaced by some $\Psi_{iH} + \tilde{\varepsilon}_{ij}$, where $\tilde{\varepsilon}_{ij}$ is now distributed independently and identically extreme value. The random error, Ψ_{iH}, would be defined across whatever grouping structure one might wish, where the subscript H is an index of sets of alternatives grouped together. Ψ_{iH} would be constant within any group of alternatives but would vary across groups, for any individual i. The error structure in the mixed logit can mimic a variety of nesting structures and correlation patterns, including overlapping nesting structures (Train, 1999; Herriges and Phaneuf, 2002). A second useful role arises when the researcher has panel data–that is, data on decisions made by a cross-section of individuals on multiple choice occasions. This is the setting of the repeated logit model, in which individuals' choices over a season are modeled. The mixed logit model can be set up so that a component of the error is constant across time for any individual, but varies over individuals (see the repeated mixed logit model of Herriges and Phaneuf).

Motivating the model from an error components or random parameters perspective would have been feasible years ago, but computation of (5.33) would not have been. Because the probability does not have a closed form solution, parameters must be estimated by simulation. (Again, see Train, 2003, for details.) Improvements in methods for estimation by simulation and faster computer speeds make the estimation of the model practical on a routine basis today and is incorporated as a standard procedure in a number of econometric packages.

Welfare measurement for the mixed logit model and the simple random utility model are based on the same principle, but the former requires more work in practice. For the mixed logit model specified in (5.32), the conditional indirect utility function is now based on iterated expectations:

$$\begin{aligned} \tilde{V} &= E_{\tilde{\alpha}} E_{\boldsymbol{\varepsilon}} \max_{j \in J} \{\tilde{\alpha} q_{ij} + \beta(y_i - p_{ij} + \varepsilon_j)\} \\ &= E_{\tilde{\alpha}} \ln \left\{ \sum_j \exp(\tilde{\alpha} q_{ij} + \beta(y_i - p_{ij})) \right\} + \bar{C}. \end{aligned} \tag{5.34}$$

Even when the preference function is linear in income, the expectation over the random parameter has no closed form solution and so must be solved numerically or by simulation. To get a sense of this, suppose that the $\tilde{\alpha}$ is distributed normally. Then $E_{\tilde{\alpha}} \ln\{\sum_j \exp(\tilde{\alpha} q_{ij} + \beta(y_i - p_{ij}))\}$ is of the form $\int \ln\left(\sum_j \exp(\tilde{\alpha} A_j + B_j)\right) f(\tilde{\alpha}) d\tilde{\alpha}$ where $f(\tilde{\alpha})$ is a normal density. When preferences are linear in income and the coefficient β is assumed constant, the calculations are considerably simplified because the $\beta(y_i - p_{ij})$ term can be factored.

5.5 The Larger Consumer Choice Problem

Despite the careful construction of the random utility model for discrete decisions, this model is not part of a larger model of consumer choice. In the standard model of consumer choice, outlined in Chapters 2 through 4, a consumer chooses the amount of each good purchased at fixed prices subject to a known budget constraint. The fact that the random utility model does not treat this full problem leaves two loose ends. One is that of modeling the *quantity* of the quality-differentiated good that is actually consumed. In the random utility framework this translates into making the number of choice occasions endogenous. The other involves the role of income.

5.5.1 The Role of Income

Allocation of a fixed income over a set of goods is the classic way of stating the general problem of consumer choice. In the random utility model, income has no effect on choices when the marginal utility of income is constant across alternatives. This is the implicit assumption when the $\beta(y - p_j)$ term enters the conditional indirect utility function linearly. Because only differences in the conditional indirect utility functions (the v^*'s) matter, income disappears from the problem and, conveniently, need never be defined or measured. As we have seen in previous sections, having conditional utility be linear and separable in income is necessary to derive the log-sum expressions (5.17) or (5.31) for expected compensating variation.

Including income nonlinearly raises practical and conceptual issues. For one thing, the appropriate measure for income must now be defined. The decision problem is a *choice occasion*, not a time period within which the individual can repeatedly make purchases of an array of consumer goods.[14] Presumably the

[14]The difficulties we allude to are obvious in most recreation demand models. The random utility model has been used in other settings, however, including models of firm decisions. In such cases, the choice occasion may align itself well with a period in which returns are normally measured. This would be true, for example, if the discrete decision was among production technologies.

income available for a choice occasion would come from an allocation function, a sort of two stage budgeting that accounts for the effects of the prices and qualities of other goods and provides a proportion of income as a sub-budget for the choice occasion. But even this is awkward and difficult to define convincingly. For the most part, studies that incorporate income effects typically use some rough measure of household income as a proxy for the available budget.

There are also calculation problems to overcome. When income enters the preference function nonlinearly, a closed form solution for expected CV does not exist. One apparent way to resolve this problem is to compute (by iterative methods) the CV that equates *expected* maximum utility before and after the change:

$$E[\max_{j \in J}\{v^*(y - p_j^0, \mathbf{q}_j^0, \varepsilon_j)\}] = E[\max_{j \in J}\{v^*(y - CV - p_j^1, \mathbf{q}_j^1, \varepsilon_j)\}].$$

This is the approach adopted in Morey, Rowe and Watson (1993), but McFadden (1999) has since shown that the CV estimate obtained in this way will in general be biased. McFadden's alternative is computationally costly, requiring repeated draws from a generalized extreme value distribution and iterative solutions to an implicit function.

Given the problems posed by including income non-linearly, it is worth investigating more carefully the meaning of the income effect for a choice occasion, and worth considering the likelihood that it will be significant. What does an 'income effect' mean in the context of a random utility model? In conventional demand analysis, a positive income effect implies that as real income rises, *ceteris paribus*, an individual can be expected to consume more of the good. But there is typically no quantity dimension in the random utility framework. Here, an income effect means that the same individual facing the same alternative set and alternative attributes would make a different choice at different levels of his real income. Intuitively we might expect income effects to be less likely or less significant in a random utility setting than in a conventional model, for the simple reason that individuals are constrained in the random utility model to select one and only one alternative from the choice set.

Instead of relating to a quantity decision, the income effect in the random utility context is tied closely to the nature of the choice set. From the individual's perspective, real income is unlikely to change dramatically with either a change in a quality characteristic or the elimination or introduction of an additional alternative, unless alternatives are of dramatically different costs. For example, if the behavior is recreational beach use and the choice set includes the array of sites available to the individual for a day trip to the beach, then we would not usually expect the recreationist's income to have a major effect on the choice of site. However, if the choice set included both local beaches as well as an array of beaches in the Caribbean, for example, then the choice may well be affected by the individual's income. But the structure of this latter problem is troubling on other grounds, since multiple day trips to the Caribbean are substantively different from day trips to local beaches, and this raises among other problems a violation of the independence of irrelevant alternatives property.

Some support for the conjecture that income effects tend to be negligible in recreation demand models is provided in Herriges and Kling's (1999) application in which they calculate welfare effects from a linear model and two nonlinear models, solving for CV using both the Morey, Rowe and Watson (1993) and the McFadden approaches. They also investigate different nesting structures (i.e. different error structure assumptions). In the application, southern California anglers choose among four sportfishing modes (beach, pier, private boat and charter boat) and the welfare effects are calculated for three scenarios: doubling price for all modes, doubling catch rates for all modes, and eliminating the beach and pier modes. The first two scenarios represent substantial changes, relative to most policy effects. And the third scenario eliminates the two low cost alternatives. For such extreme scenarios, one might expect to find income effects, if there are any to be found in recreational applications. However, for the price change and the elimination of alternatives scenarios the authors find no substantive differences in CV estimates between assuming a linear model and using either of the methods for approximating CV when utility is nonlinear in income. For the quality change case they do find non-zero differences in CV due to the linearity assumption, but these differences are smaller than those that arise due to different nesting structures.

Empirically, the only way to learn about the effect of income on choices is to look at the behavior of different people with different incomes, as we never observe choices by the same individual when in possession of different income levels. This introduces the possibility of significant bias–and misinterpretation of the results. By allowing income to enter the utility function non-linearly and by capturing changes in income by looking across people with different incomes, we can easily pick up what *looks* like an income effect, but is not. Any measure of income will be correlated with socioeconomic status and education, and thus may be a proxy for preferences rather than a budget constraint. If so, then finding that nonlinear income arguments succeed econometrically may not be evidence of a true income effect. When the prices of different alternatives do not vary dramatically, it is difficult to see why this effect *should* be due to an income effect–especially since the choice occasion is conditioned on each individual choosing no more nor less than one alternative. Admittedly, the existence of an appreciable income effect should be an empirical question, but uncertainty regarding how income should be measured and how it can be disentangled from differing preferences that are correlated with income complicates any empirical test.

Morey, Rowe and Watson provide an example where income influences choices but does not imply a gap between CV and EV for the given policy change being evaluated. In estimating a repeated nested logit model, they find that income, as it varies over individuals in the sample, is a significant variable in explaining which alternative gets chosen. However, they also show that for any given individual the difference between the EV and CV measures for a hypothesized doubling of catch rates is quite small, typically in the 2% range. Hence, while there is evidence that income is an important determinant of choices across individuals, the income effect is not large enough to induce differences in CV and EV for the particular change being evaluated. In this empirical role, any socioeconomic characteristic highly correlated with income could serve as well.

A plausible way of allowing people with different socioeconomic status (i.e. different incomes) to have different preferences, without encountering the difficulties of incorporating income in a nonlinear fashion, is to include a series of dummy variables for different income ranges that interact with the characteristics of alternatives, including price. This is the approach suggested by Morey, Sharma, and Karlstrom (2003). They specify conditional indirect utility as a piecewise linear spline function with the result that the marginal utility of income becomes a step function. The simple logit formulation for a choice probability becomes

$$\Pr(k,i) = \frac{\exp[\alpha q_{ki} - \sum_{h=1}^{H} \beta_h D_{hi} p_{ki}]}{\sum_j \exp[\alpha q_{ji} - \sum_{h=1}^{H} \beta_h D_{hi} p_{ji}]},$$

where i indexes the individual and j and k index alternatives. $D_{hi} = 1$ if household i's real income falls in range h of H possible real income ranges and zero otherwise; and β is allowed to vary over these H income groups. While an appealing approach in many respects, the usual expected compensating variation measure of expression (5.17) is now only approximately correct. $E(CV)$ has an exact specification only when β is constant. Depending on the size of the change being evaluated and the particular random drawn of ε_{ij}, an individual could, with some positive probability, move from one income range to another when the change takes place.

Morey, Sharma and Karlstrom apply this model to the choice of health providers among malaria victims in Nepal, using only two categories of income–below and above the poverty level. In their application, the errors from using the expected CV formula in (5.17) are simulated to be less than 1% for 95% of the population and less than 3.3% for everyone. In contrast, the gains from the piecewise linear model are shown to be considerable, and the households below poverty level estimated to have a substantively higher marginal utility of income than others.

5.5.2 The Frequency of Choice

In many applications of random utility models, researchers analyze the choice among alternatives on a given choice occasion, but not the number of choices. In one of the earliest applications of the random utility model–the commuter's transportation mode choice–the assumption that the frequency of choice is exogenous seems perfectly reasonable. Frequency of choice is given by the number of days of work.[15] The same cannot be said for recreational demand. When the random utility model was first used in valuing environmental and natural resources, researchers recognized its power to explain choices among alternatives, but also understood the problem posed by repeated choices. There seemed no way to make the model consistent with the more familiar model of choice in which the consumer selects the number of units of a good by maximizing utility subject to a budget constraint. The development of an empirically tractable and utility theoretic model that incorporated both the choice among alternatives and the frequency of choice appeared daunting. In this section we review the progress that has been made on this problem, as well as alternative approaches that researchers have adopted.

Consider the beach example once again. A household typically choosing a large number of beach visits in a given season might curtail its beach use in the face of substantial deterioration in the quality of all beaches or even of its favorite beach. For example, accumulation of *PCB*'s along the coastline of an area might reasonably be expected to lead to a precipitous decline in the total number of beach visits by local residents, effectively a decline in the frequency of choices. A framework that only models the choice among alternatives, conditional on choosing some beach, will obviously not lead to good welfare measures of the damage from PCB's because it will miss the most important dimension of behavioral change. In initial attempts to apply the model to this sort of problem, the researcher would typically calculate the welfare measures in (5.17) or (5.31), depending on the type of random utility model employed, and these would be estimates of the compensating variation of some change per choice occasion. But to obtain a seasonal or annual benefit measure, the researcher would have little choice but to multiply the per choice occasion benefit measure by the actual number of beach trips taken by each user, ignoring the possibility that the degradation of beach quality might

[15] Even in the model of transportation choice, the number of choice occasions can be thought of as endogenous if the changes in characteristics are sufficiently extreme. Very large increases in the cost of commuting could lead to a reduction of the number of work days, although a more likely response might be a job change.

decrease the number of beach trips.

Researchers initially attempted to incorporate changing numbers of trips into the model based not on a revision of theory but on establishing an empirical connection between a demand model for the number of trips taken and the characteristics of the alternatives available. Two approaches that give almost identical results but appear to be completely different modeling approaches are the linked model first developed by Bockstael, Hanemann, and Kling (1987) and the repeated nested logit model of Morey, Rowe and Watson (1993).[16] The linked approach, modified by Hausman, Leonard and McFadden (1995), is a two step approach in which one first estimates the random utility model (in either the nested or simple conditional logit form). With the estimated parameters of the logit in hand, one then calculates the value of the expected maximum utility, as in equation (5.12) or (5.29), where the expected maximum utility is

$$\tilde{V} = \ln\left(\sum_j \exp(v(y - p_j, \mathbf{q}_j))\right) + \bar{C}.$$

Once $\tilde{V}$ (exclusive of $\bar{C}$) is calculated for each individual, it is then included as a regressor in a model that attempts to explain the total number of trips the individual takes.

A linked model explaining the number of trips by individual i might have the form:

$$z_{Ri} = f(\tilde{V}_i, \mathbf{s}_i), \tag{5.35}$$

where z_{Ri} is the total number of recreational trips by individual i and $\mathbf{s}_i$ represents individual characteristics such as income, age and education that would not easily be included in random utility models.[17] The idea here is that $\tilde{V}(\mathbf{p}, \mathbf{q}, y)$, representing as it does the expected maximum utility possible under the price and quality conditions existing for available attributes, can be used as an index of the desirability of beach use to the individual. If a dimension of q were to change at one or more sites, then the calculated value of $\tilde{V}$ would change, indicating a change in the expected maximum utility that the

---

[16] These approaches are analyzed in detail in Parson, Jakus and Tomasi (1999).

[17] As we saw earlier, any variable that enters linearly into the conditional indirect utility function and does not change over alternatives does not affect the choice among alternatives. Unless interaction terms are entered between individual characteristics and alternative characteristics, individual characteristics will cancel out. Binary random utility models often include individual characteristics. These variables do not vary over alternatives, but they are incorporated implicitly as individual specific variables crossed with alternative specific dummy variables. Because only $n - 1$ such dummies can be included in a random utility model, where n is the number of alternatives, the individual characteristics appear associated with only one of the alternatives in the binary model.

individual facing these new circumstances could enjoy.

Bockstael, Hanemann and Kling estimated such a model and then approximated total annual (or seasonal) net benefits for a given beach user by multiplying the usual formula for expected CV, equation (5.17) or (5.31), times the predicted trips averaged over the before and after scenarios. With this approach the annual CV for individual i is estimated to be $E(CV_i) \times [(\hat{z}^1_{Ri} - \hat{z}^0_{Ri})/2]$ where $\hat{z}^0$ and $\hat{z}^1$ are predicted using the estimated parameters from (5.35) and before and after values of $\tilde{V}_i$. Other approximations are also clearly possible, such as $(\hat{z}^1_{Ri}\tilde{V}^1 - \hat{z}^0_{Ri}\tilde{V}^0)/\beta$.

Hausman, Leonard and McFadden made two modifications to this model. First, they normalized $\tilde{V}$ by $(-\beta)$, calling the resulting term a 'price index'. This is a simple rescaling of the argument used by Bockstael, Hanemann and Kling, as the marginal utility of income, β, is constant. They also estimated the 'demand' model as a count model–specifically as a Poisson, so that the expected total number of trips, z_R, becomes

$$z_{R_i} = \exp(\gamma(-\tilde{V}_i/\beta) + \boldsymbol{\delta}\mathbf{s}_i). \tag{5.36}$$

The second approach to modeling the total trip choice, the repeated nested logit model of Morey, Rowe and Watson, looks altogether different. However, as Parsons, Jakus and Tomasi (1999) show, it is almost identical in structure to the Hausman, Leonard and McFadden model and when identical functional form and error distribution assumptions are made, produces exactly the same welfare measures. The repeated nested logit expands the choice set to include not only the choice among alternatives, but also the decision of whether to choose any of the alternatives (often called the 'participation decision'). A choice occasion is now any opportunity to decide whether to go on a trip and, if so, which site to choose. In this expanded problem, the *choice occasion* cannot be made endogenous, but the outcome of each choice occasion is endogenous, including whether or not to take a trip. Making the outcome of the participation decision a function of the characteristics of alternatives makes the *frequency* of the conditional choice among alternatives a function of these characteristics as well. The model generally takes the form of a nested structure such as in (5.25): at one level the individual decides whether to participate and at another level, given participation, the individual chooses among site alternatives. To make this work, the researcher must posit a given number of (potential) choice occasions or choice 'opportunities' per year.

Because this 'participation' decision is embedded in the model, the expected number of actual trips emerges as part of the results. The number of visits will be the number of choice opportunities times the probability of taking a beach trip on a given choice opportunity, where the latter depends on the log-sum formula (the inclusive value term) calculated from the site choice decision. This is the equivalent of calculating $\tilde{V}$ for the site choice level of the nested structure only. The expected number of trips to all sites by individual i is then given by

$$Z_{R_i} = \Upsilon_i \cdot \Pr(participate_i), \tag{5.37}$$

where Υ_i is the exogenously given number of choice opportunities for the individual.

The probability of participating on a choice occasion is often estimated as a binary choice logit of the general form:

$$\begin{aligned} \Pr(participate_i) &= \frac{\exp(\eta\tilde{V}_i)}{\exp(\eta\tilde{V}_i) + \exp(\boldsymbol{\tau}\mathbf{s}_i)} \\ &= \frac{1}{1 + \exp(\boldsymbol{\tau}\mathbf{s}_i - \eta\tilde{V}_i)}, \end{aligned} \tag{5.38}$$

where i indexes the individual or household, $\mathbf{s}$ is a vector of household characteristics, $\tilde{V}$ is the inclusive value term calculated from the lower stage of the nested logit, and η and the vector $\boldsymbol{\tau}$ are parameters to be estimated. If the expression in (5.38) is to be consistent with a random utility view of the world, it begs further explanation. Perhaps the easiest way of motivating (5.38) is to assume that individual $i's$ indirect utility conditioned on participation is given by $\alpha_{1i}+ \eta\tilde{V}_i$ where α_{1i} and η are unknown and $\tilde{V}_i$ is dependent on the quality and prices of alternatives should he participate, while his indirect utility conditioned on non-participation is completely unknown and equal to α_{0i}. This yields a participation decision that looks like:

$$\begin{aligned} \Pr(participate_i) &= \frac{\exp(\alpha_{1i} + \eta\tilde{V}_i)}{\exp(\alpha_{1i} + \eta\tilde{V}_i) + \exp(\alpha_{0i})} = \\ & \frac{1}{1 + \exp(\alpha_{0i} - \alpha_{1i} - \eta\tilde{V}_i)}. \end{aligned} \tag{5.39}$$

Define $\Delta\alpha_i \equiv \alpha_{0i} - \alpha_{1i}$, which is unknown but likely varies over individuals. To attempt to capture some of this variation, we make $\Delta\alpha_i$ a function of characteristics of the individual or household, $\mathbf{s}_i$. With a linear function, (5.39) is equivalent to the last term in (5.38).

The repeated nested logit and the linked model are mathematically quite similar as can be seen by substituting (5.38) into (5.37). This gives a trip model based on the repeated logit of the form:

$$z_{R_i} = \Upsilon_i \cdot [1 + \exp(\boldsymbol{\tau}\mathbf{s}_i - \eta V_i)]^{-1}.$$

Compare this to the trip specification for the linked model (5.36). The two models capture the same type of behavioral response, but with slightly different functional forms and implied error distributions.

Parsons, Jakus and Tomasi find a difference of only about 5% in welfare estimates obtained from the two models when evaluating the closure of a prominent site, using a sport fishing data set. They also show that welfare estimates are exactly the same if the two different approaches are applied using consistent underlying functional forms and error distributions.

The Hausman, Leonard and McFadden linked model was originally asserted to be utility-consistent, in the sense that both decisions were derived from a common underlying utility specification and error structure, but the assertion has since been shown not to hold in general (Smith, 1997a, and also Herriges, Kling and Phaneuf, 1999). The utility-consistent claim is based on interpreting equation (5.36) as the first stage in a two stage budgeting problem, where the following relationship is assumed:

$$z_R = \sum_{j=1}^{J} z_j = \frac{\sum_j p_j z_j}{(-\tilde{V}_i/\beta)}. \tag{5.40}$$

In the second stage of the budgeting process, the individual distributes the total number of trips decided upon across alternative sites according to the random utility model. Now consider the assumptions implied by the random utility model in its simplest form. When the conditional indirect utility function equals $v^*(y-p_j, \mathbf{q}_j, \varepsilon_j) = \beta(y-p_j) + \boldsymbol{\alpha}\cdot\mathbf{q}_j + \varepsilon_j$, then $\tilde{V} = \ln(\sum \exp(\boldsymbol{\alpha}\mathbf{q}_j + \beta(y-p_j))$ and the Hausman *et al.* 'price index' is given by

$$-\ln(\sum_{j=1}^{J} \exp(\boldsymbol{\alpha}\mathbf{q}_j + \beta(y-p_j))/\beta.$$

As Herriges, Kling and Phaneuf (1999) note, this should fill the role of a price index in a two stage budgeting sense. Numerically it works that way because it is the negative of the inclusive value at the baseline arguments of utility, and when this 'price index' goes up, the inclusive value goes down, leading to a reduction in trips. But this is, in general, inconsistent with (5.40) because it requires that

$$z_R = \frac{\sum_{j=1}^{J} p_j z_j}{-\ln[\sum_{j=1}^{J} \exp(\boldsymbol{\alpha}\mathbf{q}_j + \beta(y-p_j))]/\beta}, \tag{5.41}$$

which will not be true except under extremely restrictive conditions. In general, the right hand side of (5.41) will sum to the total number of trips only strong restrictions. Suppose that $J = 1$ and $v^*(y-p_1, \mathbf{q}_1, \varepsilon_1) = -\beta p_1 + \varepsilon_1$. Then

$$z_R = \frac{p_1 z_1}{-\ln[\exp(-\beta p_1)]/\beta} = z_1$$

so that in these circumstances, the aggregation works.

The repeated nested logit model also looks to be utility theoretic, but the implied utility function is not the sort we would typically specify to capture the quantity dimension of the trip decision. From (5.30) the conditional indirect utility function of a nested model might appear as

$$v_{nj_n} = \alpha q_{nj_n} + \gamma b_n + \beta(y - p_{nj_n}) \tag{5.42}$$

where b varies only over the upper level alternatives and q and p vary over all alternatives. In the repeated nested model, this expression needs to be modified to reflect the special nature of the upper nest–which is a participation decision. Now, the conditional indirect utility function on a choice opportunity equals

$$v_{1j} = \boldsymbol{\alpha}\mathbf{q}_{1j} + \beta(y - p_{1j}) + \varepsilon_{1j} \tag{5.43}$$

if a trip is chosen on the choice opportunity and

$$v_0 = \boldsymbol{\tau}\mathbf{s} + \varepsilon_0 \tag{5.44}$$

if no trip is taken. Note that the utility derived from not taking a trip cannot easily be characterized as it depends on all of the other opportunities available to an individual. Typically socioeconomic variables are used to determine whether an individual takes a trip. For example, Morey, Rowe and Watson use age, fishing experience, income, and fishing club membership to explain the decision of whether to participate on the choice opportunity. This implies a puzzling form for the conditional indirect utility function in equation 5.42 such that b_n equals a vector of socioeconomic variables when $n = 0$ (non-participation) and $b_n = 0$ for $n = 1$ (participation).

In concept the repeated logit model requires the researcher to pre-specify an exogenous number of choice opportunities–something that will in general be difficult to do. The concept of a choice opportunity is not well-defined nor easily observable, and it is likely to vary over individuals. In practice, the decision of how many choice opportunities to assume makes little difference as long as it exceeds the maximum number of trips taken by anyone in the sample. The estimated parameters in the upper nest (the binary participation decision) will simply be rescaled depending on the number of choice opportunities assumed.

By writing conditional indirect utility functions as we have in (5.43) and (5.44), it becomes painfully obvious that something is awry–the multiple choice occasions for each individual are treated as independent. In the recreation context, a person might decide whether to go to a beach on a sunny day in June and choose the beach to visit based on attributes of available beaches, including the distance she must travel to each. A few weeks later, the choice is made again. What is the relationship between these choices? In the original version of the repeated logit model, presented above, there is no relationship.

The two choice opportunities are independent. Unless characteristics have changed, the deterministic portion of the choice is identical over opportunities. The random portion is drawn from an identical distribution but is a completely independent draw.

To address at least one troubling aspect of this independence, the mixed logit model of Train has been applied in the context of the repeated nested logit model (Herriges and Phaneuf, 2002). This produces a model analogous to a random effects model using panel data. An individual specific error component is introduced that varies over people but is constant over choice opportunities for any one person. The unobserved part of preferences now includes a component that 'sticks' to the individual over time.

Provencher, Baerenklau, and Bishop (2002) introduce a temporal stochastic connection in a somewhat different way. They use a finite mixture logit to model the binary decision of whether to take a fishing trip on each day of the fishing season. This is a choice of whether to fish on each day, and not a choice of alternatives of which sites to fish, given fishing. Utility for individual i conditioned on a trip taken in period t is $v_{it} = \beta X_{it} + \varepsilon_{it}$, and zero otherwise. To make the model dynamic, in the sense that decisions at different dates will be made differently, the model is temporally evolving in two ways. First the random component of preferences is serially correlated: $\varepsilon_{it+1} = \rho \varepsilon_{it} + \eta_{it}$, where η_{it} is i.i.d. extreme value. Some unobserved components influence choice over time. For example it may take a month–several fishing occasions–to repair a boat. Further, the variables X_{it} evolve also, depending on current changes such as weather, and past decisions. Hence both parts of the preference function evolve over time. Each individual's evolution is different, partly because of different histories and partly because of different random components. A random event for individual i influences current behavior, which is then transmitted to the future.

Both of these approaches represent improvements in the random utility model by incorporating correlation in the stochastic portion of the model. But another aspect of temporal interdependence consistent with conventional demand analysis, that of diminishing marginal utility from increased consumption of trips, does not convincingly emerge from a correlated error structure. The Provencher, Baerenklau, and Bishop model incorporates an additional element of interdependence but in the systematic portion of utility. The variable, the time elapsed since the last trip, is meant to capture a dynamic analog to diminishing marginal utility. In a static model, one would expect that utility would increase with elapsed time. The results are disappointing in that the time elapsed since the last trip is found to have a 'corrosive' effect on utility from taking a trip. The authors interpret this as an indication of habit formation, but it may be due to their inability to correct for the fact that more avid fishermen take more frequent trips.[18]

Many valiant attempts have been made to address the shortcomings in an otherwise desirable choice model. But, at the end of the day, none of them is entirely satisfactory at explaining both elements of the decision–the choice among alternatives on a choice occasion and the selection of the number of choice occasions–in a way that is consistent with how we view choices to be made. Researchers have settled for practical solutions to the problem for the past 20 years. Until increases in computing power and advances in estimation by simulation, the estimation of a 'generalized corner solution model' remained unattainable.

5.5.3 The Generalized Corner Solution Model

A more comprehensive solution to the problems set out in the previous sections lies in a model that more closely matches conventional demand analysis.[19] The 'generalized corner solution' or 'Kuhn-Tucker' model was developed and made operational by Phaneuf (1997) and Phaneuf, Herriges, and Kling (2000) but has a history going back to Hanemann (1978) and Wales and Woodland (1983). In setting out the framework below, we continue to use the terminology of a recreational demand model, as that is the context in which the authors developed the model, but its applicability is clearly more wide ranging.

In this model, the individual chooses the number of visits, z_j, to each of $j = 1, .., J$ sites. The preference function in this model is given by

$$u(z_1, ..., z_J, y - \mathbf{z} \cdot \mathbf{p}, \mathbf{q}_1, ..., \mathbf{q}_{J+1}, \varepsilon), \tag{5.45}$$

where $\mathbf{q}_j$ is a vector of quality characteristics of site j, and $\mathbf{q}_{J+1}$ represents a quality vector associated with the composite commodity. This looks like a model we might have encountered in Chapter 3. But the generalized corner solution model advances conventional analysis by allowing for zero consumption of some commodities. That is, expression (5.45) is maximized subject to the *non-negativity* constraints: $z_j \geq 0$ for the first J goods; the composite commodity is treated as essential and therefore strictly positive. A solution that allocates all one's income to recreational trips would be nonsensical.

---

[18]The authors recognize this problem, which arose in an earlier paper (Provencher and Bishop, 1997) in which they sought to capture this same effect. In the later paper, (Provencher, Barenklau and Bishop) an attempt was made to introduce heterogeneity in preferences that would presumably control for avidity, but it is not clear that the problem can be entirely mitigated in this framework.

[19]Some might argue that conventional demand analysis has problems too. For example, how does one define the time period over which the model's quantity decisions are analyzed and diminishing marginal utility is expected to be exhibited?

The generalized corner solution is necessary to mimic the nature of recreational demand, but proves useful in any analysis where highly disaggregated data makes dealing with zero consumption of some commodities imperative. This model allows any number of z's to be zero or positive. These non-negativity restrictions require that the first order conditions be written as complementary slackness conditions for trips to all J sites:

$$\begin{aligned} \partial u(\mathbf{z}, y-\mathbf{z}\cdot\mathbf{p}, \mathbf{q}, \varepsilon)/\partial z_j - p_j\lambda &\leq 0 \\ z_j[\partial u(\mathbf{z}, y-\mathbf{z}\cdot\mathbf{p}, \mathbf{q}, \varepsilon)/\partial z_j - p_j\lambda] &= 0 \\ z_j &\geq 0 \\ \partial u(\mathbf{z}, y-\mathbf{z}\cdot\mathbf{p}, \mathbf{q}, \varepsilon)/\partial y &= \lambda. \end{aligned} \tag{5.46}$$

Given certain properties of the random vector and given a feasible utility function, Phaneuf *et al.* were able to write the likelihood function that permitted estimation of the parameters embedded in (5.46). This, of course, requires specifying a utility function. Phaneuf *et al.* chose the modified Stone-Geary function discussed in the last chapter:

$$u(z_1, ..., z_J, y-\mathbf{z}\cdot\mathbf{p}, \mathbf{q}, \varepsilon) = \sum_{j=1}^{J} \Psi(\mathbf{q}_j, \varepsilon_j)\ln(z_j+\Omega)+\ln(y-\mathbf{z}\cdot\mathbf{p}, \mathbf{q}_{J+1}) \tag{5.47}$$

where

$$\Psi(\mathbf{q}_j, \varepsilon_j) = \exp(\sum_h q_{hj}\delta_h + \varepsilon_j).$$

In this specification, the ε_j are random to the researcher but known to the agent, a distinction that matters when welfare is calculated. The parameters to be estimated are Ω and δ_h, where the subscript h indexes different quality characteristics. (Phaneuf, *et al.* also include an intercept). The composite commodity is represented by $y - \mathbf{z}\cdot\mathbf{p}$.

This is really the complete version of the quality-differentiated goods model explored in Chapter 3. It allows choices of no trips and any number of trips to one or multiple sites. It allows for diminishing marginal utility of trips and can accommodate weak complementarity. The marginal utility of quality h at site j is

$$\partial u/\partial q_{hj} = \delta_h \exp(\sum_h q_{hj}\delta_h + \varepsilon_j)\cdot\ln(z_j+\Omega).$$

Weak complementarity implies that this term should be zero when $z_j = 0$, a condition that can be imposed by setting $\Omega = 1$. In their paper, Phaneuf *et al.* estimate Ω rather than constrain it to one. Specifically Phaneuf *et al.* estimate a value of Ω equal to 1.76 (Table 3, page 89) which is significantly different from zero and from one.

Testing for the parametric restrictions required by weak complementarity reverses the classical approach to hypothesis testing. Typically in statistical tests we seek to reject the null hypothesis in favor of the alternative hypothesis. In such tests, the null hypothesis is given a very precise form–often a zero value for a parameter. In the case of weak complementarity for the modified Stone-Geary, the *alternative* hypothesis has a very precise form–in this case the parameter of the Stone-Geary equals one–and the null hypothesis has a diffuse form. In the Stone-Geary case, the diffuse form of the alternative hypothesis is that the parameter is not equal to one. This reversal of tests by itself would be manageable. But the difficulty comes with the implications of not rejecting the null. The conceptual problem is the nature of the *null* hypothesis, which is that the public good affects some behavior but not in the principal commodity investigated. When q is a publicly supplied characteristic of a private good, the failure of weak complementarity implies the presence of a pathway whereby individuals respond to changes the quality characteristic of a good that is not consumed. This may be feasible but it requires a good story. Without strong motivation, rejecting weak complementarity on the basis of a test on a single parameter cedes far too much power to the econometrics in determining the preference structure. The potential econometric causes of the test failure are large: a different preference function, errors in measurement and specification, aggregation errors, etc. Nevertheless, this is a critical issue, one that will be more easily addressed as more options for estimating the generalized corner solution model become available.

The type of top-down modeling facilitated by the generalized corner solution model would appear to make welfare measurement transparent. Welfare measurement in principle relies on the indirect utility function, and top-down modeling allows one simply to write down the indirect utility function given the estimated parameters. In practice, however, random elements are part of the preference function in equation (5.47), so that the indirect utility function must once again be written as an expected maximum utility:

$$\tilde{V}(\mathbf{p},\mathbf{q},y) = E_{\varepsilon} \max u(z_1, ..., z_J, y - \mathbf{z} \cdot \mathbf{p}, \mathbf{q}, \varepsilon).$$

Welfare calculations are further complicated by the fact that income effects are non-linear, so that the welfare measure cannot be solved explicitly. Phaneuf *et al.* devise an algorithm for calculating welfare effects that handles this non-linearity.

Although not obvious on the surface, the generalized corner solution model requires a certain amount of non-linearity in preferences to produce feasible solutions. This is consistent with most conventional demand analysis which requires for its solution that utility be nonlinear in the quantity of a good. This often induces non-linearity in income in the indirect utility function, implying non-zero income effects. The non-linearity may be necessary for solutions but its intuitive appeal depends on the importance of the recreational budget in the household's budget. It is quite feasible that income influences choices, as many researchers have discovered. But typically one cannot discriminate between the effects of income as a proxy for socioeconomic status and income as a constraint on spending.

The generalized corner solution model has all the desirable properties one would want for the study of recreational behavior. It incorporates site choice and frequency of choice within a single optimization. It can accommodate quality variables for different sites. And one can impose weak complementarity in the original utility function. Of course, the rather parsimonious preference functions used to date restrict the range of preferences that can be represented. It is different from the typical logit model, but in many ways not less restrictive. Early versions of the model were difficult to estimate but growth in computing power and experience with the model will probably result in a significant expansion of its applicability. A few attempts have been made to compare the welfare effects estimated from this model with those from the more loosely constructed linked models (Herriges, Kling and Phaneuf, 1999). Additional models have been estimated with larger datasets than originally used by Phaneuf *et al.* The most extensive analyses can be found in Von Haefen, Phaneuf, and Parsons (2004). Their comparisons across models have tended to illustrate the value of the logical consistency of the generalized corner solution model. (See also Von Haefen and Phaneuf, 2003).

While the generalized corner solution model will not be suitable for all discrete choice problems, it offers a setting for modeling choices involving a moderate number of alternatives when the choice is somewhat important in the consumer's budget allocation decisions. It represents one of the most significant advances since the conditional logit model. It moves beyond the logit framework by using the basic model of consumer choice. Another alternative, the hedonic travel cost model, uses the hedonic idea to build a competing model.

5.6 The Hedonic Travel Cost Model

The hedonic travel cost model (Brown and Mendelsohn, 1984) has been offered as an alternative to the random utility model.[20] This approach blends aspects of the random utility model and Rosen's hedonic property value model (see Chapter 6). It takes the setting and type of data used for the former and the modeling strategy of the latter. All three models use information on how people choose among alternative packages that cost different amounts and embody

[20]See Pendleton (1999) and Pendleton and Mendelsohn (1998, 2000).

different levels of attributes. Despite the similarities, the underlying story is far more difficult to rationalize in the hedonic travel cost model.[21]

5.6.1 The Structure of the Model

The random utility and hedonic travel cost models view environmental quality and other site attributes as a set of vectors, one for each site. The vector of environmental qualities at different locations is exogenous to the individual and affected by policy or accident, but the individual chooses the location and hence the vector of attributes he is exposed to. In the random utility model setting these choices have the potential of revealing how the individual values environmental quality. The question is whether they have the same potential in the hedonic travel cost model.

The modeling strategy begins with an individual maximizing a utility function that depends on the 'effective' vector of attributes, $\mathbf{q}$, and a numeraire z. The 'effective' levels of these attributes are the levels present at the chosen site. Unlike the random utility model, however, the hedonic travel cost model skips the discrete choice among sites and deals directly with the implied choices of the $\mathbf{q}$'s.

In the hedonic travel cost model, the individual is believed to

$$\max_{\mathbf{q},z} u(\mathbf{q}, z, \mathbf{s}) \textit{ subject to } y - C(\mathbf{q}) - z = 0 \tag{5.48}$$

where $\mathbf{s}$ is a vector of individual characteristics, y is income, and $C(\mathbf{q})$ is the hedonic travel cost function, which describes the costs of acquiring any bundle of attributes $\mathbf{q}$. Substituting the budget constraint into the objective function and differentiating gives the first order conditions

$$\frac{\partial u(\mathbf{q}, z; \mathbf{s})/\partial q_k}{\partial u(\mathbf{q}, z; \mathbf{s})/\partial z} = \partial C(\mathbf{q})/\partial q_k;\ k = 1, ..., K, \tag{5.49}$$

where the subscript, k, denotes one of the K dimensions of site quality considered. Note that $\mathbf{q}$ is no longer subscripted by alternatives because this model does not view the problem as one of discrete choice. Hence site distinctions disappear.

The hedonic cost function represents the costs of obtaining different levels of the K dimensional quality vector, given the location and configuration of site alternatives. If the hedonic cost function is linear in all attributes, demand functions for *attributes* can be derived from the first order conditions, where

[21] There is some discussion in the literature about the coherence of the hedonic travel cost model. See Smith and Kaoru, 1987; Bockstael, Hanemann and Kling, 1987; and Smith, Palmquist and Jakus, 1991.

the constant marginal cost of an attribute is treated as a parametric price. In the following, c_k is the constant marginal cost associated with characteristic k. The demand for attribute q_k is

$$q_k = f_k(c_1, ., c_k, .., c_K, \mathbf{s}, y); \; k = 1, ..., K,$$

but it is the inverse function

$$c_k = f_k^{-1}(q_1, ., q_k, .., q_K, \mathbf{s}, y); \; k = 1, ..., K \tag{5.50}$$

that is usually estimated.

The intuitive appeal of the hedonic travel cost model comes from the analogy with the hedonic model of housing characteristics. People who want more of any given characteristic make the trade-off—higher costs for more of the characteristic. The trade-off is common in the housing market. More rooms and better views increase the price of a house. Markets sometimes provide such trade-offs in recreation. One can pay for professional guide services and probably be more successful in hunting and fishing enterprises. The question is whether the usual circumstances in which travel cost models are applied align themselves with the basic workings of the hedonic model.

5.6.2 The Hedonic Cost Function

The cornerstone of the hedonic travel cost model is the cost function. It plays an analogous role to the hedonic price function in conventional hedonic analysis. Understanding the distinctions between these two functions gives insight into the hedonic travel cost approach.

Rosen (1974) provides the theoretical basis for the hedonic price function for housing. The underlying mechanism is a simple one. Buyers compete for houses through price. The fact that a given property can be purchased by only one household causes the price of properties possessing more of a desirable attribute to be bid up. In the course of estimation, when a statistically significant negative marginal price occasionally emerges for a desirable attribute, it generally signals a specification problem. Most likely, levels of a desirable characteristic included in the regression may be negatively correlated with levels of another desirable characteristic omitted from the regression. Hedonic housing prices are, however, generally quite systematic in showing positive marginal prices for important, desirable attributes.

Hedonic travel cost applications presume a similar hedonic price function for characteristics of recreational sites. This presumption implies some underlying process by which the cost depends positively on the levels of the characteristics. In the recreational case, however, there is no mechanism to allocate attributes by price. An individual faces an array of alternative sites with different quality characteristics and different costs of access, where cost of access is determined by how far from a site an individual happens to live. Consequently, if we plot travel costs against attributes for all sites available to the individual, there is no reason for the resulting scatter of points to be a function at all, let alone a well-behaved one where price is monotonically increasing in environmental quality. Nature organizes the spatial distribution of sites and characteristics. Some people may happen to live quite close to the environmentally most desirable sites, for example.[22] If nature arranges sites such that the site with the most desirable attributes is the closest one for an individual, then his choice of site conveys no information about how he values site attributes.

Dependence on a relation between costs and attributes is required whether one uses the random utility model, the hedonic travel cost model or any other model that attempts to extract information from site choice. When the hopeless case exists–everyone lives closest to the best site–random utility models will fail to reveal anything about the value of attributes. However, as we look across individuals in the sample, if there is an opportunity for individuals to trade money for quality between any pairs of alternatives, this can be exploited by the random utility model. If too little of this sort of trade-off is visible in the data, then the model will return insignificant coefficients. In contrast, the hedonic travel cost model *requires* the estimation of a monotonic cost *function* and then assumes equality between the slope of this function and marginal value. This places a heavy burden on 'nature' to produce systematic variation.

5.6.3 Making Sense of the Story

Smith, Palmquist and Jakus (1991) make explicit the need for a regularity constraint on the choice set and use information on technology (cost and attribute information) to identify a subset of sites that posses the needed properties. In taking a technical approach by formulating a Farrell frontier, Smith *et al.* avoid confounding preferences and costs. In practice, most hedonic travel cost applications constrain the choice set by grouping individuals by origin zone and then estimating separate hedonic cost functions for each origin zone using only those sites that are actually chosen by recreationists from that zone.

Researchers using the hedonic travel cost model argue that while we may

---

[22]While nature dictates the location of the recreation resource, in some special cases the individual's residential location may be endogenous to the problem. For example, retired couples may purchase property close to their favorite recreational spots. For these special cases the residential location choice becomes part of the problem. Taking residential location as exogenous in this case will confound the difficulties already embedded in the hedonic travel cost, because the travel cost to the most desirable site will be lower, not higher, than to other sites.

not know anything about the configuration of the choice set, it must be true that wherever an individual ends up, at that site his marginal value for each attribute will be equal to his marginal cost of obtaining it (e.g. Englin and Mendelsohn, 1991, p 277). This is not in general true, unless the choice set is approximately continuous. The assumption that the *slope* of a function, econometrically fitted to a set of chosen sites, equals the individual's marginal value for q becomes more tenuous the sparser or more irregular the frontier or constraint set. If the individual cannot marginally adjust his chosen level of q but must decide among a finite set of discrete choices, his optimum will have the nature of a corner solution; it will be characterized by a set of inequalities, not the equality of marginal costs and values.

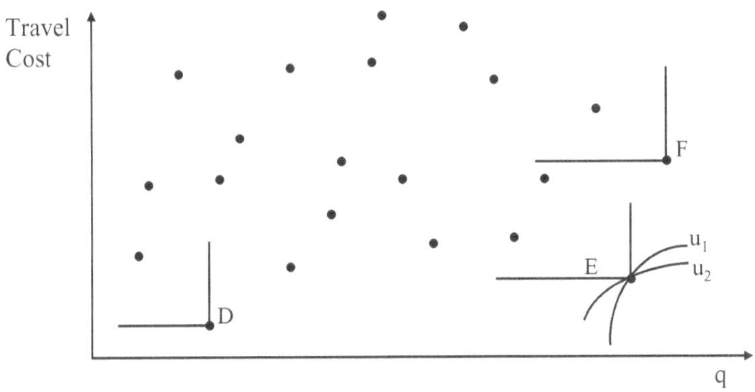

FIGURE 5.2. One Possible Configuration of Quality and Cost Pairs

To illustrate, Figures 5.2 and 5.3 represent two different potential configurations of quality-travel cost pairs, where we restrict the problem to a single quality characteristic for ease of illustration. Both dispersions are realistic sets of quality-travel cost pairs. If preferences for attributes are not satiated, any alternatives that lie northwest of another available choice will not be chosen. The combinations lying northwest cost at least as much but provide no more q and would never be chosen when individuals have positive marginal values for the attribute. Given the domination by any southeast point, Figure 5.2 provides only three non-dominated alternatives, denoted D, E, and F in the graph. If the set of non-dominated choices constitutes the basis for determining $C(q)$, then $C(q)$ is based on a sparse set indeed. Indifference curves with all sorts of slopes will be consistent with the choice of alternative E as the optimal alternative, where u_1 and u_2 are examples. Any functional form fit to this scatter of points will be tangent to an individual's indifference curve only by chance. The constraint set is better behaved in Figure 5.3, but the results still yield a hedonic cost function with a slope quite different from marginal value,

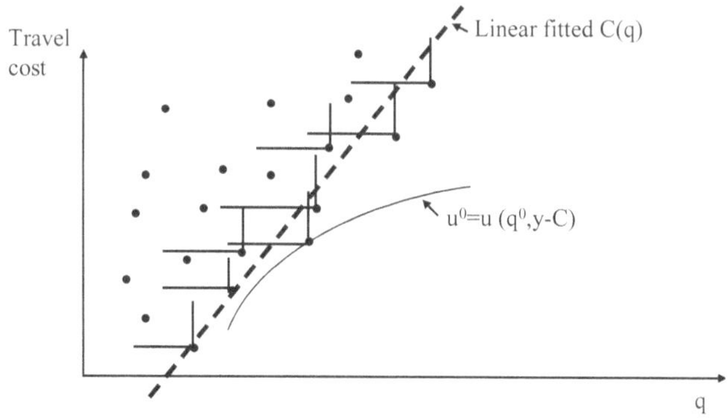

FIGURE 5.3. Another Possible Distribution of Travel Cost-Quality Points

as implied by the indifference curve (u^0).

Our examples have adopted the usual procedure of estimating a linear cost function, and this will clearly exacerbate the problem since fitting a linear-in-attributes function to the complex and irregular shape that the cost frontier is likely to exhibit will usually yield estimated attribute prices that bear no resemblance to actual marginal costs. However, if the hedonic cost function is not linear in attributes the marginal prices are no longer exogenous to the individual in the first stage of estimation. Further 'prices' in the second stage will not be parameters, making it difficult to define conventional demand functions.

An empirical manifestation of irregularity of the hedonic travel cost surface is the negative marginal price problem.[23] Negative 'marginal prices', obtained by the empirical application of the hedonic travel cost approach, could occur in problems where individuals are really making discrete choices among bundles rather than acting as though the first order conditions in equation (5.49) hold for all attributes. Given that the problem is really a choice among discrete alternatives whose cost frontier is likely to be characterized by irregular curvature, these first order conditions may not be part of the optimization problem at all.

Englin and Mendelsohn rationalize negative marginal prices as a consequence of 'over-satiation'. To use their example, an individual might value some Dou-

[23]Negative coefficients for desirable attributes can result from the estimation of random utility models as well as hedonic models for housing attributes. Collinearity among attributes is the most likely cause for these unexpected signs. This occurs most often when the number of attributes is large and the attributes are not central to the decision process. In random utility models one rarely finds a negative sign on the travel cost parameter because it is central to the decision process, and in hedonic housing price models, the number of rooms or square feet of the house, arguably the most important attributes, typically have positive signs.

glas fir along a hiking trail, but might actually value a mix of tree species rather than all Douglas fir–implying that preferences can exhibit satiation in the amount of Douglas fir. If, in addition, the cost frontier is continuous and declining in this attribute, it is possible for the individual to optimize at a point where the value for more Douglas fir is declining.

Distinguishing between satiation and an ill-conceived optimization problem may not be easy. Both conceptual and econometric problems could lead to negative prices, and it will be difficult to rule one or the other out. Admitting satiation further unravels the hedonic travel cost model, however. It means that we must relax our requirements on the constraint set. Now all we can require is that no site be included that would involve paying more than necessary for any level of q; we cannot rule out ranges in which less of the attribute costs more.

5.6.4 Welfare Measures in the Hedonic Travel Cost Model

We abstract from the difficulties in the previous sections and assume that the condition that marginal value equals marginal cost in fact characterizes the individual's choice. Assume also that we have accurately recovered these marginal costs and, using this information, we have successfully estimated demands for attributes, such as in (5.50). Note that welfare measurement will be fundamentally different if the cost function is non-linear because the standard Marshallian demand functions will not exist in that case. Hence we also assume the usual practice of estimating hedonic cost functions that are linear and separable in attributes.

Suppose that a linear hedonic travel cost function has been estimated. The usual procedure is to use the constant marginal costs ($\mathbf{c_q}$) implied by this linear function to estimate a system of inverse demand functions of the form:

$$\mathbf{c_q} = \mathbf{f}^{-1}(\mathbf{q}, y, \mathbf{s}). \tag{5.51}$$

We consider whether the results from these steps can be used for welfare estimation.

In the context of environmental valuation, the goal of welfare measurement is to determine the compensation required for an exogenous change in one or more elements of the vector of characteristics, $\mathbf{q}$. In the random utility model this is a well-formed welfare question because the vector of characteristics at each site is exogenously determined and the choice is among sites. It is not such a well-formed question in the hedonic travel cost framework, which skips over the site choice and models the choice of the $\mathbf{q}$'s directly. The difficulty is in knowing how to reflect an exogenous change in one or more elements of $\mathbf{q}$ in the context of the hedonic travel cost structure. Because the structure of the model requires linking levels of $\mathbf{q}$ with costs of access, changes in $\mathbf{q}$ must be represented as changes in the cost function, which turns out to be both conceptually and empirically difficult.

Begin with the question: what is the welfare effect for an individual if q_1 changes at one or more sites? If we accept the hedonic travel cost framework then, in concept, the welfare effect of a change in a quality characteristic, q_1, at one or more sites would need to be given by

$$CS_{\Delta q_1} = [\int_0^{q_1^1} f^{-1}(q_1, \mathbf{q}_{-1}, y, \mathbf{s})dq_1 - c_{q_1}^1 q_1^1] \tag{5.52}$$

$$-[\int_0^{q_1^0} f^{-1}(q_1, \mathbf{q}_{-1}, y, \mathbf{s})dq_1 - c_{q_1}^0 q_1^0] \tag{5.53}$$

$$= \int_{q_1^0}^{q_1^1} f^{-1}(q_1, \mathbf{q}_{-1}, y, \mathbf{s})dq_1 - c_{q_1}^1 q_1^1 + c_{q_1}^0 q_1^0.$$

The expression on line (5.52) is intended to represent total consumer surplus after the change, with the analogous expression on line (5.53) reflecting the same measure before the change. Each of these expressions includes the total area under the estimated inverse demand function for the quality characteristic, q_1, evaluated at the *chosen* level of this characteristic minus that chosen level times the constant marginal cost of q_1. If f^{-1} is indeed the inverse demand for the quality characteristic, then the first part of each expression makes sense. A problem arises in knowing what to use for $c_{q_1}^1$. The only way in which exogenous changes in levels of q_1 can be captured in this problem is if the cost function changes. Otherwise, the individual would not make any change in his choice of q_1 *and* there would be no change in costs, resulting in no welfare change. So $c_{q_1}^1$ must be different from $c_{q_1}^0$ for the problem to make sense. Yet we have no way of knowing what $c_{q_1}^1$ is. The cost function is, in a sense, a figment of the researcher's imagination. It is a frontier, estimated by the researcher and based on the observed choices of individuals from any given origin zone. Without the benefit of observing choices already made, its nature is unknown; we do not know $c_{q_1}^1$, and as a result, we can not determine what level of q_1 is chosen after the change. *A priori*, there would seem no way to complete the valuation task.

Pendleton and Mendelsohn (2000), using linear cost functions, give consumer surplus in the hedonic travel cost framework as

$$CS_{\Delta q_1} = \int_{q_1^0}^{q_1^1} f^{-1}(q_1, y, \mathbf{s})dq_1 - C(q_1^1) + C(q_1^0), \tag{5.54}$$

where we alter the notation so as to match our own.[24] The second and third

---

[24]The authors also include an error term whose distribution is discussed extensively. We

terms in (5.54) appear to be the same cost function evaluated at different levels of q_1. Note that the entire cost function is included as the authors appear to simplify the problem to one in which only a single quality dimension is relevant. The motivation is of course to measure the welfare effects of a change in q_1. The difficulty is that this variable is endogenous, and according to the model, won't change unless costs change. The absence of a clear idea about how the attributes can change when they are endogenous prevents a plausible welfare story. A subsequent statement that 'the HTC [hedonic travel cost model] suffers from the restriction that it can estimate welfare only for those who choose the same bundle no matter the level of attributes' (Pendleton and Mendelsohn, 2000, p 103) attempts to rectify the situation. Presumably the 'same bundle' means the same site. This restriction makes the approach of limited value in welfare measurement. Also, it reminds us that, however we may attempt to estimate a cost function for quality characteristics, the travel cost to each site will remain constant even though site quality changes. Therefore, it is possible to know total costs if one knows the chosen site. In particular, if the chosen site does not change, then the two cost terms in (5.54) cancel, leaving only the area under the inverse demand function. All this makes even a stronger argument for modeling the problem as one of discrete choice rather than a continuous choice of attribute levels embedded in this artificially contrived hedonic framework.

On the surface, it might appear that the attractiveness of the hedonic travel cost model turns on the completeness of the recreationist's choice set. With large numbers of sites, the choice of attribute levels looks more continuous and the hedonic travel cost model might seem increasingly appropriate, although welfare analysis remains problematic. This is the usual argument in the housing literature where thousands or even tens of thousands of distinct alternatives are not uncommon. However, without market competition for attributes there is no guarantee of a well-behaved cost function, because no mechanism exists to ensure that even the boundary of the 'technology' set is a well-behaved function with appropriate curvature. Consequently, there is no reason to argue that marginal value and marginal cost are equated, and no way to rationalize a new cost function and a new quality choice when exogenous changes occur.

5.7 Conclusion

Modeling discrete choices has proven to be one of the most effective ways of valuing resources, especially when individuals choose among alternatives with many attributes. McFadden's random utility model emerges when individuals choose among discrete options with different bundles of attributes and the

suppress this to make the conceptual argument clear.

researcher's uncertainty about preferences takes the form of an additive extreme value random element. Discrete choice models of consumer goods and modes of transportation are quite common in the literature, but the applications to recreational resources have taken full advantage of the utility-theoretic property of the model.

In recent years, the conditional logit model has been extended to incorporate more complex deterministic preferences and greater generality in the random part of preferences. The random parameters model, for example, allows for a richer characterization of uncertainty. But the principal outstanding challenge, that of consistently modeling frequency of choice occasions together with the discrete choice, has been addressed by the generalized corner solution model of Phaneuf, Herriges and Kling, which offers a solution for problems of small to moderate size. Extensions of this approach that incorporate more flexibility in the structure of preferences are already being made.

Chapter 6

Hedonic Models of Heterogeneous Goods

6.1 Introduction

More often than not economists treat marketed goods as homogeneous and estimate demand curves for goods with homogeneous quality. An economist might be interested in estimating the demand for water from a public water supply, where public water is viewed as a homogeneous good whose quality declines as a result of contamination. Bottled water would normally be considered a separate although related good. A market demand curve would exist for each, although each demand would be conditioned on the price and quality of the other.

The degree of heterogeneity varies with commodities. A book may sell in both hard cover and paperback with no other substantive difference among copies of the same book. Some commodities, like electrical appliances, vary across makes and models with different units of the same model effectively identical. Some commodities are so heterogeneous that each *unit* is essentially a different good. The prime example is real estate or, more specifically, housing. Location is critical in housing. No two properties can occupy the same location, so no two properties can be identical however similar their structural characteristics. Whether the heterogeneous good is an automobile or a house, markets will force higher quality units to sell at higher prices. Better cars or houses may be more expensive to produce. These more expensive goods are produced because people are willing to pay more for them. Economists have long believed that the trade-offs households make between price and quality characteristics should reveal something about the value they place on changes in those characteristics.

Since *in situ* environmental quality can sometimes be an important characteristic of housing, economists have attempted to use the market relationship between housing price and environmental quality to obtain welfare measures of changes in the quality. This chapter reviews the theory of hedonic models as applied to housing and evaluates what can and cannot be said about welfare effects of quality changes using this construct. In Chapter 7, we explore a separate body of hedonic literature that uses wage/job risk trade-offs to deduce willingness to pay for changes in risk. In the second part of that chapter, efforts to combine wage and property hedonics to value such regionally varying amenities as climate will be explored.

Our ability to evaluate welfare effects in the context of hedonic markets is limited in part because it is difficult to recover enough information about preferences. Welfare evaluation is further complicated because the level of the public good is not imposed on households. They can adjust their consumption of it by changing their residence or job. By such adjustments, they may face a different price or wage. With a sufficiently large change in the public good, and consequent adjusting by households, the entire price/wage schedule may change. These issues were not encountered in Chapters 3 and 4 because no markets intervened and the public good was seen as being imposed on the individual. Chapter 5's treatment allowed individuals to make discrete choices among 'packages' of cost/quality combinations and therefore to adjust the level of the public good that affected them, but these decisions did not induce changes in the price schedule. In this and the next chapters, the action takes place in a market setting with the consequence that policy changes or exogenous events will quite often lead to changes in price.

6.2 The Theory of Hedonic Models

The most popular model of quality differentiated goods is the hedonic model, whose theoretical underpinnings were developed by Rosen (1974) in the context of the housing market. Rosen initially constructed the model in terms of both consumers and producers, incorporating consumer preferences and producer decisions to add to the housing stock or renovate existing structures. However, it is easier and more intuitive to tell this story for a given stock of housing and housing characteristics, assuming a perfectly inelastic supply at a point in time. The existing housing stock typically dominates the housing market so that producer decisions in any one year may add no more that one or two percent to the stock. Most researchers take the stock of housing and housing attributes as fixed and model prices as determined by the distribution of these attribute bundles within the housing stock and the distribution of preferences for these attributes within the population. The difficulties that arise in welfare measurement can be found in this simpler story, so we present the problem in this setting and modify Rosen's original model by taking the stock of housing as given. Palmquist (1991, 2005b) provides a complete exposition of the hedonic problem, and much in this chapter is indebted to his work.

In the context of Rosen's hedonic model, the household, choosing a housing unit, is viewed as maximizing utility subject to a budget constraint that contains a price *schedule* rather than a parametric price for housing:

$$\max\ u(z, \mathbf{q}; \delta) \text{ subject to } y = z + P(\mathbf{q}). \tag{6.1}$$

Here z is the amount of the numeraire good consumed by the household, $\mathbf{q}$ is a K-dimensional vector of housing characteristics chosen, y is the household's income, and δ characterizes the household's preferences. Utility maximization models are rarely explicit about differences in household preferences. However, as we will see below, the hedonic price function depends on the distribution of preferences among individuals in the housing market. The vector of characteristics of the house, $\mathbf{q}$, typically includes structural characteristics such as square feet, number of bedrooms, presence of a swimming pool, and neighborhood characteristics such as crime rate, school quality, and pollution levels. For the time being, we will concentrate on the relationship perceived by individuals in the market, the function $P(\mathbf{q})$. It represents a relationship between housing prices and characteristics, and is known as the hedonic price function.

We will have more to say about how this hedonic price function comes about shortly. First there are elements of this formulation that require discussion. Housing is a durable good and, for some, a once in a lifetime purchase. As such, the temporal dimension of equation (6.1) is confusing. Viewing $P(\mathbf{q})$ as the equivalent rental price or monthly payment for a housing unit resolves the confusion, at least in part. With costless moving and no transactions costs, households could be viewed as making this periodic payment for housing, even if they own their residence, and re-optimizing the housing choice in each period. In reality of course, there are large costs to changing one's residence. The existence of transactions and moving costs will cause us to qualify some of our welfare measurement results later on.

Assuming an interior solution for all quantities, we write the first order conditions that characterize the household's solution to the problem as

$$\partial u/\partial q_k = \lambda \partial P(\mathbf{q})/\partial q_k \quad k = 1, ..., K \tag{6.2}$$

$$\partial u/\partial z = \lambda \tag{6.3}$$

$$y - P(\mathbf{q}) - z = 0. \tag{6.4}$$

The final condition, equation (6.4), is the budget constraint which, as we will shortly see, need not be linear in the attributes that enter the utility function. The set of $K+2$ first order conditions in equations (6.2) and (6.4) can be collapsed into the K conditions:

$$\frac{\partial u(\mathbf{q}, y - P(\mathbf{q}); \delta)/\partial q_k}{\partial u(\mathbf{q}, y - P(\mathbf{q}); \boldsymbol{\delta})/\partial z} = \partial P(\mathbf{q})/\partial q_k \quad k = 1, ..., K, \tag{6.5}$$

where we have substituted the budget constraint for the numeraire and eliminated the multiplier by forming ratios of marginal utilities. The right hand side of equation (6.5) is the slope of the hedonic price function with respect to the kth attribute and the left hand side is the marginal rate of substitution between the kth attribute and the numeraire.

6.2.1 Rosen's Bid Function

Understanding Rosen's bid function is central to understanding the theoretical literature on hedonic models. The bid function describes the maximum an individual would be willing to pay for a house with a particular set of characteristics given his income, preferences, and a baseline level of utility. To define this concept more precisely, consider the indifference surface over which utility is held constant at u^0 while z and the elements of $\mathbf{q}$ are allowed to vary

$$u(z, \mathbf{q}; \delta) = u^0. \tag{6.6}$$

In this expression, z is the composite commodity with normalized price. It represents the amount of the composite commodity consumed and the amount of money left over from income, after the heterogeneous good (e.g. the house) is purchased, that can be spent on all other goods. Given this definition, it must be true that

$$z(\mathbf{q}, y, u^0; \delta) = y - \theta(\mathbf{q}, y, u^0; \delta), \tag{6.7}$$

where θ is Rosen's bid function. It is obvious that θ must be a function, since it will vary with the housing attribute bundle, $\mathbf{q}$, as well as with household income and the base utility level.

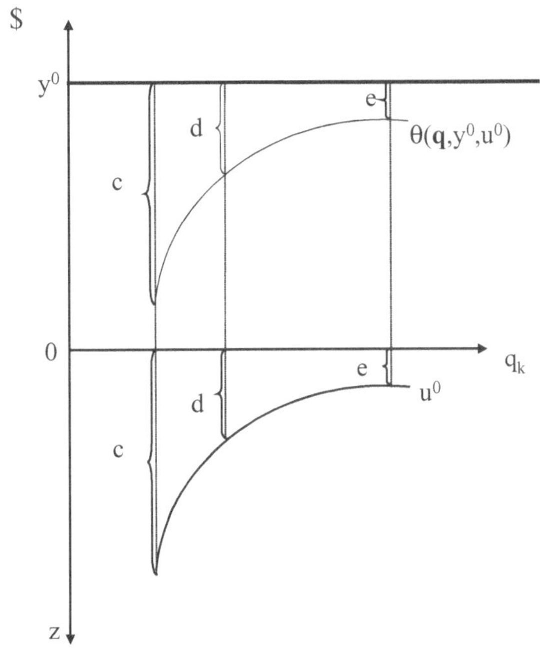

FIGURE 6.1. Transforming Bid Functions into Indifference Curves

Because utility is monotonically increasing in z, equation (6.6) can be solved for z as an inverse function of utility:

$$z = u^{-1}(u^0, \mathbf{q}; \delta).$$

From equation (6.7)

$$\theta(\mathbf{q}, y, u^0; \delta) = y - u^{-1}(u^0, \mathbf{q}; \delta), \tag{6.8}$$

which leads to an explicit definition for the bid function and to the surprising result that this function is additively separable in income. That is, a dollar increase in income increases the bid function by one dollar.

Rosen's bid function is easily drawn in income-characteristic space. θ can be plotted as a function of one of the elements of $\mathbf{q}$, holding other elements of $\mathbf{q}$, income, and utility constant, as illustrated in the upper quadrant in Figure 6.1. This figure also shows how easily the bid function can be translated into an indifference curve in (z, q_k) space. Substituting (6.7) into (6.6) yields:

$$u(y - \theta, \mathbf{q};\delta) = u^0, \tag{6.9}$$

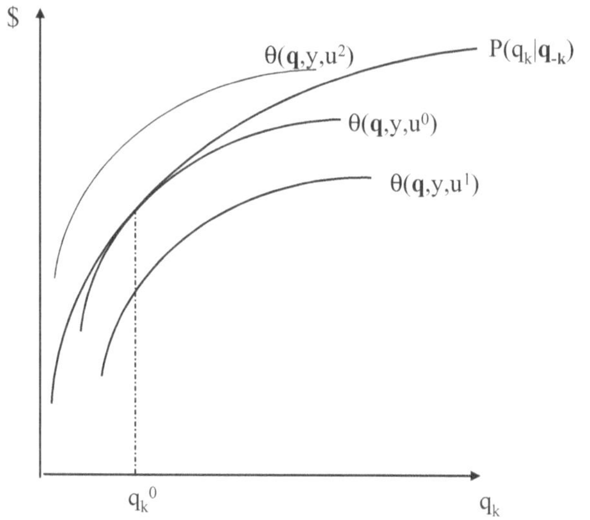

FIGURE 6.2. The Hedonic Price Function and Bid Functions

which implies the correspondence between an indifference curve in (z, q_k) space and the bid function depicted in this same graph.

Given that $u(y - \theta(\mathbf{q}, y, u^0; \delta), \mathbf{q}; \delta) = u^0$, then

$$\frac{du}{dq_k} = -\frac{\partial u}{\partial z}\frac{\partial \theta}{\partial q_k} + \frac{\partial u}{\partial q_k} = 0 \tag{6.10}$$

which implies that the slope of the bid function with respect to q_k is

$$\frac{\partial \theta}{\partial q_k} = \frac{\partial u(\mathbf{q}, y - \theta(\mathbf{q}, y, u; \delta))/\partial q_k}{\partial u(\mathbf{q}, y - \theta(\mathbf{q}, y, u; \delta))/\partial z} > 0. \tag{6.11}$$

This result is not surprising, given that the right hand side of the expression is the slope of the indifference curve drawn in the lower quadrant of Figure 6.1. It is an important result, however, because from (6.5) we know that at the optimum the individual will equate the slope of the indifference curve to the slope of the hedonic price function. This leads to the usual figure depicted in the hedonic literature–one in which the individual's bid function is tangent to the hedonic price function as illustrated by the tangency of bid function $\theta(\mathbf{q}, y, u^0)$ in Figure 6.2. While we see only a cross section of each of these functions the 'tangency' is assumed to exist in all K dimensions.

Figure 6.2 also includes two other bid functions conditioned on two other levels of utility, where $u^1 > u^0 > u^2$. The bid function reflects the maximum an individual would be willing to pay for a bundle of attributes to attain a given level of utility. In contrast, the hedonic price function represents the amount the household would *need* to pay if attempting to purchase this bundle in the market. At the individual's optimum, the slope of the two functions must be equal, a result that must be true by combining the results from (6.5) and (6.11) and the budget constraint.

6.2.2 The Hedonic Price Function

Many economic concepts are abstract constructs with little appeal outside the discipline. Not so the hedonic price function. Even though continuous variation in all attributes may be rare and some markets thinner than others, a systematic relationship between price and characteristics is taken for granted by participants in housing markets. No one doubts that, all else equal, houses with more square footage, on larger lots, with more amenities, and easier commutes will command higher prices. Buyers will bid up the price of houses with more desirable characteristics and houses with relatively poor amenities will sell at a discount. The hedonic price function is a characterization of this multi-dimensional relationship between price and characteristics.

In Figure 6.2, $P(q_k|\mathbf{q}_{-k})$ is a cross section of the hedonic price function that all participants in the market are assumed to perceive. Drawn in this space, it illustrates the market price of housing that the household faces for each level of q_k, holding other characteristics constant. The household maximizes utility by choosing the highest bid contour that is feasible, given the market hedonic price schedule. At the optimal solution the household will be at a position such as q_k^0 in Figure 6.2. This will be true in all K attribute dimensions if continuous levels of attributes are available over a sufficient range. The optimal level of z is given by $y - P(\mathbf{q})$, evaluated at that point.

Different households will have different bid functions, $\theta(\mathbf{q}, y, u^0; \delta)$, that depend on their income and their preferences for attributes. Although each household takes $P(\mathbf{q})$ as given, the hedonic price function emerges as the result of an equilibrium process that allocates each unit of housing to the highest bidder. The equilibrium can be modeled as an assignment problem in which available housing units are allocated to households according to their preferences and income, such that once the allocation is accomplished no household would willingly outbid other households for any other house. In a market with a large stock of housing and virtually continuous variation in attributes, a hedonic price function would emerge that represents an envelope of relevant household bid functions such as depicted in Figure 6.3. At equilibrium, expression (6.5) must hold for all households and all attributes where the right side of (6.5) is the partial derivative of the hedonic price function.

We have denoted the hedonic price function as $P(\mathbf{q})$. This notation obscures the function's origins. Underlying the market function $P(\mathbf{q})$ is an equilibrium *process*. Given the process by which the function emerges, it must change

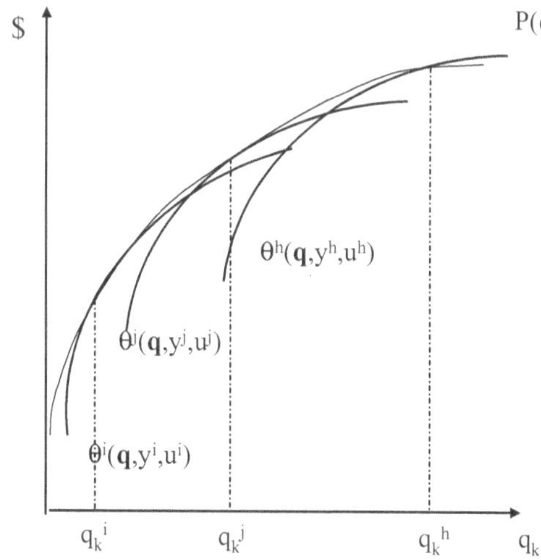

FIGURE 6.3. Hedonic Price Function Formation

with changes in the distribution of people's tastes and incomes. For example, if households systematically lose interest in fireplaces or pools, then bid functions would change shape in this dimension and the hedonic price function would change. The structure of the hedonic price function also depends on the distribution of attributes in the stock of housing. For example, if most neighborhoods in a housing market are plagued by congestion and traffic noise, then houses in quiet neighborhoods will command a premium. The more common are quiet neighborhoods, however, the lower the implicit price for that attribute. Changes in either the distribution of household tastes or of attributes in the housing stock lead to shifts in the hedonic price schedule. This property of the hedonic price function will have bearing on welfare measurement.

In concept, if all households were to have identical incomes and preferences, then the hedonic price function would be identical to the common bid function held by all households. The hedonic price function is an envelope of individual bid functions. When all individuals have the same bid function, the hedonic price function must converge to a common bid function. That is, the hedonic price function simply replicates the common bid function over the range of attribute levels found in the existing housing stock. In this case, houses would be allocated arbitrarily across households, with every household paying a different amount for its housing unit and therefore having a different amount of money left to buy z. At the same time each household would be acquiring a different set of attributes such that the varying combinations of $(z, \mathbf{q})$ would make each indifferent to every other household's situation.

Of course, in reality households will differ in income and tastes so the allocation process will not be an arbitrary one. Using simulation, Cropper, Deck and McConnell (1988) provided an illustration of the allocation process with varying household income and preferences and showed that if a stock of housing is allocated among buyers by a maximum bid rule, a solution will be consistent with the equilibrium conditions in (6.5).[1] One element of the allocation process is particularly important. All other things equal, households with more income and/or stronger preferences for an attribute will chose more of the attribute and will therefore be instrumental in determining the marginal price at higher levels of that attribute. Likewise, households with lower income and/or weaker preferences will be found at lower levels of the attribute in question and will affect the hedonic price function's shape in that region.

Except that a hedonic price function should be increasing in amenities, little *a priori* information about its functional form can be deduced from theory. The shape will be induced by the distribution of attribute bundles in the housing stock and the joint distribution of income and preferences in the population of households. Because the price function is an envelope, it should be true that bid functions are more concave (from below) than the price function. The price function can be concave or convex in attributes. There is certainly no reason to expect $P(\mathbf{q})$ to be linear in all elements of $\mathbf{q}$, and as we will see this causes problems.[2]

Estimation of hedonic price functions for housing is a common practice in applied econometrics.[3] Moreover, there are many papers that relate prices and attributes of other products such as agricultural goods, automobiles, computers, etc. The hedonic papers of interest to us come from the subset that attempts to measure the welfare effects of changes in exogenously determined attributes. The attributes most commonly valued are environmental amenities or public goods. In the sections that follow, we consider whether a hedonic model that includes as an attribute some environmental amenity (such as air quality at the housing location) can help us value a change in that amenity. As we mention above, the econometric question of whether any information

---

[1] While the conditions described here must be true of the hedonic price function, there is nothing in the above discussion that defines this function uniquely. Returning to the case of the common bid function, one of these will exist for each different level of utility. In order to define $P(\mathbf{q})$ uniquely, one must incorporate further considerations. A general equilibrium model would provide closure of the model, but at the least this would require a treatment of housing production. The conditions set forth above must still hold however, and they are sufficient for the purposes here.

[2] Linearity implies perfect repackaging and makes little sense for many attributes. This is best illustrated using structural attributes. For example, it makes no sense to assume that each additional bathroom in a house adds the same amount to the price of the house (especially holding other attributes constant, such as square footage and number of bedrooms.)

[3] An ECONLIT search yielded almost 600 papers.

about preferences can be recovered from information about households' housing choices has plagued researchers since Rosen's seminal paper, but it turns out that recovering this information is sometimes not necessary, and at other times insufficient, to assess the welfare consequences of a change in amenities.

6.3 Welfare Measures in Hedonic Markets

In previous chapters the definition of the welfare measure has been quite clear: the change in income necessary to return the individual to his initial utility level after some change in the level of the public good. Compensating variation has been defined implicitly by the equality $v(\mathbf{p}^0, \mathbf{q}^0, y) = v(\mathbf{p}^0, \mathbf{q}^1, y - CV)$, where the amounts of the consumption goods are optimally chosen under each regime.

In the context of the hedonic property model the definition of welfare can become muddled. What we typically want to value is an exogenous change in the level of some attribute (specifically an environmental amenity or disamenity) at one or more locations. This attribute will be provided by nature or policy, not by private actions. For example, it might be a reduction in airborne pollutants in a neighborhood due to new environmental standards, the introduction of a health risk from a new hazardous waste site, or an increase in odors from an expanding hog industry. Several things make this type of change confusing. First, while the change in an attribute at a location is exogenous, the change is not, strictly speaking, an exogenous change imposed on the household if the household is free to change its housing location. This is not so very different from previous welfare measurement stories. We have, all along, recognized that individuals adjust their behavior in response to changes in external stimuli. The hedonic story is especially similar to the circumstances modeled in Chapter 5 where the individual chose the level of 'exposure' to an environmental amenity or disamenity by selecting one of a finite set of alternatives. But unlike the discrete choice setting in Chapter 5, the hedonic problem is one in which a market intervenes. This means that if the level of the exogenous attribute changes at a given housing location, the price of that housing unit can be expected to change as well. Finally, if the change is sufficiently wide-spread the distribution of the attribute in the housing stock will change, possibly causing a shift in the entire hedonic price function. This may lead to changes in prices at all locations–even those not experiencing an attribute change.

Before proceeding, we make two distinctions. The first, and one we continue to highlight, is the distinction between a marginal valuation and a non-marginal one. As is usual, a household's value for a marginal change in an attribute will be relatively easy to estimate, but marginal changes do not typically characterize proposed policies or natural resource damage cases. We need a way to evaluate the welfare effects of discrete changes in attributes.

The second distinction arises in the context of non-marginal changes, but is a subtle distinction and often goes unmentioned. Some researchers are interested in using hedonic models to capture, *in general*, how people value reductions in pollution or increases in environmental amenities. We will call this the 'pure willingness to pay' measure. This is the answer to a question such as: how much would the household be willing to pay to experience an improvement in ambient environmental quality at their location of residence, all other things held constant? And given the theoretical development, we know that pure willingness to pay must be the change in the value of the bid function with a change in the public good.

Often researchers have been interested in estimating the welfare effect of an actual (or proposed) change in pollution or health risk due to some policy or event that has taken place or will take place. Such a question might be: what are the losses or gains to a household from a policy that actually alters ambient air quality in a certain area or that sites a hazardous waste landfill at a particular location? The distinction between this type of question and the one asked above may seem moot at this point, but it becomes important as we attempt to define welfare measures.

As we shall see, the real welfare effect of a policy, project or event generally differs from the 'pure willingness to pay' measure. The difference arises in part because, when faced with changing environmental quality at their current residence, households can adjust. As always the proper welfare measure must take into account behavioral adjustments. But what makes the problem even more complicated is that prices of affected houses will also change, changing the implicit cost to residents of remaining in the same location, and if the change in the public good is sufficiently widespread, it may lead to changes in prices for all houses, through a shift in the entire hedonic price function.

In the following sections marginal changes and discrete changes of both types will be treated. For the time being we will ignore the serious econometric difficulties that arise in attempting to recover the information needed. The treatment will be purely conceptual at first, with operational difficulties postponed until later in this chapter.

6.3.1 Defining 'Pure Willingness to Pay'

We have defined our 'pure willingness to pay' measure, loosely, as the answer to the question: how much would a household be willing to pay to experience a change in an exogenously determined characteristic (an element of the $\mathbf{q}$ vector) holding all else constant? This is not a sufficiently precise definition. For one thing, we must investigate further what we mean by 'holding all else constant'. In the conventional welfare economics problem, the welfare measure we sought equaled the change in income necessary to return the individual to the original utility level when one or more prices changed, holding all other prices constant. Implicit in the definition was the notion that the individual would re-optimize with regard to his choice variables (the quantity of each good consumed) when he faced the new price(s) and the compensated change in income. Prices are clearly parameters to the individual in such a problem and quantities consumed are choice variables. When we first considered the welfare effect of a *quality* change in Chapters 3 and 4, quality was viewed as a parameter also. The compensating variation of a quality change implicitly held prices constant but, once again, allowed the individual to re-optimize with regard to all quantities consumed, except for the exogenously imposed quality characteristic.

In the hedonic model, as with the discrete choice model in Chapter 5, we no longer have a straightforward optimization problem in which the exogenous and endogenous variables are clearly defined. In the hedonic model the individual chooses among heterogenous goods that embody different levels of quality characteristics. There are a number of ways to characterize theoretical welfare measures in this context. Palmquist (2005b) provides a discussion of the alternatives. Economists have opted for basing the definition on the bid function. Recall that θ is the maximum amount the household with income y could pay for attribute bundle $\mathbf{q}$ so as to have just enough income left over to buy enough z to achieve utility level u^0. This seems rather convoluted, but because one unit of the heterogeneous good (e.g. a house) is consumed, the bid function reduces to a simple and useful concept. The *change* in θ associated with a *change* in one or more elements of $\mathbf{q}$ represents the extra amount the individual would be *willing to pay* for the heterogeneous good (holding utility constant). It is the amount of the numeraire that the household is willing to give up in order to get the added amount of the attribute. Because only one unit of the heterogeneous good is chosen, and that does not change irrespective of the level of attributes, a change in θ is equivalent to a change in exogenous income, and therefore has a willingness to pay interpretation.

From here on, we consider an exogenous change in one characteristic which we label, q_a, where q_a is an attribute of location that is determined by nature or public policy. The 'pure willingness to pay' for a change in quality from q_a^0 to q_a^1 in the context of the hedonic model is defined as the difference in the bid functions evaluated at the initial and final levels of q_a:

$$WTP_{\Delta q_a} = \theta(q_a^1, \mathbf{q}_{-a}, y, u^0) - \theta(q_a^0, \mathbf{q}_{-a}, y, u^0). \tag{6.12}$$

This expression represents the maximum additional amount the individual could pay for the house after the quality change, and remain at the original utility level achieved before the quality change. The individual is at the original location (with no behavioral adjustments) and the only change that has taken place is the change in the quality at that location. By defining willingness to pay in this way, we allow the household to freely choose its quantity of the numeraire, but the elements of the vector $\mathbf{q}$ and any changes in those elements are taken as exogenous to the household, because we hold it at its current location. In this usual definition of the 'pure willingness to pay' measure for a change in q_a, all the other attributes are assumed fixed. An implicit mathematical definition is given by:

$$\tilde{v}(q_q^0, \mathbf{q}_{-a}^0, p_z, y) = \tilde{v}(q_a^1, \mathbf{q}_{-a}^0, p_z, y - WTP_{\Delta q_a}) \tag{6.13}$$

where p_z is simply the price of the numeraire (equal to 1) and is included to make clear that the numeraire is the choice variable but its price is exogenous. The function, $\tilde{v}$, is analogous to an indirect utility function that treats attributes as exogenous.

Expression (6.12) can be rewritten to illustrate just how simple a concept this is. Remember that θ can be written as $\theta = y - z(\mathbf{q}, u)$, where $z(\mathbf{q}, u)$ is simply the indifference surface solved explicitly for the numeraire. Since y is the household's initial income, a change in θ with a change in the attribute, q_a, can be expressed as minus the change in the expenditures on the numeraire holding utility constant:

$$\begin{aligned} WTP_{\Delta q_a} &= \theta(q_a^1, \mathbf{q}_{-a}, y, u^0) - \theta(q_a^0, \mathbf{q}_{-a}, y, u^0) \\ &= \{y - z(q_a^1, \mathbf{q}_{-a}, u^0)\} - \{y - z(q_a^0, \mathbf{q}_{-a}, u^0)\} \\ &= z(q_a^0, \mathbf{q}_{-a}, u^0) - z(q_a^1, \mathbf{q}_{-a}, u^0). \end{aligned} \tag{6.14}$$

This is the change in the amount of money spent on z that holds utility constant when q_a changes. This $WTP_{\Delta q_a}$ is fundamentally different from previously defined welfare measures, as it partially circumscribes the household's ability to adjust its behavior. We will soon see that when households are allowed to adjust, the welfare measure changes.

6.3.2 Revealing 'Pure Willingness to Pay'

Throughout this book, our aim is not just to define a welfare measure but to discuss how that measure might be revealed empirically. Having defined our 'pure willingness to pay' effect in expression (6.14), we are faced with the task of estimating it using information on household behavior. In earlier chapters, the first empirical step has been to estimate one or more Marshallian demand functions and then consider how to use these to move to compensated demands or to calculate approximations of compensating variation directly. In the hedonic framework, a version of this same problem is encountered, as well as a further problem–a nonlinear budget constraint.

Rosen argued that inverse demand functions for attributes could be estimated by using marginal prices calculated from an estimated hedonic price function. He suggested that the effective price of an attribute, calculated as $\partial P/\partial q_k$, could be regressed on the observed level of the attribute, q_k, to obtain an (uncompensated) inverse demand function. Ignoring the econometric difficulties with this procedure, there should indeed exist an observable relationship between any q_k and the corresponding slope of the hedonic price function, $\partial P/\partial q_k$. The data generating process that produces observations on $\partial P/\partial q_k$ and q_k is characterized by the full set of first order conditions presented in equations (6.2)–(6.4) and is represented by the system of K conditions in (6.5) where the budget constraint has been substituted for the numeraire.

If $P(\mathbf{q})$ could be assured of being linear in attributes, then the budget constraint would look exactly like a conventional one. One could solve for the endogenous attributes as a function of the constant marginal prices and income (as well as other exogenous variables) and estimate a conventional system of Marshallian demand functions. In such a case, exogenous income and exogenous implicit prices would explain the observable choices of attribute levels. Unfortunately, linearity is not always assured and in fact is unlikely because buyers cannot separately purchase housing attributes and combine them costlessly (they can not 'repackage' attributes) nor are some sets of attributes logically additively separable. A linear budget constraint has the added drawback of generating no variation in marginal prices over individual observations within the same housing market. In the absence of multiple hedonic price functions from different housing markets, estimation of demand functions would still be infeasible.[4] Hedonic price functions are almost never estimated in linear form whether they are used for welfare measurement or not.

When the hedonic price function is nonlinear in attributes, marginal attribute prices will be functions of $\mathbf{q}$. To show the nature of solutions, adopt the notation $P(\mathbf{q}) = P(\mathbf{q};\boldsymbol{\gamma})$ where $\boldsymbol{\gamma}$ is a vector of parameters of the hedonic price function. Look again at the individual's first order conditions:

$$\begin{aligned} \partial u/\partial q_k &= \lambda \partial P(\mathbf{q};\boldsymbol{\gamma})/\partial q_k \;\; k = 1, ..., K \\ \partial u/\partial z &= \lambda \\ y - P(\mathbf{q};\boldsymbol{\gamma}) - z &= 0. \end{aligned}$$

These conditions do not lead to standard Marshallian demand curves, because implicit prices are not parameters and therefore are not constant. The choice variables ($\mathbf{q}$) appear on both sides of the first K equilibrium conditions, so that the marginal prices of the q's vary with the levels of the q's. With non-constant marginal prices, one can still solve for the endogenous choice variables but not as functions of parametric prices. Writing the demands for the characteristics,

[4] Parsons (1986) assumed a linear hedonic price equation and estimated an almost ideal demand system for housing characteristics across housing markets in 14 cities.

we have for each attribute:

$$q_k = f_k(\gamma, y). \tag{6.15}$$

These are equivalent to Marshallian demand functions, but only in the sense that choice variables are solved as functions of exogenous variables. No parameter is present that plays the role of price and the marginal cost of an attribute will be endogenous.

Consider the simple case in which the hedonic price function is a quadratic in lot size, D, so that $P = \beta_0 + \beta_1 D + \beta_2 D^2 + \ldots$. The marginal price of lot size is $\beta_1 + 2\beta_2 D$, so that by choosing D the household also determines the price of that attribute. Further, as Epple points out, substitution of the nonlinear budget constraint for z implies that the observable relationship must include the full term, $y - P(\mathbf{q})$, and not just income. Under these circumstances it is still possible to specify a relationship among the marginal prices, quantities, and $y - P(\mathbf{q})$, but usual duality results break down.[5] Note also that $y - P(\mathbf{q})$ is endogenous, depending on the choice of $\mathbf{q}$.

To resolve this difficulty, two alternatives to exact welfare measurement have emerged from the literature. One requires first choosing a form for the underlying utility function. With functional form in hand, one can solve the $K+2$ first order conditions from equations (6.2)–(6.4), yielding a system of relationships that can be estimated. From this estimation one can recover estimated values for the parameters of the utility function, and then calculate welfare estimates using expression (6.13).[6]

Chattopadhyay (1999) implements this approach using Chicago housing market data. He compares results from twelve models–combinations of six different variants of a Box-Cox functional form combined with two alternative functional forms for utility (the Diewert and the translog). Welfare estimates for both marginal and non-marginal changes in particulate matter and sulphur dioxide are found to be remarkably stable over different specifications.

The second approach that suggests itself would require estimating an observable demand function and integrating back. Given that conventional Marshallian demand functions do not exist when the budget constraint is nonlinear, this approach would appear to be infeasible. Palmquist (1988) resolved these difficulties by employing the notion of a linearized budget constraint. In Fig-

---

[5]With this result of Epple's it becomes clear that if one attempts to estimate the behavioral function, then $P(\mathbf{q})$ must represent some periodic payment for housing rather than the sales price of the house. Different authors have used different concepts for the periodic payment, including an annualized value and a pseudo-mortgage payment.

[6]Choice of functional form is critical for identification of preference parameters. Brown and Rosen (1982) show that when both the marginal value function and the marginal hedonic price function are linear, no information about preferences can be derived by estimating parameters in first order conditions.

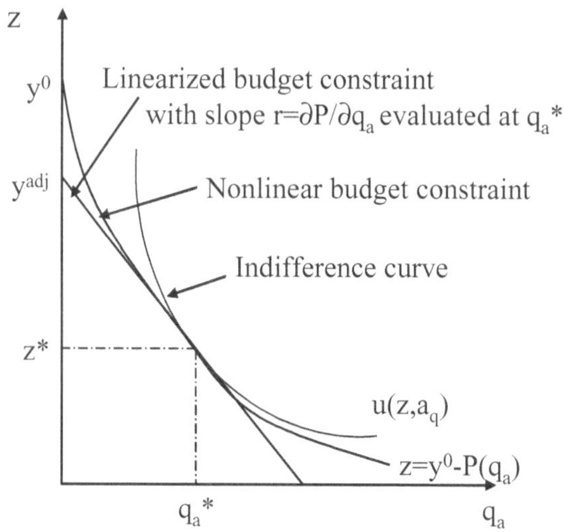

FIGURE 6.4. Linearizing the Budget Constraint

ure 6.4, the indifference curve drawn in the southeast quadrant of Figure 6.1 is redrawn in more conventional fashion and all attributes other than the one of interest are assumed away. The actual budget constraint appears as a nonlinear function tangent to the indifference curve at the optimum choice (z^*, q_a^*). Actual income is plotted as the intercept on the numeraire axis; at that point none of the housing attribute is consumed and all income is spent on z. A linear budget constraint, $z = y^{adj} - rq_a^*$, is also drawn on this graph, with slope, r, equal to the slope of the nonlinear budget constraint evaluated at the optimum choice, and with an intercept equal to an adjusted income term. This adjusted income can be calculated with knowledge of actual income, the hedonic price function, and the household's actual housing choice, because the adjusted budget constraint must pass through $z^* = y - P(q_a^*)$. As a result, the adjusted income is defined as: $y^{adj} = y - P(q_a^*) + rq_a^*$.

As Palmquist explains, a household facing this linearized budget constraint would choose the same optimum solution as a household facing the actual nonlinear budget constraint. Consequently, the solutions to the two problems contain equal information about the underlying utility function. Of course the linearized budget constraint varies over households, because its location is determined by the household's income and its choice of location on its nonlinear budget constraint. An application can be found in Boyle, Poor, and Taylor (1999) who estimate consumer surplus measures for improvements in water clarity using hedonic models of prices for properties along Maine lakes.

While the above approaches are possible, several econometric problems make them challenging, as we will discuss later in this chapter. To avoid these, many authors have restricted themselves to calculating marginal values. Figure 6.2 illustrates that a household's optimum will be characterized by equality between the slope of the bid function and the slope of the hedonic price function for that attribute. Mathematically, this follows from condition (6.5) that requires the marginal rate of substitution relative to the numeraire for each attribute to equal the slope of the hedonic price function for that attribute, and from condition (6.11) that requires the marginal rate of substitution to equal the slope of the bid function. At the optimum, it must be true that

$$\frac{\partial \theta}{\partial q_a} = \frac{\partial u / \partial q_a}{\partial u / \partial z} = \frac{\partial P}{\partial q_a}. \tag{6.16}$$

This relationship has been used to show that the slope of the hedonic price function at the optimum equals the household's willingness to pay for a marginal change in the environmental amenity. The result follows from the interpretation of the marginal bid function, $\partial\theta/\partial q_a$, as the increase in the amount of money the household would be willing to pay for the house (so as to keep utility unchanged) if the attribute, q_a, were marginally increased.

As long as the hedonic price function can be estimated, information about any household's *marginal value* for an attribute can be recovered. It will simply be the slope of the hedonic price function at the level of the attribute enjoyed by the household. For example, if the *log* of house price were regressed on a vector of attributes including airborne particulate matter measured at each location, then the estimate of household $i's$ valuation for a marginal change in particulate matter would be $\hat{\beta}_{q_a} \cdot P(\mathbf{q}_i)$, where $\hat{\beta}_{q_a}$ is the estimated coefficient for the air quality variable in the hedonic price function, and $P(\mathbf{q}_i)$ is household $i's$ house price.

Exactly what constitutes a marginal change is open to debate. Many researchers, recognizing the difficulties in non-marginal welfare measurement, have used marginal values–sometimes for changes that would be difficult by any standard to justify as marginal. Others have used marginal values to approximate the welfare effects of admittedly discrete changes by multiplying the discrete change in q_a by the marginal value of q_a:

$$\frac{\partial P}{\partial q_a} \cdot (q_a^1 - q_a^0) = \frac{\partial \theta}{\partial q_a} \cdot (q_a^1 - q_a^0). \tag{6.17}$$

This can be illustrated by drawing a tangent to the bid function at the optimally chosen level of q_a. The tangent line, which is the linear projection of the slope, will lie above the bid function and so the approximation will overestimate the pure willingness to pay for an increase in q_a and underestimate it for a decrease. An alternative approximation to the pure willingness to pay measure is

$$P(q_a^1, q_{-a}) - P(q_a^0, q_{-a}), \tag{6.18}$$

which will also overestimate the pure willingness to pay for an increase in q_a and underestimate it for a decrease. Expression (6.18) will be a better approximation than (6.17) if the hedonic price function is concave in q_a.

Brookshire, Thayer, Schulze and d'Arge (1982) use this upper bound measure to draw inferences about 'pure willingness to pay', ultimately using this bound as a means of verifying results from a contingent valuation experiment. Observations on Los Angeles housing prices and housing attributes including air quality (q_a) allowed the authors to estimate a hedonic price function and calculate $\Delta P/\Delta q_a$ at different levels of q_a. In a companion analysis, a stated preference experiment elicited willingness to pay bids for changes in air quality that were intended to reveal on what terms households would be willing to trade q_a for money along the bid function, θ. The finding that changes in the hedonic price function exceeded willingness to pay bids from the stated preference experiment was taken as corroboration of the latter, as this is the relationship that should exist between $P(\mathbf{q})$ and $\theta(\mathbf{q})$ as portrayed in Figure 6.3.

6.3.3 Welfare Effects of Exogenous Events

While the 'pure willingness to pay' effect is of interest, it is rarely the answer to the question: what is the ultimate welfare effect on households of an actual (or proposed) change in an environmental amenity at one or more locations? To help sort out the various effects caused by the market's intervention we use an abstraction, introduced by Lind (1973) and adapted by Bartik (1988b), that has become common in the hedonic literature. We divide society into two groups: renters and landlords. In practice, many renters will also be landlords–those who own the houses they live in. But we make this conceptual distinction because the welfare effect on households in their role as residents at particular locations is substantively different from the welfare effect on households in their role as owners of assets whose value may have changed due to external forces. The same household may lose in its role as renter but gain in its role as landlord, for example. Ultimately we will be interested in the sum of effects over the two groups so that the distinction is one that helps the thought process but does not alter the outcome of the analysis.

Localized Changes in Amenities

The distinction between 'localized' and 'non-localized' changes in attributes has been important in the hedonic literature since Polinsky and Shavell (1976) and Freeman (1979). Localized changes are exogenous changes in an environmental amenity that occur only at a limited set of housing locations. Too few sites are affected to cause a change in the hedonic price function. Freeman argued that when this characterizes the policy, project or event that causes an environmental change, then welfare measurement can be quite simple. In fact, in this case, no information about households' preferences is needed. This seems odd at first blush. Why should we be able to calculate welfare effects without any information about preferences? The reason is simply that in the absence of moving costs and transactions costs, the entire effect of the environmental change can be measured as a 'windfall' gain or loss in asset value to landlords.

A simple way to motivate this story is the following. There is a continuum of renters in the market, initially allocated to the stock of houses arrayed along a continuum of one or more attributes.. In this stylized story, we ignore all other attributes of houses except the environmental attribute we are interested in and order the houses according to the level of q_a embodied in them. The households will also be ranked according to their willingness to pay for q_a which will be determined by their income and preferences. The allocation process will accordingly match the ordered renters with the houses in a way such that no renter can outbid any other renter for a given house. As described earlier, a hedonic house price function will emerge as an envelope of bid functions.

Now, imagine that the environmental quality at some small number of houses is improved for some reason. Perhaps these houses are on an inlet that has been polluted by the effluent from a nearby plant. The plant has now been forced by regulation to clean up its activities. To make the exposition easier, assume that the small number of affected houses is really just one house. How do we measure the welfare effect of the non-marginal improvement at that one house?

Initially, the renter living at this location experiences an increase in welfare due to the discrete jump in quality associated with the change in water quality. But very quickly the environmental improvement is coupled with a rise in the house's rent because quality has improved. Although experiencing an increase in q_a, this household is worse off because it is paying more in extra rents than it is willing to pay for the change that has taken place in the attribute. Figure 6.5 depicts the household's disequilibrium position.

If it is costless for households to move, this household will seek to return to equilibrium by moving back along the hedonic price function to a house embodying an environmental level approximately equal to its original position at q_a^1. In a static world, this relocation will be possible by marginally adjusting positions of other households along the attribute continuum. The hedonic price function will remain stable, as too few houses are changing to make a difference in the location of $P(q_a)$. The result is no real change in welfare for any households in their roles as renters. The renting household at the affected location finds a house embodying no more than a marginal change in quality from its initial house. Since marginal willingness to pay equals the marginal change in rent, the household is no better or worse off. The full effect of

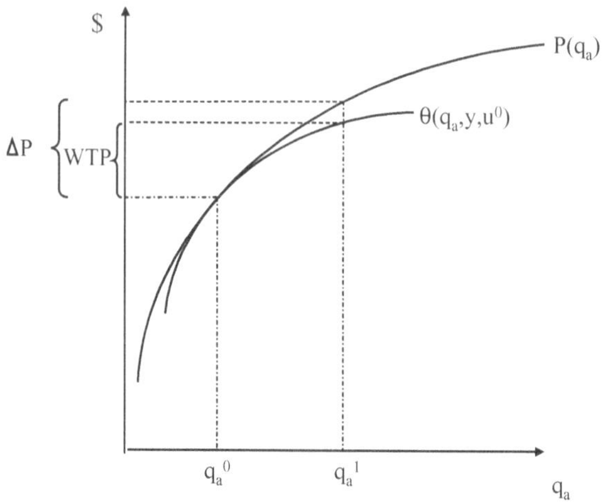

FIGURE 6.5. Disequilibrium Position of Affected Household

the change accrues to the *owner* of the one house that has made a discrete jump along the hedonic price function. The increase in value is reflected by the movement of that house along the stationary hedonic price function and is equal to:

$$P(q_a^1) - P(q_a^0). \tag{6.19}$$

This is equivalent to an exogenous change in income for the owner of the improved house. The exogenous income change in the form of increased rents equals the total compensating variation for the environmental change.

This story can hold approximately for an environmental change that affects a small number of houses in a relatively dense market. If the environmental change is a negative one, households in the role of renters are still left unaffected because of their ability to move, while households in the role of asset owners suffer a loss in welfare equal to the decline in the value of the asset as it moves along the hedonic function. All of this makes sense as long as moving is costless. In reality, moving and transactions costs can be substantial. If these costs are substantive but not so large as to outweigh the gains of moving, then households at affected locations will attempt to return to equilibrium but will incur costs in doing so. Transactions costs cause the owners' windfall gain or loss to be an upper bound (along the real number line) on the total welfare effect to society. A more accurate measure would deduct transactions costs from (6.19) but these are difficult to estimate.[7]

[7]Introducing moving costs into this stylized story complicates matters. The 'device' of having other households adjust marginally to 'let in' the affected household no longer makes

If transactions costs are sufficiently large to outweigh the gains from moving, then renters will be stuck at their old locations but with the new levels of q_a. Landlords will still enjoy the gains represented by (6.19), but renters will suffer losses because the increase in rents they now must pay exceeds the value of the environmental improvement to them. Summed over both renters and landlords, however, the net effect of the environmental improvement is positive. When no one moves, the change in rents can be treated as a transfer payment between owners and renters, resulting in no net welfare change. What is left is the gain to renters due to the environmental improvement (or loss due to an environmental decline) at their residence. This can only be valued with knowledge of preferences as it is the 'pure willingness to pay' effect discussed in the previous section. With no knowledge of the magnitude of transactions costs or their size relative to the gains from moving, one can only say that in the presence of transactions costs the net welfare effect associated with each affected property will lie somewhere between the property's changing price along the hedonic price function and the corresponding change along the bid function.

Non-Localized Changes in Amenities

Non-localized changes imply environmental quality changes at sufficient numbers of properties to effectively change the 'stock' of environmental quality available in the market. With non-marginal and wide-spread changes such as this, the hedonic price function can be expected to shift, but predicting this shift *a priori* is near to impossible.[8] In principle we know the measure we want, although we need to be careful about notation.

The renter's compensating variation, $CV_{renters}$, will be measured as the willingness to pay for the change in housing characteristics that the household ultimately consumes minus the actual change in the rent paid, once again temporarily ignoring transactions costs. This is given by

$$CV_{renters} = [\theta(\mathbf{q}_n^1, y, u^0) - \theta(\mathbf{q}_o^0, y, u^0)] - [P^n(\mathbf{q}_n^1) - P^o(\mathbf{q}_o^0)]. \tag{6.20}$$

where we use the notation as follows. The household lives in houses denoted by the attribute vector $\mathbf{q}_s^t$ where $s = o$ denotes the original house, $s = n$ the

sense, since other households will not incur moving costs when they have nothing to gain. The static story above now becomes too limiting. In reality, there will be continual movement of households in and out of the market for a variety of reasons, as well as new additions to the housing stock.

[8]The urban equilibrium sorting models predict a new hedonic price equation following a non-localized change (Bayer and Timmons, forthcoming and literature cited therein). With the new hedonic price equation, these models give the complete welfare measures.

new house, $t = 0$ the initial attribute vector, and $t = 1$ the attribute vector after the change. The change in rent will be sufficiently large and pervasive to change the hedonic price function, with $P^o(\mathbf{q})$ denoting the original hedonic function and $P^n(\mathbf{q})$ denoting the new function. The pure willingness to pay for the change in characteristics is given by the first two terms and represents the fact that the renter will move from location o to location n as part of the adjustment process. The change in rent paid is calculated as the price on the newly evolved hedonic price function (P^n) for the newly chosen house under the new quality conditions minus the price of the original house on the original hedonic price function before the quality change.

In contrast, the welfare effect for landlords, $CV_{landlords}$, is given by the change in rents:

$$CV_{landlords} = [P^n(\mathbf{q}_n^1) - P^o(\mathbf{q}_o^0)].$$

Summed over the two groups, the term $[P^n(\mathbf{q}_n^1) - P^o(\mathbf{q}_o^0)]$ is clearly a transfer payment, leaving $\sum[\theta(\mathbf{q}_n^1, y, u^0) - \theta(\mathbf{q}_o^0, y, u^0)]$ as the ultimate welfare effect. However, without knowledge of the bid function, the new hedonic price function, and the resulting new equilibrium positions, there is no obvious way to obtain the exact welfare measure from a non-localized change.

Bartik (1988b) provides a means of arguing a lower bound on the true welfare measure, an argument which we present here in a simpler form. Bartik's story relates to the benefits of households and profits of producers. We pare down the story to apply to households only, assuming away any new construction activity. To establish bounds, the adjustment process is divided into three artificial steps. We consider three phases of adjustment that would move renters and landlords from their initial position to their final position, but not necessarily in the order in which these steps would take place in the real world. The reordering of the adjustment steps makes it possible to sign the outcome.

In the first of these artificial steps, an exogenous change in an amenity level at several (or all) locations takes place. Renters at all affected locations experience a welfare change that is measured in money terms as

$$\theta(\mathbf{q}_o^1, y, u^0) - \theta(\mathbf{q}_o^0, y, u^0). \tag{6.21}$$

If the change is an improvement in quality at several sites then renters at these housing locations enjoy positive benefits.

In the second of these artificial steps, all price changes take place but all households remain in their original houses. The hedonic price function shifts and all houses are associated with potentially different rents as they move from their initial rental price before any change in quality to what will be the final rental price after their house has moved along a new, shifted hedonic price function. In reality the shift in the hedonic price function will not take place until renters start making behavioral adjustments. However in this artificial step, we are constraining renters to remain in their original houses as we change the hedonic function to what it will be after all adjustments take place. The change in rents that each household pays is given by $P^n(\mathbf{q}_o^1) - P^o(\mathbf{q}_o^0)$. The household has not moved, but the new hedonic price function and the new level of quality determines the rental price of the house. Likewise the landlord of each house receives $P^n(\mathbf{q}_o^1) - P^o(\mathbf{q}_o^0)$. In step 2 there are no net social gains, only transfers between renters and landlords.

Finally, in step three we allow renters to adjust. They are allowed to move to what becomes their final locations (the locations that support the new hedonic price function that we artificially allowed to emerge in step 2). The potential moves are not restricted to renters in originally affected houses, because all renters may now be out of equilibrium given the price shift. In any event, moves will only be made by households if they can make themselves better off by moving. Households that were previously restricted (in our story) to remain in their original houses are now free to change housing locations. By Le Chatelier's Principle, they can be made no worse off in this step. Therefore, the net effect at this stage is a non-negative welfare change for renters, with no effect on landlords.

The aggregation of these three steps provides Bartik's result. In step one, the net gain to society of improvements at a set of locations is given by the pure willingness to pay measure in expression (6.21) and is summed over households affected by the amenity change. Step 2 yields no net welfare change, while step 3 produces an unmeasurable improvement. This leads to the result that the pure willingness to pay measure must be a lower bound on the welfare effects of an improvement. If the initial change is a degradation in quality, expression (6.14) will be a negative number, but the rest of the argument holds. There will be no welfare effect in step two and step three still produces a non-negative change in welfare. The 'pure willingness to pay' measure, $\sum[\theta(\mathbf{q}_o^1) - \theta(\mathbf{q}_o^0)]$, is therefore a lower bound on compensating variation along the real number line. It underestimates gains and overestimates losses.

To the extent that moving and transactions costs prevent some households from re-optimizing, the pure willingness to pay lower bound becomes an increasingly better estimate. Failure to re-optimize affects only the third stage of the artificial story. It may also affect the extent to which the hedonic price schedule shifts if few households actually make adjustments, but that makes no difference since step two involves only pecuniary transfers. The presence of moving and transactions costs *will* reduce gains from environmental improvements and exacerbate losses from environmental degradation, however, because it will either prevent adjustments that would make renters better off or charge a 'price' for this adjustment.

General Equilibrium Approaches to Welfare

Researchers have long recognized that environmental regulation can be the source of non-localized changes in the housing market and can change the hedonic relationships in various communities. The idea is very simple, and nicely illustrated for Los Angeles by the analysis of Smith, Sieg, Banzhaf and Walsh (2004). Suppose a policy intervention changes the level of air quality in a variety of housing districts. Household equilibrium will be disturbed not just in one place, but perhaps in all housing districts, and we know that if the displacement is large enough, a new hedonic price function will emerge. Under those circumstances, as we have seen, it will not be possible to use the old hedonic price function for welfare analysis. Smith *et al.* have confronted this problem by constructing a general equilibrium model of location and housing markets that lets one assess the welfare effects by looking at the *ex post* equilibrium. This represents a significant improvement in the use of hedonic models. In their study of the prospective benefits of ozone changes from Clean Air Act Amendments in southern California for the period 2000 to 2010, they find that partial equilibrium effects, the kind discussed in this chapter, and general equilibrium effects can be substantially different. Assessing per individual benefits of control technologies by county, they find that for the year 2010 the partial and general equilibrium welfare estimate for Los Angeles county differs by less than $1. In contrast, the partial equilibrium welfare estimate for Ventura county is estimated to be $21, while the general equilibrium estimate is $539. Differences will naturally vary by application, but this study provides empirical evidence of what we know intuitively, that when there are large changes the conventional hedonic approach to welfare measurement may be seriously deficient.

6.4 Some Econometric Issues

In the previous section we found that welfare measures or bounds on these measures depend on knowing about the hedonic price function, the bid function, or both. These are the only functions of consequence in welfare evaluation in the context of hedonic property markets. In general the latter is much more difficult to obtain than the former. Calculating what we have called the 'pure willingness to pay' effect is particularly difficult because it requires knowledge of the bid function, and therefore of preferences. In a previous section we discussed how, in principle, the researcher could recover this information. In practice the statistical problems that arise in attempting to do so are formidable. A complete treatment of the econometric issues is beyond the scope of this book on welfare measurement, but we sketch out the issues here and point to important papers and applications.

6.4.1 Estimating the Hedonic Price Function Only

Before proceeding to what has been called 'the identification problem' in the hedonic literature, we consider first the issues that arise in obtaining $P(\mathbf{q}^1) - P(\mathbf{q}^0)$, the change in price as measured along the hedonic price function when one or more attributes change and the hedonic price function does not shift. This requires only the estimation of the hedonic price function itself which may appear straightforward but is well known to present a number of practical challenges. The ones mentioned in most reviews of hedonic analysis are multicollinearity among attributes (especially structural attributes, but also neighborhood and proximity measures), spatial correlation in the errors and arbitrariness in the choice of functional form. These problems usually have acceptable, although not perfect, resolutions, and statistically significant relationships between important attributes and price can usually be found.

Almost all actual applications of hedonic analysis implicitly take the view that any econometric problems that do arise in the estimation of the hedonic price function itself are those listed above and are really of the nuisance variety. Yet Epple (1987) and others argue that except under fairly strict conditions, OLS estimation of the hedonic price function yields inconsistent results. How do we reconcile these different views?

When the inconsistency issue is discussed, it is often attributed to the fact that the property characteristics are endogenous–in the sense that households choose the bundle of characteristics jointly with the price they pay. This source of endogeneity is certainly grounds for concern when estimating the inverse demand function for attributes as represented in expression (6.5). But since the hedonic price function is a locus of equilibrium points and not a behavioral function, this does not seem a sufficient argument to establish inconsistency in the hedonic price equation. And in fact, it is not sufficient. If all property characteristics were measured,[9] then a regression of price on property characteristics would produce unbiased estimated coefficients. The estimated equation would describe the relationship between price and characteristics that emerges as a result of the allocation of properties with different characteristics to the highest bidders.

The inconsistency problem 'only' arises if some property characteristics are omitted and are correlated with included characteristics. The previous statement is obvious–this will be true of any regression. If omitted variables are correlated with included ones, then estimated coefficients of the included variables will be biased. The likelihood that many attributes will be highly correlated is also well-known. In fact researchers regularly struggle with the trade-off be-

[9] Strictly speaking we require that they are all measurable and measured without error, but we will concentrate on the case of omission, as it is the easiest to explain.

tween omitted variable bias and multicollinearity in hedonic models, especially with regard to structural characteristics.[10]

Most researchers view this as a serious problem, however, only if the focus of the study is a variable with highly correlated but omitted attributes. In the studies of interest to us, the key question is whether the environmental amenity or disamenity of interest is correlated with omitted variables. Some authors have worried about correlation among pollutants. In one of the earliest studies of air pollution and housing prices, Harrison and Rubinfeld (1978) find that air pollution variables are so highly correlated that they are forced to use nitrous oxides as a proxy for all pollutants (p 86). Others have considered the omitted variable bias that arises if the sources of pollution are also undesirable amenities, such as smoke stacks or automobile congestion (Leggett and Bockstael, 2000). In both these cases omission of important variables can bias upward the effect of changes in pollution on housing prices because of functional relationships between the target variable of interest and the omitted variables. Once recognized, the problem can often be addressed by controlling for these correlated effects.

Epple's concern extends beyond this simple result, however. He argues that the omitted variable bias is of special importance in the context of hedonic models because of the way in which the hedonic price function emerges. As we outlined earlier, the formation of the hedonic price equilibrium entails a type of endogenous sorting. Holding income constant, households with stronger tastes for air quality, for example, will choose to consume relatively more of it, and so it will be these households that are found in positions such as q_k^h in Figure 6.3, while households with relatively weak preferences for air quality will be found at positions such as q_k^i. Household preferences for various attributes will tend to be correlated, so that households choosing to be at high levels of one environmental attribute are more likely to be at high levels of other environmental attributes (and possibly other amenities) as well. Allowing for variation across households in income only further magnifies this correlation. Wealthier households, all other things equal, will tend to consume larger amounts of all desirable attributes. This, in itself, does not cause any bias. However, it increases the likelihood that attributes omitted from the hedonic price function will be correlated with included attributes. This source of omitted variable bias no longer depends on some external functional relationship between attributes (such as square footage and number of rooms or air pollution and proximity to smoke stacks) but is likely to occur simply because household preferences over attributes will be correlated and this will be exacerbated by the income

[10]A larger than average number of bathrooms will be found in conjunction with a larger than average number of bedrooms, for example. And houses with more rooms are more likely to contain more square footage.

effect. Consequently, households that are observed buying high levels of air quality will likely also be buying better views, quieter or safer neighborhoods, and any number of other unmeasured attributes. Overcoming or avoiding this type of omitted variable bias is much more difficult, as it is difficult to reason out which omitted variables may cause bias.

Epple describes conditions by which the hedonic price function parameters can be consistently estimated, but these conditions require estimation of the entire system described in (6.5). This negates any of the practical gains from obtaining the simple measure, $P(\mathbf{q}^1) - P(\mathbf{q}^0)$, from estimation of the hedonic price function alone, and leaves some skeptics to argue that without an appropriate strategy to eliminate the problems caused by endogenous sorting, the hedonic price function does not fully reflect what we want and expect. As a consequence, virtually all applied papers ignore this source of bias, and little is known about whether the bias thus generated is of a significant magnitude.

6.4.2 Recovering Information on Preferences

In his seminal paper on hedonic models, Rosen suggested that inverse demand functions for attributes could be obtained in a second stage estimation using information from the estimated hedonic function.[11] Specifically, one could estimate the hedonic function:

$$P = f(\mathbf{q}) + \varepsilon \tag{6.22}$$

and then calculate the marginal price paid for a relevant attribute, i.e. $\partial P/\partial q_k$ evaluated at the chosen level of attributes, for each household. This value could then be used as the dependent variable in an inverse demand function regression. Epple has since shown that the problem is one of estimating the system of $K+1$ equations:

$$\begin{aligned} P &= f(\mathbf{q}) + \varepsilon \\ \frac{\partial P}{\partial q_k} &= g_k(\mathbf{q}, y - P(\mathbf{q}), \mathbf{s}) + \upsilon_k \quad k = 1, ..., K, \end{aligned} \tag{6.23}$$

where $\mathbf{s}$ is a vector of household characteristics and ε and $\boldsymbol{\upsilon}$ are stochastic terms. The function $g_k(\cdot)$ is simply a marginal value function for attribute k defined as in (6.5).[12] Because the household's chosen $\mathbf{q}$ is clearly endogenous,

[11] To be precise, the functions suggested by Rosen are not inverse demand functions but rather marginal value functions as described in equations (6.5). Inverse demand curves with the properties of standard demand systems do not exist when the budget constraint is not linear.

[12] In Epple's presentation, there is also an inverse supply function for each attribute, as he treats the general case in which the heterogeneous good is produced by firms. However, he also treats the case of interest here–when supply is exogenous.

the inverse demand functions cannot be estimated consistently without the use of instruments, but exogenous instruments are difficult to come by.

In markets for homogeneous goods, where a single equilibrium price is determined by the intersection of supply and demand, econometric identification of the demand function is achieved by having sufficient systematic variation in supply. Heuristically, shifts in supply trace out a demand function. This concept of identification cannot be carried over to the hedonic model. Contrary to initial beliefs, the identification problem in hedonics is not a matter of distinguishing between supply and demand. In fact, even in the usual case where the supply of housing is assumed fixed, the identification problem still exists. The problem arises because of the difficulty in differentiating the inverse demand function, which provides the necessary information about preferences, from the hedonic price function.

The graph in Figure 6.3 illustrates the problem. In this graph, the bid functions of three households are depicted, all tangent to the hedonic price function but at different points. These households optimize at different points because of different preferences and incomes. Having information on each household's marginal attribute price and quantity gives us only one point on each of the household bid functions, however. It tells us nothing about the shape of any bid function, and it is the shape of this function (which is related so directly to the indifference curve) that captures information about preferences. The identification problem with respect to the inverse demand functions arises even if the hedonic price function were measurable without error, because there can be an infinite number of bid function families consistent with the same hedonic price function.

Brown and Rosen (1982) were the first to demonstrate the identification problem, but did so using a very specific example–one in which the hedonic price function was estimated as a quadratic and the demand function as a linear function. If a quadratic hedonic price function given by

$$P(\mathbf{q}) = \delta_0 + \Sigma_k \delta_k q_k + 0.5 \cdot \Sigma_k \Sigma_j \delta_{kj} q_k q_j + \varepsilon$$

is estimated, then the marginal price of q_k for household i is calculated as $\hat{P}_{q_k i} = \hat{\delta}_k + \Sigma_j \hat{\delta}_{kj} q_j^i$. When the second stage regression of this constructed price on attribute quantities is linear in attributes (as well as any number of socio-economic variables) it will merely return parameter estimates that are simple transformations of the $\delta' s$ that could have been derived exactly, without any estimation. In fact, the second stage will yield no new information and we will learn nothing about preferences. In retrospect, Brown and Rosen's choice of functional forms forced this result and is overly restrictive. However, the example still helps to illustrate the problem.

What has emerged from the several papers are two alternative modelling strategies. These strategies turn out to be two ways to effect identification.

Kahn and Lang (1988) present a clear statement of the problem, rewriting (6.23) as

$$\begin{aligned} P &= f(\mathbf{q}, \mathbf{W}) + \varepsilon \\ \frac{\partial P}{\partial q_k} &= g_k(\mathbf{q}, y - P(\mathbf{q}), \mathbf{s}) + \upsilon_k \quad k = 1, ..., K. \end{aligned}$$

In this system, $\mathbf{W}$ is a vector of exogenous variables that do not appear in the inverse demand functions. Kahn and Lang show that one identification strategy depends on the existence of measurable elements of $\mathbf{W}$.

Under what conditions will a vector such as $\mathbf{W}$ exist? In a single market, there can be no exogenous variables that affect the hedonic price function that are not inherent in the inverse demand functions, since it is these demands that determine the hedonic function. Variables can affect P and not g_k only through the matching process. If multiple distinct markets for the housing good exist, however, they will likely contain different distributions of attribute bundles and different joint distributions of incomes and preferences over their populations. Even if demand functions are the same over all households, irrespective of market, the matching process will be different because of the different bundles of attributes in the stock of housing. These different markets provide elements of $\mathbf{W}$ that can be used for identification of the inverse demand functions.

Intuitively, multiple markets allow us to observe more than one point on any one type of household's bid function, as similar people are presented with different shaped hedonic price functions. As Kahn and Lang point out, dummy variables for different markets (and possibly cross products of the dummy variables and elements of the $\mathbf{s}$ vector) can be used as instruments. One important requirement for identification is that buyers cannot have self-selected into different markets on the basis of the attributes of interest. Applications of the multiple market identification strategy include Palmquist (1984), Parsons (1986), and Boyle, Poor, and Taylor (1999).

A second strategy involves specifying functional forms for the $P(\mathbf{q})$ and $g(\mathbf{q}, y - P(\mathbf{q}), \mathbf{s})$ functions that mathematically force identification. Kahn and Lang show that parameters of the marginal value function can be recovered if elements of $\mathbf{q}$ appear in the marginal price function ($\partial P/\partial q_k$) with a higher power than in the inverse demand function. Some view this approach as not very satisfying since the choice of functional form is made, not on the basis of goodness of fit or economic theory, but because it permits a solution to the identification problem. However, if sufficiently general functional forms can be found that allow identification, then this criticism loses some of its force. Chattopadhyay (1999) offers some hope. As we mentioned earlier, he employs two functional forms for utility and six for the hedonic price function and finds stable results across specifications.

Both identification strategies require considerably more work than the estimation of the hedonic price function itself (ignoring Epple's concerns about the identification of the latter function's parameters). They also require more data. The estimation of the hedonic equation only is quite straightforward. Researchers generally estimate $P(\mathbf{q})$ using sales and accompanying housing attribute data obtained from Multiple Listing Services or tax assessment files, both of which are reasonably easy to acquire. Large datasets are widely available because real estate transactions are so common. In order to recover preference parameters, the researcher needs not only house price and attribute data (and, sometimes, for more than one market), but also socioeconomic data (elements of $\mathbf{s}$) for the buyers. Such data are not readily available. (The same demands on data are made by discrete choice models, as we discuss below.) It is not surprising that most hedonic analyses stop at the estimation of the hedonic price function and use the results as bounds on welfare measures.

Recently Ekeland, Heckman and Nesheim (2004) have revisited the identification of preferences problem in the hedonic model. They demonstrate that with appropriate instruments–that is, instruments that have the required independence from stochastic errors in preferences–it is feasible to identify preference parameters. Such instruments are difficult to find. Indeed, the endogenous sorting posited by Epple suggests that there may be no such instruments.

6.5 The Housing Choice as a Discrete Choice

Setting the house choice problem in a discrete choice setting was initially suggested by McFadden (1978). The appeal is a practical one. Housing attributes are not always available continuously over a wide range and in all dimensions, even in large markets with heterogeneous housing stocks. The discrete choice setting emphasizes the heterogeneity of alternatives in the market but does not assume that individuals can optimally adjust at the margin in all attribute dimensions. It also provides a direct means of revealing something about preferences.

Random utility model results from simulation experiments have produced *marginal* welfare measures similar to those from a corresponding conventional hedonic model using the same data (Mason and Quigley, 1997), but more accurate *non-marginal* welfare measures than those generated in a two-stage hedonic model (Cropper, Deck, Kishor and McConnell, 1993). While a number of discrete choice hedonic applications can be found in the general housing literature, few environmental applications of this approach have been published. Palmquist and Israngkura (1999) compare a conventional two-stage hedonic analysis using 13 housing markets with a discrete choice model estimated with the same market data. Chattopadhyay (2000) compares a two-stage hedonic where identification is achieved by assuming appropriate functional forms with a *nested* random utility model, using data from Chicago.[13]

The decision being modeled by the discrete choice version of the hedonic problem is each household's choice of house among all available houses ($m \in M$, where M is the available set), based on the house attributes, $\mathbf{q}^m$, and household characteristics, $\mathbf{s}$. The behavioral model is one in which the household is observed to choose house j, if

$$u(z_i, \mathbf{q}_i^j | \mathbf{s}_i) \geq \max u(z_i, \mathbf{q}_i^m | \mathbf{s}_i) \; for \; all \; m \in M_i, \tag{6.24}$$

where i denotes the household. The amount of the composite commodity purchased by household i, denoted z_i, cannot be observed, but a functional relationship exists between z_i, observable income, and house price: $z_i = y_i - P(j)$, where $P(j)$ is the price that prevails for a house with attribute bundle $\mathbf{q}^j$. In the discrete choice model this notation is preferred to $P(\mathbf{q}^j)$, signifying that price need not be a smooth function of quality characteristics. It also emphasizes the fact that the empirical model depends on the actual housing price, not some estimated function.

Suppose utility in equation (6.24) can be written $u(z, \mathbf{q}^j | \mathbf{s}) = v(z, \mathbf{q}^j | \mathbf{s}) + \varepsilon_j$ and ε_j has a type-I extreme value density function. The probability of household i choosing house j can be written explicitly as

$$\Pr(i \text{ chooses } j) = \frac{\exp[v(y_i - P(j), \mathbf{q}_i^j; \mathbf{s}_i]}{\sum_m \exp[v(y_i - P(m), \mathbf{q}_i^m; \mathbf{s}_i]}. \tag{6.25}$$

In effect, the model in (6.25) attempts to explain *which* household chooses *which* house. If all households were alike, then the random utility model would fail, because by its nature the random utility model attempts to explain the allocation process based on differences in households. If all households were alike the random utility model could contribute nothing to our understanding of how much households value different elements of $\mathbf{q}$, but estimating the hedonic price function would tell us everything. This is because the $P(\mathbf{q})$ function would then be synonymous with the common household indifference curve. But people are not alike, and the discrete choice approach avoids some of the difficulties that arise and assumptions required when using the conventional hedonic framework.

This approach is appealing for conceptual as well as practical reasons, particularly when the goal is to estimate pure willingness to pay. The indirect utility function, $v(\cdot)$, is exactly the function we wish to recover. The random utility framework reveals it in one step rather than through the convoluted

[13] Both studies rely on Federal Housing Authority (FHA) data, which contains information about the socioeconomic characteristics of the buyers.

and econometrically questionable route required by the conventional two-stage hedonic analysis. But the strongest argument for using the random utility model approach is still that it does not require a differentiable relationship to exist between price and attributes. Approaches to recovering preferences in a conventional hedonic setting rely heavily on the accuracy of the slopes of the estimated hedonic model as they represent attribute prices used in the second stage. While it is certainly the case that an active market will generate housing prices that are increasing in amenities, the idea that a well-defined, differentiable hedonic price function exists is largely a fiction.

Even if such a function did exist, errors in estimating it, including errors in choice of functional form, could generate considerable measurement error in the price of attributes used in the second stage of analysis. Banzhaf has provided evidence from simulations of the difficulties in recovering accurate marginal willingness to pay measures, even when there is no measurement error and no omitted variables, especially for those attributes of secondary importance to households. These problems grow with either discreteness or correlation in the attributes. In contrast, the random utility model requires that we be able to measure the house price, but not marginal prices, accurately. The random utility framework would also seem to avoid the difficulty of finding adequate instruments for the endogenous levels of the attributes in the second stage inverse demand function. However, it is not clear that the discrete choice framework avoids endogeneity altogether, as we will discuss below. What this approach does not avoid is the necessity of imposing considerable structure on the functional form of the indirect utility function. As we will note in a moment, some non-linearity will be necessary, but estimating generalized functional forms, especially non-linearities in parameters, makes the discrete choice problem more difficult to solve.

Violations of the independence of irrelevant alternatives property of the simple logit random utility model will obviously arise in the housing choice. But introducing nesting structures can resolve this problem and, in doing so, allows the researcher to provide a potentially more realistic characterization of the choice problem. In fact, it was in the context of housing choice that McFadden first illustrated the nested logit model. Typical applications (such as the early empirical application by Quigley, 1985) consider the housing decision as the choice of dwelling, conditioned on choice of neighborhood, which itself is conditioned on choice of city, town, county or other local governmental district. This model approximates a decision process in which the best house in each possible neighborhood and town is first evaluated based on structural and lot attributes. Then the best neighborhood is determined using the housing information from the first stage (as represented in an inclusive value) together with neighborhood attributes such as crime, accessibility, and environmental quality (e.g. air quality, proximity to hazardous waste site, etc.). Finally, the best town is determined based on information from the second stage (as represented in an inclusive value) together with town varying characteristics such as school quality, tax rates, etc. While most nesting attempts are structured along geographic lines, nests could be constructed according to the importance of attributes in households' preferences. For example, school quality may take precedence over everything else for some families, followed by size of house, etc. Environmental quality may matter, but only in differentiating among the houses that are identified as preferred by other criteria.

6.5.1 Drawbacks of Discrete Choice Housing Models

Although nesting based on preference hierarchies might better approximate decisions, it is precisely the importance of attributes in households' preferences that we are attempting to estimate. Typically we have little *a priori* knowledge, except that gained by introspection. Chattopadhyay (2000) compares welfare measures from four different nesting structures with those obtained in a conventional two-stage hedonic. He finds that non-marginal welfare estimates based on nested random utility models tend to be much lower than corresponding estimates from traditional hedonic models, but that different nesting schemes can also generate quite different non-marginal welfare estimates. Chattopadhyay's different nesting structures simply define the groupings of city-type and neighborhood-type differently. For example, one nesting strategy groups cities by property tax ranges while another groups on the basis of per-capita municipal spending ranges. Neighborhoods are grouped on the basis of different configurations of racial composition and income. More dramatic redefinitions of nesting structures might lead to increasingly unstable welfare estimates. The finding that nesting structure affects welfare measurement is not new. It has been found to be true in the recreational demand literature as well (see Kling and Thomson, 1996).

Other problems carry over from the recreational demand experience with random utility models–such as the definition of the set of available alternatives facing the household. In the recreation setting, much has been written about the problems that arise when households are ignorant of some of the alternatives that the researcher includes in the set (e.g. Haab and Hicks, 1997; Hicks and Strand, 2000). In the hedonic framework, the Multiple Listings Service (MLS) and increasingly available internet resources make information widely available, suggesting that potential buyers know what is available in the market. But the timing of the set of available alternatives relative to the timing of any household's search process is difficult for researchers to ascertain. Ideally, we would want to include only those houses that were on the market at the time a household was searching, but such information is not typically available. One strategy is to include all houses that sold in a given time period (one or more years, for example) and all households that purchased a house in that same time period. The conventional hedonic model uses this approach, but does not attempt to explain precisely the same behavior as the discrete choice model. Including all houses on the market in a given time frame will typically produce a very large number of alternatives, much too large to be empirically tractable. The difficulty is overcome using McFadden's (1978) solution of including the chosen alternative and limited random draws from other alternatives, a procedure that produces consistent estimates of the unknown parameters.[14]

In one important way, the housing choice problem differs from the recreational demand application. In the recreation application, site quality characteristics often do not vary over households, unless perceived rather than objective measures of quality are used, but travel costs do vary over households, and this variation is key in explaining choices. In the housing model, the only things that vary over a set of households facing the same alternative set are the characteristics of the households themselves. Even the prices of alternatives do not change across households. Add to this the fact that every house is chosen by some household and we realize that the entire burden of explaining *which* household chooses *which* house is borne by variation in household characteristics.

This feature of the housing choice–that each house has only one household–imposes limits on the ability of the RUM approach to estimate preferences. By way of illustration, consider a problem in which the researcher ignores the composite commodity and attempts to estimate the random utility model expressed in (6.25) as a function solely of housing attributes, $\mathbf{q}$. To make the point even clearer, suppose the elements of $\mathbf{q}$ are all discrete variables–for example, whether the property is on waterfront, whether the house has air conditioning, whether it is served by public sewer and water, whether it is located in an area of air quality attainment, etc. Estimation will produce results that reflect only the prevalence of these characteristics in the housing stock currently on the market and will reveal nothing about preferences. The coefficient on waterfront will be negative for the simple reason that more households will be observed choosing non-waterfront homes because there are more of them on the market than waterfront houses.[15] The relationship between parameter estimates and the composition of the choice set is a reflection of the role

---

[14]Palmquist and Israngkura use another approach. They include for each household, the house actually purchased, as well as the ten houses that sold just before and the ten that sold just after that purchase.

[15]In the simplest case, where there is a single trait–say being on the waterfront–the proportion of waterfront houses in the sample completely determines the coefficient on the waterfront dummy variable.

of sampling in estimating random utility models, as discussed in Manski and McFadden (1981).

The point is that inclusion of household characteristics is important in explaining why a given household (or household type) makes a particular choice. Introducing cross-product terms of household characteristics and housing attributes allows preferences and not just income to be different across households and helps explain what is being attempted in (6.25). Chattopadhyay crosses all housing attributes with family size and race and interprets his results in terms of the relative preference of one group over another for certain attributes.

Even when other household characteristics are included, but especially when they are not, the inclusion of the $y - P$ term holds a critical place in the model. The v function clearly cannot be linear in this term, as is usually assumed in recreation applications, or y will cancel out making choices independent of income. Including $y - P$ nonlinearly makes economic sense because, unlike the usual recreation case, the difference in prices across housing alternatives will be sufficiently great to generate income effects. To see how the model works, consider the simplest of models in which there is one property attribute, q_a, and the only household characteristic is income. Then $(y_i - P(j))$ is implicitly being traded off against q_{aj}; each household is trading the amount of money left over after buying a house against the characteristic that the house has to offer and choosing the house that provides the most utility. The term $(y_i - P(q_a))$ must be included non-linearly so as to retain income in the model but also because the level curves of $v(\cdot)$ are indifference curves and will not be linear in housing and the numeraire. Unfortunately household socio-economic variables, so important in estimating discrete choice models for housing, are not typically available in housing sales datasets. For example, data from real estate listing services such as MLS do not include buyer characteristics.

In the recreational model as many households as want can choose any given alternative, subject only to congestion effects. In the housing choice problem, there is an implicit one-to-one matching between households and alternatives. If two households attempt to choose the same house, the price will be bid up until one prospective buyer wins. Unlike the homogeneous good case, households do have an influence on prices in the housing market and it is within this bidding framework that the hedonic prices are determined. Hence the model in (6.25) is not a complete description of equilibrium. Prices are endogenous and there is a side condition that requires that one and only one household can choose each alternative. The new equilibrium sorting models (Bayer and Timmons, forthcoming) solve this problem through a multiple step estimation process. In addition to providing a useful approach to managing unobserved housing characteristics in the estimation of the discrete choice models, this framework imposes the condition that fixed supply equals demand, thus enforcing the one house-one household constraint.

A different description of the equilibrium process is given by Ellickson's random bid model (Ellickson, 1981; Lerman and Kern, 1983). In this model the probability that a household ends up in a given house is estimated as a function of household and housing attributes. Specifically, the probability that a household of type h outbids households of other types for house j is given by

$$\Pr\{\theta(\mathbf{q}^j, y_h, \mathbf{s}_h, u_h) + \varepsilon_h^* > \theta(\mathbf{q}^j, y_n, \mathbf{s}_n, u_n) + \varepsilon_n^*\} \text{ for all } n \neq h,$$

where ε_h^* represents the maximum of the ε_{i_h}'s over all households $i \in h$ and ε_{i_h} is the unobserved part of preferences for household i in group h. With the appropriate assumptions about the error structure, the probability that a household in group h chooses house j becomes

$$\frac{\exp(\theta(\mathbf{q}^j, y_h, \mathbf{s}_h, u_h))}{\sum\limits_{n \in N} \exp(\theta(\mathbf{q}^j, y_n, \mathbf{s}_n, u_n))}.$$

In this form, the parameters need to be normalized on one consumer type. Lerman and Kern have suggested a modified approach that makes use of the information provided by the sales price. They note that $\theta(\mathbf{q}^j, y_h, \mathbf{s}_h, u_h) = P^j$, while $\theta(\mathbf{q}^j, y_n, \mathbf{s}_n, u_n) \leq P^j$ for $n \neq j$. This model implies a different likelihood function that can be maximized using iterative methods. Using the Chicago housing data, Chattopadhyay (1998) compares the Lerman and Kern model with that from a standard hedonic application and finds that results do not differ markedly.

6.6 Conclusions

Hedonic housing models provide one of the few settings in which observations of *market* behavior can provide information about willingness to pay for environmental and publicly provided amenities. This has enormous appeal. Although housing markets are not perfect, no one who has bought or sold a house doubts the connection between attributes and house price. The argument that variation in housing prices might contain some information about the values people place on attributes can easily be made convincing to non-economists.

By using housing prices to value environmental attributes, researchers have come to understand that recovering *marginal* values of amenities may not pose huge problems. The recovery of preference schedules is made a great deal more complex, however. We have explored how the hedonic model works to provide estimates of the marginal values of amenities directly from fitting functions to observable data, and how researchers have attempted to recover preference schedules that allow estimates of non-marginal changes in amenities from observations on hedonic markets. Knowing about preferences is necessary if we want to estimate 'pure willingness to pay' measures. However, 'simply' knowing the hedonic price function is sufficient to analyze the ultimate welfare consequences of a localized change in environmental quality. Once moving and transactions costs are introduced, available answers tend to take the form of bounds, but the good news is that these will sometimes depend on information from the hedonic price function rather than preference schedules.

Although easier than recovering preferences, estimating the hedonic price function poses problems as well. It is the nature of hedonic housing markets that bundles of attributes are often highly correlated. Omission of correlated variables causes bias and inclusion affects precision. Yet accurate estimates of environmental attribute coefficients are key to environmental valuation. All this means that the central tendency and precision of parameter estimates for attributes that are of secondary or tertiary importance in the housing decisions are likely to be unstable, changing substantially with different specifications. The practice of reporting values for environmental attributes based on a significant regression parameter, regardless of their likely importance in the house purchase process, is questionable.

To date the case that environmental amenities affect housing prices has been made for some attributes, such as air quality in cities with particularly steep quality gradients, noise pollution from air traffic, and proximity to hazardous waste and landfill sites. The case is notably more difficult to make for water quality where variation within a market is more difficult to find. Attempts to value rainfall or sportfishing quality using housing hedonics are far less plausible. To have confidence in even marginal welfare effects of attributes, one must be confident that the attribute has some salience in the purchaser's mind.

Chapter 7

Hedonic Wage Analysis

7.1 Introduction

For at least two hundred years, economists have argued that a competitive labor market will generate higher wages in return for less desirable working conditions, such as hazardous conditions or poorer on-the-job amenities. This expectation has lead to the development of the theory of compensating wage differentials and the estimation of hedonic wage models, the second type of hedonic model that has engaged environmental economists. In this chapter we investigate how wage differentials have been employed in valuing changes in environmental amenities.

We consider the hedonic wage model in a separate chapter because its literature differs significantly from the hedonic housing model. Further, as we explain below, empirical evidence based on hedonic wage equations provides substantial support for the benefits of the Clean Air Act, yet this evidence remains controversial. Because of the inordinate leverage of estimates from hedonic wage models, we explore the estimation process and data sources to a much greater extent than for other methods.

Differences arise between property and wage hedonic models for several reasons. For one thing, property is a durable asset and jobs are not. In the housing market, resales of already produced homes dominate the market, leaving the role of the producer of houses as largely peripheral to the analysis. In the job market, the decisions of firms that supply jobs are potentially as important in sorting out the underlying hedonic story as the decisions of those who seek employment. A second difference is that the wage schedule a worker faces depends on the characteristics of the job and the characteristics of the worker. A given worker's productivity in a job will be determined in part by his/her education and experience. Hence the wage schedule can be expected to differ among individuals in different skill groups. Indeed some workers simply will not be qualified for some jobs.[1] The importance of these two features will become clear as we discuss the assumptions generally made in estimating the hedonic wage model.

Hedonic wage analysis has played a role in two different environmental valuation literatures. The first has to do with valuing workplace amenities. In theory it should be possible to use hedonic wage analysis to value exogenous changes in these job attributes, but empirical evidence of compensating wage differentials has been difficult to uncover. In practice, the amenity that has received almost all the attention is workplace safety. As we will see, the value of safety is estimated in the context of the labor market with the purpose of inferring the value of reducing risk in other, often environmentally-induced, health risk settings.

The second type of application of hedonic wage analysis is quite different and is really an extension of the hedonic property analysis. Most property value studies have considered price variation along an amenity gradient within a metropolitan housing market. When the public good of interest varies only across broad regions but not within metropolitan areas, intercity wage variation has seemed intuitively to provide a better, or at least additional, source of information about preferences. Climate variations clearly fall into this category. Recent applications have attempted to measure the effect of climate amenities and disamenities on wages, spawned by the increased interest in global warming and climate change.

Because the underlying modeling issues are very different in these two wage hedonic settings, we develop the analytical models separately, beginning with the valuation of workplace amenities.

7.2 Hedonic Wages in Theory

The setting is a market in which workers have preferences for job amenities and firms supply those amenities and hire workers in making production decisions. Firms differ in the costs of providing the amenities and workers differ in their preferences for the amenities. These amenities could be any non-wage attributes of the job, such as the level of comfort (perhaps in terms of temperature control) of the work place, the existence and quality of on-site child care facilities, or the level of workplace safety. As in Chapter 6, attributes are

[1]It is possible that a similar situation could arise in a housing market if there existed 'red-lining' or other severe forms of discrimination based on race, ethnicity, age or other characteristics of potential buyers.

designated by a vector, $\mathbf{q}$, and are measured such that higher levels of any q are valued more highly by workers and may cost firms more to supply.

The marketplace interaction of workers with different preferences for amenities and firms with different cost structures generates a locus of equilibrium wage-amenity combinations which we will call the hedonic wage function. Abstracting from reality for a moment and treating q as a scalar, we denote the hedonic wage function as $w(q)$. This is a menu of wage-amenity pairs that workers and firms can choose from. The hedonic wage function emerges as the result of concerted actions by agents in the labor market. However each agent takes the function as given.

7.2.1 The Simple Model

Consistent with almost all the literature in this area, the worker's labor supply decision is ignored.[2] The worker is assumed to choose one job, with the hours worked at that job predetermined. The focus of the analysis is the choice of job, defined by the combination of wage and workplace amenity levels. In this exposition we will assume only one such amenity to keep the notation simple. In order to see how the market wage hedonic emerges, first consider the trade-off between wage and amenity levels that firms and workers are willing to make. The worker, facing the hedonic wage schedule $w(q)$, seeks to maximize utility

$$\max_{z,q} u(z, q; \alpha) \text{ where } \bar{y} + w(q) - z = 0 \tag{7.1}$$

where z represents the composite commodity, $\bar{y}$ is non-wage (exogenous) income normalized on the price of the composite commodity, w is the normalized annual wage and α is a parametric index of the worker's taste for the desirable amenity q. The distribution of α across the population of workers describes the distribution of preferences for q in that population. To make the theoretical exposition transparent, we assume for now that all workers are identically productive, not differing in their human capital or level of effort. In addition, the worker's taste parameter is assumed not to affect his productivity, so that the parameter α is of no consequence to the employing firm.

Substituting $z = \bar{y} + w$ into the utility function yields $u(\bar{y} + w, q; \alpha)$, where utility increases in both w and q. By substituting for the composite commodity, we can view the household as choosing among amenity-wage combinations. We may use the notation u_w from this formulation because the substitution makes clear that $u_w \equiv u_z$. At the utility maximizing solution the worker sets his marginal rate of substitution between q and income equal to the marginal

[2]An exception is Smith, Pattanayak, and van Houtven (2003) who merge a labor supply model with a model of work-place risk.

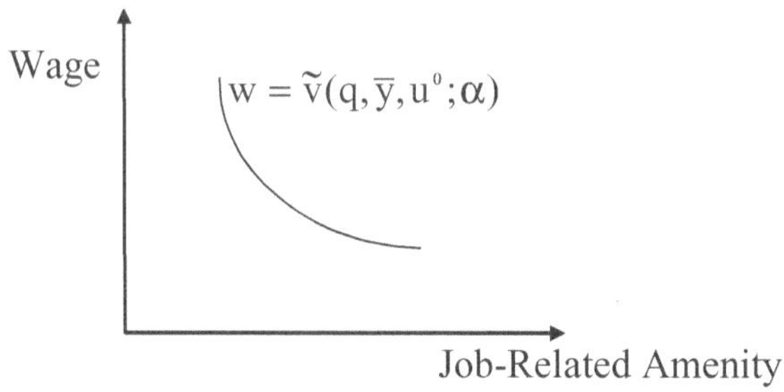

FIGURE 7.1. Indifference Curve of Worker

trade-off between wages and the amenity in the work place:

$$-\frac{u_q}{u_w} = \frac{dw}{dq}. \tag{7.2}$$

The second term in expression (7.2) is simply the slope of the hedonic wage function that the worker faces, while the first term is the slope of an indifference curve between money and the workplace amenity. Because u_w is the marginal utility of income, the first term can be interpreted as (minus) the marginal value of q. This marginal rate of substitution will be a function of the amenity preference parameter, α, and exogenous income, $\bar{y}$. When the amenity is desirable, the implied indifference curve will naturally slope downward in q, w space, so that $-u_q/u_w$ will be negative. Whether drawn in terms of (q, z) or (q, w), the indifference curve will be convex to the origin as long as preferences for q and z are well-behaved. In Figure 7.1, the function $w = \tilde{v}(q, \bar{y}, u^0; \alpha)$ denotes one such indifference curve between w and q, holding utility constant at u^0.

From the firm's perspective, the job-related amenity is likely costly to supply. For example, firms will incur costs if they attempt to control ambient temperatures in a manufacturing plant, adopt improved safety methods, or supply day care for workers' children. Thus firms possess isoprofit functions reflecting various combinations of wages and levels of q that hold profits at a constant level. Firms can be viewed as hiring workers to produce some output, z, and supplying q of the job-related amenity at some cost.

The firm, also facing the market hedonic wage function, maximizes profits:

$$\pi = \max_{L,q} pz(L) - w(q)L - c(q;\mu)L - c^0. \tag{7.3}$$

In this expression p is the output price, L is the amount of labor hired, w is wage per worker, c^0 is fixed costs, $c(q;\mu)$ is the cost of providing different levels of the work-place amenity per worker, and μ is a parameter indexing the efficiency of the firm in providing the amenity. Workers have no interest in the efficiency of firms and so are indifferent to different values of μ.

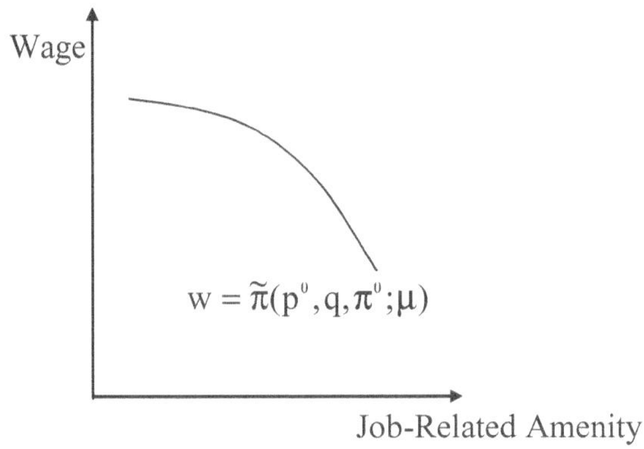

FIGURE 7.2. Isoprofit Curve of Firm

The profit maximizing firm will set the marginal cost of producing q equal to the market trade-off between wages and amenities, so that

$$-\frac{\pi_q}{\pi_w} = -c_q = \frac{dw}{dq}.$$

The slope of the isoprofit function equals (minus) the marginal cost of providing q and will be a function of the amenity efficiency parameter, μ. The isoprofit function will be downward sloping and concave to the origin as long as costs are rising at an increasing rate in q. As with the worker's indifference curve, it will not depend on the shape of the hedonic wage function but will be tangent to it. Figure 7.2 portrays an isoprofit function, labelled $w = \tilde{\pi}(p, q, \pi^0; \mu)$.

Each agent in this market takes the hedonic wage function as given. This function will emerge as the locus of equilibrium points in the market. The solution of the problem would be simple if all workers had the same preferences and exogenous income, and if all firms were identical. Then all workers would have a common indifference surface and all firms a common isoprofit curve. Add the simplifying assumption we have been making, that all jobs require the same skills and all workers possess the same levels of those skills, and an equilibrium would arise that consisted of one point–i.e. one wage-amenity combination (w^*, q^*). A hedonic wage *schedule* emerges only if there is heterogeneity in firms or workers or both.

If all workers were the same and firms differed in the cost of providing safety, the hedonic wage function would trace out the common indifference curve of workers (as in Figure 7.3). If all firms were the same and workers' preferences differed, the hedonic wage function would trace out the common isoprofit function of firms (see Figure 7.4). In practice, there will be heterogeneity among firms and among workers. When both workers and firms vary in type, there exists an exogenous distribution of workers induced by the joint distribution of

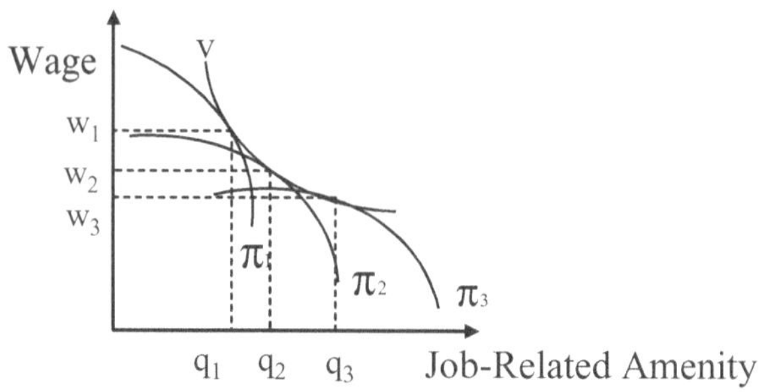

FIGURE 7.3. Hedonic Wage Function as Worker Indifference Curve

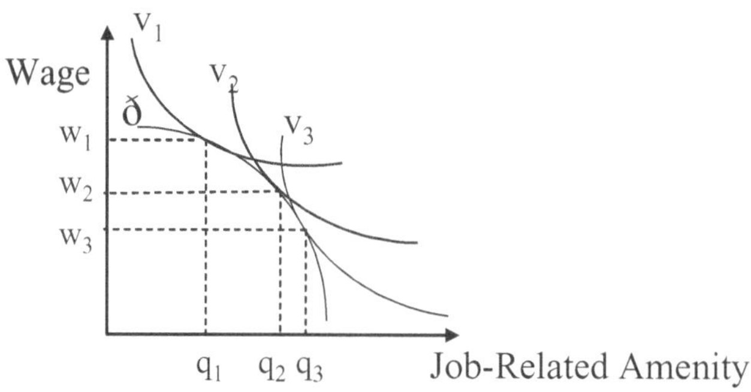

FIGURE 7.4. Hedonic Wage Function as Firm Offer Curve

α and $\bar{y}$ and an exogenous distribution of firms induced by the distribution of μ. Equilibrium requires transforming these distributions into a common set of (w, q) endogenous solutions by matching firms and workers. Firms supply jobs with different levels of the amenity and different wages; workers with different tastes for the amenity will sort themselves accordingly. If too few workers choose jobs at low levels of q, then supply and demand are not equated and wages must adjust to entice more workers.[3]

A hedonic wage schedule, such as the one depicted in Figure 7.5, emerges as an envelope of mutual tangencies. Note that a kind of matching is taking place in this graph. At a point like (w_3, q_3) workers with higher than average preferences for q end up matched with firms (or jobs) where the costs of supplying high levels of q are relatively low. Likewise, workers with relatively weak preferences for the amenity will match with firms that find that amenity relatively costly to supply.

[3]Kniesner and Leeth (1988) develop a numerical simulation of the equilibrating process in the labor market.

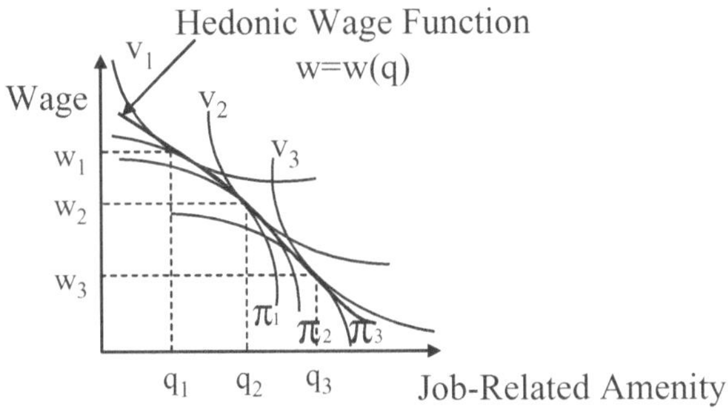

FIGURE 7.5. Hedonic Wage Function with Heterogeneous Workers and Firms

7.2.2 Revising the Model: The Wage vs Risk Trade-off

The most common workplace amenity treated in the hedonic wage literature is safety or its converse, risk. The essential idea is that a worker must be paid a wage premium to accept a job with higher risk of on-the-job injury or fatality. This literature investigates the trade-offs workers make between wages and risk.

Jones-Lee (1976) formalized the model that is the basis of the literature. The nature of the model is expected utility maximization over state dependent utility functions, where the two states are health and death or injury. The crucial element in this type of model is that the individual's decision affects the likelihood of the different states. We will modify the Jones-Lee model slightly to reflect both our notation and the nature of the compensating wage differential problem, as the Jones-Lee's model is stated in terms of wealth rather than wages. This distinction between wealth and annual income is potentially important but receives little attention in the literature.

We denote $u_0(y)$ as the utility in the state of no accident and $u_1(y)$ as the utility in case of an accident where y represents income from wage and non-wage sources. The two functions, u_0 and u_1, are usually expected to differ, and most certainly must differ in the mortality risk case for the problem to make sense. When the accident involves mortality, one may wish to specify the utility in the state of an accident as equal to zero, although it could alternatively be viewed as the utility one has for the income one leaves to one's heirs.

The maximization problem in (7.1) is now replaced by

$$\max E[u] = \max_{\rho} [(1-\rho) \cdot u_0(\bar{y} + w(\rho)) + \rho \cdot u_1(\bar{y} + \tilde{w})], \qquad (7.4)$$

where $\bar{y}$ still denotes exogenous income. $w(\rho)$ is the hedonic wage function that emerges from the equilibrating process described in previous sections and is taken as given by the worker. Now the attribute of interest is risk of injury or death, denoted ρ. The risk level, ρ, at which we observe an individual is often called the baseline risk, and as we shall see, this risk plays a considerable role in welfare analysis of the value of marginal changes in risk. In the hedonic literature, the risk is interpreted as occupational risk, the likelihood of job-related injury. Only a few authors have asked whether ignoring other sources of health risk biases the results. Eeckhoudt and Hammitt (2001) develop a model with other risks and demonstrate that greater background mortality and/or financial risks decrease the marginal value of job-safety. The authors conclude, however, that failure to account for background risk may not lead to substantial bias unless individuals face large risks elsewhere.

When the risk is one of mortality, u_1 reflects the worker's preferences over the death benefits ($\tilde{w}$) that are paid to heirs in the event of a job-related death. Alternatively, $\tilde{w}$ could represent the compensation payments made in the event of injury. It may be that $\tilde{w}$ will be related to $w(\rho)$ such that higher paid jobs also yield higher levels of compensation. This is true of most worker compensation plans, based as they are on the notion of lost income replacement. In any case, a good deal is 'swept under the rug' in this expression. While presumably a higher wage would generate more wealth and therefore higher exogenous income in future periods, the expression in (7.4) ignores the dynamic aspect, in keeping with common practice in the wage hedonic literature.

The first order condition can be rewritten as

$$\frac{u_0(\bar{y} + w(\rho)) - u_1(\bar{y} + \tilde{w})}{(1-\rho)u_0'(\bar{y} + w(\rho)) + \rho u_1'(\bar{y} + \tilde{w})} = \frac{dw}{d\rho}$$

which results from setting the slope of the indifference curve between w and ρ equal to the slope of the hedonic wage function. Note that this will be true no matter how we treat income in the injury or death state. Even if $\tilde{w}$ is zero or has nothing to do with $w(\rho)$ this interpretation will hold. Hence in equilibrium, the worker equates his marginal value of avoiding risk to the wage reduction he is forced by the market to accept for a marginal reduction in risk.

Jones-Lee argues that at any given level of y, we would expect $u_0(y) > u_1(y)$ and $\partial u_0/\partial w > \partial u_1/\partial w$. These inequalities are intuitively obvious when the accident outcome is death. When dealing with injury, the inequalities may not always hold. For example, if the pain and suffering from the injury are not great but the disutility associated with work is, then $u_0(y)$ might actually be less than $u_1(y)$ at any given value of y, although in this literature the risk is generally of serious injury. The second inequality is even less assured in the case of injury. A debilitated individual may have greater marginal utility of income, even though absolute levels of well-being are, at every level of income, smaller. This second condition is not required for the hedonic wage story to make sense, so we will not pursue it further.

The firm's optimization problem, when the amenity is on-the-job risk reduction, could look essentially like (7.3) if the firm is not responsible for paying death benefits or compensating for injury and if equally experienced and trained workers can be obtained to replace those who have been injured. In this case ρ simply substitutes for q and higher levels of ρ lower $c(\rho;\mu)$. However, the expression for the firm's profits can become considerably more complicated than in (7.3) depending on how one treats compensation in the injured or death state. For example, Kniesner and Leeth (1991) model the problem as one in which increased workplace safety raises the costs per worker, raises the productivity of any given sized workforce and lowers injury compensation, death benefits, and/or insurance costs. One way to represent profits is

$$\pi = \max_{L,\rho}\{pz(\tilde{L}) - \tilde{L}w(\rho) - c(\rho;\mu) - c^0 - b(\rho,\tilde{w})L\} \tag{7.5}$$

where $\tilde{L} = (1-\rho)L$ and terms within parentheses denote arguments of functions. The first four terms are identical to profits in (7.3) except that only the number of uninjured workers, $(1-\rho)L$, are paid wages and contribute to productivity. Now there are additional costs to higher risk in the workplace. The function $b(\rho,\tilde{w})$ reflects per worker insurance premiums which increase in risk and in the payout in the event of injury or death, $\tilde{w}$. Other representations of the firm's profit position are possible, depending on the nature of compensation schemes and the presence of regulations.

The first order condition for ρ for problem (7.5) is

$$\frac{\partial \pi}{\partial \rho} = -pz_L L + Lw - (1-\rho)L\frac{dw}{d\rho} - c_\rho - b_\rho L = 0, \tag{7.6}$$

where $dw/d\rho$ is the slope of the hedonic wage function in risk. Unlike the dw/dq of the previous section, this slope is positive because ρ is a 'bad'. Higher wages are needed to compensate for higher levels of risk. Expression (7.6) reflects the fact that marginal increases in risk lower profits by decreasing revenues through declines in productivity and increasing insurance costs. They reduce the costs of producing safety because the level of safety provided is lower. Higher wages also have an effect on the wage bill by reducing the number of workers (not injured) and increasing the wage per worker needed to compensate for the higher risk. Expression (7.6) can be rewritten as

$$\frac{-pz_L L + Lw - c_\rho - b_\rho L}{(1-\rho)L} = \frac{dw}{d\rho}. \tag{7.7}$$

The right hand side of (7.7) is the slope of the hedonic wage function that all firms face. The left hand side is actually the slope of an isoprofit curve drawn in w,ρ space. With this more complex profit expression, the slope of

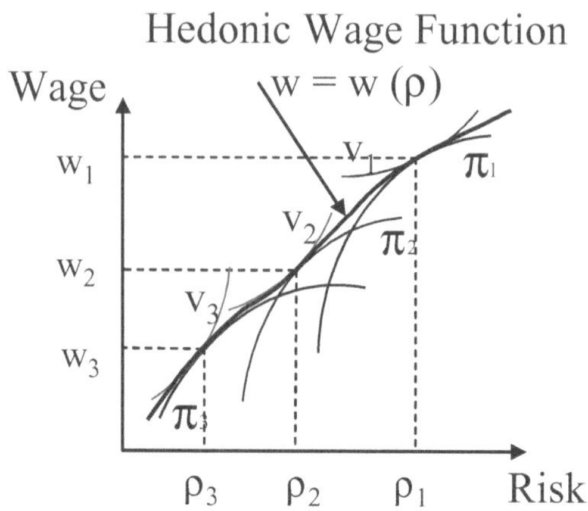

FIGURE 7.6. Hedonic Wage Function for Job Risk

the isoprofit function is not guaranteed to be positive over all levels of ρ. It is possible, for example, that the more the firm is made liable for accidents, the more truncated will be the distribution of workplace risk the firm cares to offer. However, over lower levels of risk, it is not unreasonable to assume that increasing safety raises costs. This may be especially true as we look across different types of industries and production processes as opposed to looking across firms producing the same goods and services. In fact, if firm behavior is attenuated by liability (as well as by safety regulations), it will be variation across industries or types of occupations rather than across firms in the same industry that will supply most of the variation in wage-risk pairs offered in the labor market.

Indifference curves for workers with varying preferences and isoprofit functions for firms with varying cost structures are depicted in Figure 7.6, where the envelop of tangencies traces out the hedonic wage function. Figure 7.6 illustrates the same concept as Figure 7.5 except that a) risk is a disamenity changing the slope of the relevant functions and b) the indifference curve now holds *expected* utility constant.

7.2.3 Important Underlying Assumptions

The underlying assumptions of any economic model are, of course, subject to challenge. The ones underlying hedonic wage analysis seem especially tenuous. The theoretical development outlined above argues that individual workers will choose their location on the wage hedonic function such that their marginal value of safety equals the marginal effect of safety on the wage rate. This is the key result that all subsequent welfare analysis turns on. For these results to hold, it must be true that a) workers value even small changes in workplace safety, b) they are fully informed of small distinctions in these safety levels, c) safety is provided at a cost by firms, and d) the labor market works well–it is approximately competitive so that risk can be treated as a 'transacted' attribute of a job and participants are observed at wage-risk equilibriums. As with any set of underlying assumptions, these are unlikely to hold perfectly. The question is the extent to which they are violated and the degree to which the validity of the compensating wage differential theory can then be challenged. Because the estimated *magnitude* of the wage-risk trade-off is the target of this analysis, the burden of proof is considerable.

Some of these assumptions have received considerable attention in the literature. The first of these is that workers possess accurate perceptions of risk levels in different jobs. While workers may have a good notion of the *relative* riskiness of different jobs, it is not clear that they would necessarily have a good sense of *absolute* levels. Occupational risks tend to be quite small, at least in the U.S. labor market where the mean workplace fatality risk is on the order of 4 in 100,000.[4] It is perfectly plausible that workers would view mining, which has an estimated risk of fatality rate over 25 in 100,000, as more dangerous than the finance, insurance, and real estate sector, with its 1.3 in 100,000 fatality risk. Whether workers respond to perceptions of differences among relatively safe industries (differences which may be on the order of less than 1 in 100,000) is more difficult to believe. Stated choice studies find especial difficulty in inducing systematic responses to very small changes in risks.[5] Even if workers perceive such differences in fatalities as consequential, they are unlikely to observe accurate estimates of risk. Fatalities are reported in numbers, requiring the estimation of total employment levels to construct risk rates. This leaves room for errors in inferences by even well-informed market participants who may at best be cognizant of fatality numbers rather than risks. For example in a study of industrial accidents, Viscusi and Zeckhauser (2004) demonstrate that potential jurists are quite insensitive to change in the denominator of a risk calculation, responding instead to the level of accidents.

Few papers have attempted to elicit workers' perceptions of risks. In a mail survey collected in 1984, Gegax, Gerking and Schulze (1988) asked workers to place their job on a risk ladder in which different steps were related to different numbers of fatalities per 4000 workers. The mean perceived risk of the full sample of respondents was 6.5 in 10,000, a figure much higher than what is

---

[4]This estimate is based on 1992-1997 Bureau of Labor Statistics data. See the tables at http://stats.bls.gov/iif/oshwc/cfoi/.

[5]See especially the study by Hammitt and Graham (1999), which reviews the basic issuses of perceptions of risk changes, and the contingent valuation experiments by Smith and Desvousges (1987).

commonly believed to be representative of workplace risk. These high levels are consistent with the fact that the ladder included examples of job risks obtained using Society of Actuaries estimates which have since been identified as over-estimating risk levels because they include non-work-related fatalities. While there have been other efforts using perceived risk in hedonic wage models (e.g. Liu and Hammitt, 1999), we know little about the correlation between risk levels perceived by workers and those calculated by researchers.[6] We know even less about the correlation between the perceived risk and 'actual' fatality risk, as the perceived risk poses problems in definition as well as calculation. We will return to this problem in a subsequent section.

The second assumption receiving particular attention in the literature is that workers can be found at wage-risk equilibriums. While this implies that workers are freely mobile, large transactions costs will restrict worker mobility. To understand the strength of this criticism, recall the parallel circumstances in the housing market. The hedonic housing literature recognizes that because of transactions and moving costs, households can not always be expected to attain the equilibrium described by first order conditions when attributes change. Changes in housing attributes need to be large enough to overcome these costs in order to induce relocation. Because hedonic housing price functions are usually estimated with sales data, it is reasonably safe to assume that if a household has overcome the 'inertia' due to these transactions costs and has entered the housing market, then its choice of location will reflect its marginal valuation for attributes. Transactions costs exist in the labor market as well. Individuals wishing to readjust their wage-amenity position will need to incur costs of job search to obtain equilibrium and may forfeit benefits associated with job tenure if they change positions. Herzog and Schlottman (1990) address the lack of mobility and costs of adjustment as impediments to equilibrium in the labor market. Because hedonic wage equations are estimated using samples of workers with no particular constraints on length of time in their current job, it is more difficult to dismiss the potential importance of these impediments than it is in the housing hedonic setting. In their study, Herzog and Schlottman argue that a large proportion of individuals included in most worker samples are out of equilibrium and thus are not equating their marginal value for amenities to the slope of the hedonic wage equation.

A third assumption is that labor markets are at least approximately competitive. That is, for any skill level, wages should reflect the attributes of a job, including risk, and the same wages should be available to all workers with that skill level. Dorman and Hagstrom (1998) argue that the large literature

---

[6] Slovic (1987) compares actuarial risks with perceived risks for a variety of hazards that are not typically work-related.

on inter-industry wage differentials has 'generated substantial evidence for the presence of noncompetitive forces operating within contemporary labor markets' (p 119). The 'inter-industry wage differential' to which they refer is a well-documented phenomenon in the economics literature. The term reflects the persistent empirical evidence that wage premia exist for some industries, irrespective of type of job or job attributes. Certain industries seem to pay higher wages, even controlling for observable worker characteristics, and this differential wage structure seems to have persisted over a very long time.

The literature on inter-industry wage differentials shows persistent differentials across industries and provides powerful *prima facie* evidence of the failure of wage rates to equilibrate (Dickens and Katz, 1987; Krueger and Summers, 1987). Krueger and Summers (1988) conclude 'we believe that the results here call into serious question the view that wage differentials can be rationalized with competitive textbook models. These differentials appear to be a pervasive empirical regularity' (p 280). Since the initial research, the presence of inter-industry wage differentials has been documented in numerous other countries and has been shown to be quite stable over time. Subsequent research has attempted to sort out how much of these differentials is due to heterogeneity in workers that is typically uncontrolled for by the researcher but observable to the firm and worker and how much is due to noncompetitive aspects of wage determination. While each may offer a partial explanation, the latter appears to explain the largest share in the inter-industry wage differential (see Allen, 1995; Blackburn, 1995; Blackburn and Neumark, 1992). Wage premiums are found to be correlated with product market power, rising capital intensities, and profit growth. Even if worker heterogeneity is responsible for some of this inter-industry wage differential, one needs to ask why better workers (in unobservable ways) are persistently found in certain industries. One explanation is that the industries making more than normal profits can outbid other industries for these better workers.

7.2.4 The Determinants of the Hedonic Function

In most of the wage hedonic literature the details of the firm, as opposed to the worker, are largely ignored. Yet the theory implies that the distribution of firm costs is as important in determining the shape of the hedonic wage locus as the distribution of workers' risk preferences. In fact, Rosen states that in the long run the equilibrium wage function will be the envelope of firms' isoprofit functions.

To appreciate the role of the firm side, consider a labor market that looks like Figure 7.6. If the π functions are long run isoprofit functions in a perfectly competitive labor market, then the level of 'excess' profits along each will be zero. Imposing the zero excess profit condition implies a unique envelope of isoprofit curves. In contrast, individuals' indifference curves are not uniquely determined, *a priori*. A family of such functions exists, depending on the utility level the individual is able to attain, once his constraints are considered. If safety is costly to produce and if the marginal cost of supplying it is increasing with safety–both within and across firms' cost structures, the envelope of the isoprofit functions will represent a market offer curve that workers take as given in determining the highest indifference curve they can attain (Smith, 1979, p 340).

Two sorts of changes can shift this locus. The first is a change in the cost structure of supplying safety. If technical change makes supplying safety cheaper, then firms' zero isoprofit functions will shift leftward in some pattern, allowing workers to reach higher indifference curves. The exact location of the new envelope of long run isoprofit functions will be determined by the differential impact of technical change on different types of firms. The second type of change that can shift the locus originates with the workers. As Smith points out, if many workers of type v_1 (in Figure 7.6) change their risk preferences to look more like workers of type v_3, then this will initially depress wages for this latter group. Also, there will be fewer workers taking jobs with firms denoted by π_1, increasing wages in those firms. Profits will increase for firms of type 3 and decrease for firms of type 1, a situation that can not be sustained in the long run. This will be true whether these different types of firms supply to the same output market or not. In the end, capital will move from firms of type 1 to firms of type 3. The result will be a shift in the equilibrium locus to a new distribution of firms in the industry. Given the new distribution of firms, the new envelope of firms' isoprofit functions will determine the hedonic wage equation.

This story of adjustment explains the role of workers' preferences, highlighting the ultimate importance of the structure and distribution of the costs of supplying safety. Arguing from Rosen's results, Smith (1979) maintains that as long as marginal costs of supplying safety are increasing over firms, the hedonic wage locus should be concave from below, even though it can shift as a result of both changes in the distribution of firms' costs and changes in the distribution of workers' preferences. Recognizing the role of both safety technology and risk preferences in determining the shape of the hedonic wage function, Kniesner and Leeth (1988) mimicked the process using simulation. The hedonic wage function emerged as linear or concave from below except when providing safety was extremely costly for some firms and workers with relatively low preferences for safety were rare. In any event, it is important to recognize that the shape and location (and therefore the slope at different levels of risk) will be affected not only by preferences for risk but also by the cost structures of firms. Changing technology of supplying safety will alter the hedonic wage function and alter the tangencies, and therefore will alter the wage-risk trade-offs of individual workers. If this is true, then safety regulations will affect the equilibrium locus and the wage-risk tangencies, as safety regulations alter the costs of supplying safety by raising the penalties for workplace risk.

7.2.5 The Anomaly of Safer Jobs and Higher Pay

In the housing market, where hedonic models were first used, it is common knowledge that houses with more desirable amenities sell at higher prices. In hedonic housing model applications this relationship is regularly supported by empirical findings, especially for the most important attributes such as size of house, number of rooms, lot size, commuting distance, etc. It is also supported by casual observation. Finding houses with better attributes that are also cheaper, at least within a given geographically defined market, is virtually impossible. The same consistency does not hold for hedonic wages. Jobs that are safer and pay more abound. CEO's have lower risk of accident and higher wages than construction workers. Computer programmers earn more and have safer jobs than machine operators in factories.

This anomaly arises in the labor market but not in the housing market because of differences in segmentation. Segmentation exists in the housing market, but it is based on geography. If in a geographic housing market two identical houses sell at different prices, buyers in the market will arbitrage the price difference until it disappears. The matching between buyers and houses takes place over the buyer's willingness to pay and the characteristics of the house. In the housing market, the seller has an interest in the buyer's willingness to pay, but not other characteristics of the buyer, and the buyer has interest only in the characteristics of the house but not of the seller. The labor market is characterized by considerable segmentation even within any geographical market. The buyer (the firm) has a significant interest in the characteristics of the seller (the worker). The worker's productivity (ability, education and/or skill) in the given job matters. Clearly some workers will not be eligible for some jobs for lack of appropriate skills or training, and others may be eligible but less qualified and will be hired only at different wage-risk combinations. When surgeons face higher wages and lower risks than manufacturing workers, differences in skills prevent arbitrage.

This segmentation has important empirical implications, as we will see, but it also reinforces the conceptual message of the previous section. Individuals with exactly the same risk-income preferences but different productivities will likely have different marginal valuations for risk because they will be facing different offer curves and will optimize at different wage-risk combinations. Figure 7.7 illustrates one example. In this figure, we draw two offer curves, one for workers with high productivity (π_H) and one for workers with low productivity (π_L). Workers are assumed to have identical preferences irrespective of productivity,

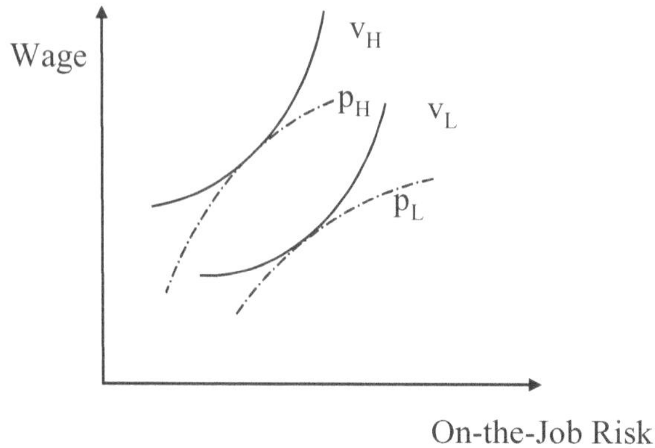

FIGURE 7.7. Market Segmentation Caused by Differences in Worker Productivity

but productivity levels will dictate the utility level each group is able to attain in the labor market. Marginal risk valuations can differ between these two types of individuals for two reasons. First, individuals optimize where indifference curves are tangent to different offer curves. Second, at higher wages an individual is likely to value risk more highly if safety is a normal good. Consequently, the slopes of a family of indifference curves evaluated at any given level of risk may be expected to be increasing in wages. This means that the indifference curves drawn in Figure 7.7 need not be vertically parallel. The relative choices of w and ρ by the two groups are impossible to determine without more information; we have represented one possible solution in which higher productivity workers choose less risk. In any event, there is no guarantee that the slopes of the indifference curves at the two optimal solutions will be the same, so we find that individuals with identical preferences can exhibit different marginal valuations for risk

Realistically, there are differences in workers and the same set of wage-risk combinations will not be available to all workers in the market. The segmentation is caused, fundamentally, by different marginal products for workers with different skills. It can be intensified by other (perhaps arbitrary) means such as licensing, institutional barriers, or accreditation. Any empirical analysis must take this into account. Smith (1979) argues for choosing a functional form that would allow for multiple offer function envelopes for workers with different productivity (signaled presumably by different amounts of human capital, such as education). This could be achieved by including such worker attributes in some non-linear form of the hedonic wage function when the sample includes heterogeneous workers. More often researchers have attempted to control for differences by estimating the hedonic wage function for a relatively homogeneous subset of the labor market, such as for blue collar workers. Perhaps the most common subgroup includes young males with limited education, as these individuals are found more often in high risk jobs. Doing so does not eliminate the problem, however. For one thing, not all productivity differences are measurable, even with proxy variables. As we will see in the section on estimation, unobserved heterogeneity in productivity causes severe problems for hedonic wage analysis. In addition, limiting the estimation to a subset of labor market participants limits the usefulness of the results in reflecting risk valuation for the population at large.

7.2.6 Endogenous Sorting

It is not always easy to determine, on the basis of observable data, whether individuals face the same or different hedonic wage functions, as not all relevant characteristics are measurable. This potential source of error takes on particular significance in the presence of Epple's endogenous sorting–an idea we introduced in the last chapter. In both the hedonic housing and wage analysis, there is some confusion in the literature between variation in preferences on the one hand and 'endogenous sorting' on the other. Varying risk preferences simply means that we can observe equilibria such as that pictured in Figure 7.6. Individual workers have different shaped indifference curves and find their optimum position along a wage-risk offer curve at different places. This variation does not cause any particular problems. In fact it provides one explanation for why individuals with similar skills can be found at different wage-rent trade-offs.

Endogenous sorting is a different phenomenon. It arises because both wage and risk are really endogenous variables, jointly chosen by the worker. Epple (1987), who raised this issue in the context of hedonic models in general, points out that OLS estimates of hedonic price function parameters are inconsistent in the presence of endogenous sorting. The endogenous sorting problem is particularly relevant in hedonic wage analysis. Since workers observed at any point on the wage hedonic will have jointly chosen both wage and risk level, unobserved heterogeneity in workers will be correlated with both wage and risk. The most troubling type of unobserved heterogeneity is differences in productivity, because these differences imply that workers are really facing different wage-risk offer curves, unbeknownst to the researcher.

In the last section we discussed the anomaly of higher pay for less risky jobs, when workers vary in productivity. We now revisit that same problem, considering the econometric implications. As we noted earlier, workers of different productivity face different wage-risk offer curves. The $X's$ in Figure 7.8 depict wage-risk pairs that might be observed in a labor market. Suppose these are really generated by differences in worker productivity that are perceived by firms. Consequently they lie along different offer curves. Increasing produc-

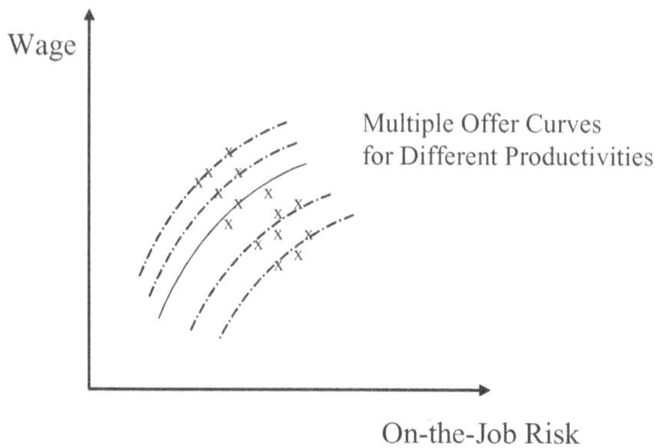

FIGURE 7.8. Firm Offer Functions for Workers with Different Productivities

tivity will allow a worker to move northwest in the diagram–that is, to move to higher wage-safety schedules and higher utility levels. But, suppose the researcher does not have the necessary information to control for heterogeneity in productivity. Given the way this diagram is drawn, any attempt at estimating a hedonic wage function will produce a negative rather than positive sign on risk. Hwang, Reed and Hubbard (1992) have shown with this kind of specification that the larger the share of worker productivity variance that is unobserved and the smaller the variation in risk preferences across workers, the larger will be the bias in the risk coefficient. This type of endogenous sorting creates a downward bias and, according to Hwang *et al.*, can easily be sufficient to reverse the sign on risk.

Garen (1988) considers another possibility. While not exhibiting higher productivity in general, some individuals may be relatively more productive in dangerous jobs. This is the 'cool-headed' effect often mentioned in the literature. Such individuals will have a higher marginal product in risky jobs than others and, as a consequence, will earn a higher wage in such jobs even though they may earn the same wage as others for normal jobs. All other things equal, such individuals will choose riskier jobs and this will bias the coefficient on risk, especially at high risk levels. This is an unobserved productivity phenomenon that is correlated with both wage and risk choice, but it is specific to risk levels.[7] For such cases, the wage hedonic model might alternatively be written as

$$\ln w_i = \beta_0 + \boldsymbol{\beta}_1 \mathbf{X}_i + \beta_2 \rho_i + \varepsilon_i + \eta_i \rho_i, \tag{7.8}$$

[7]Leeth and Ruser (2003) consider a related phenomenon. If some workers are less productive in risky situations (i.e. they are more 'accident prone'), but if those same workers are relatively less risk averse, then workers in risky situations may actually end up receiving less *total* wages than workers in safer jobs.

where for each worker w is wage, ρ is risk level, and $\mathbf{X}$ is a vector of worker characteristics. The error is composed of $\varepsilon_i + \eta_i \rho_i$, where ε_i may be correlated with ρ_i in the sense that Hwang, Reed and Hubbard describe the problem. In addition a portion of the error is not only correlated with risk but changes as a function of the risk level.

Garen attempted to resolve both problems using instrumental variables. As instruments he used non-wage income (principally the income of the spouse) and proxies for degree of risk aversion (specifically marital status, number of dependents, and value of owned home). Hwang, Reed and Hubbard question the usefulness of these instruments as all are correlated with marital status which is widely used as a proxy for unobserved human capital. Finding convincing and useful instruments in these hedonic models is notably difficult as Epple has argued and as such the endogeneity has been largely ignored in empirical applications. It has the potential of causing greater bias in the wage hedonic than housing hedonic setting, however, because worker's unobserved characteristics matter to firms.

7.2.7 Welfare with the Hedonic Wage Model

Schelling (1968) was perhaps the first to suggest using compensating wage differentials to reveal how people value safety in the workplace and to use this information in cost-benefit analyses of any policy that reduces health risks. By far the most prevalent use of wage hedonics is toward this end. The willingness to pay for safety (or in its more common form, the willingness to accept compensation for increased risk) has been estimated in the context of wage-risk trade-offs. However, these estimates have typically been used, not to value actual changes in workplace safety, but instead they have been applied generically to environmental policies that reduce health risks (and most commonly mortality risks) in other settings. Thus, the implicit argument underlying all this literature is that we learn about preferences for risk as they are *revealed* in the workplace, where the individual *can* actually choose to trade money for risk. Taking what we learn from the labor market, we apply this trade-off to cases where risk levels and changes in those levels are imposed on individuals.[8] In a subsequent section we will discuss some of the drawbacks of this 'extrapolation'.

---

[8]Wage hedonic analysis is rarely used to estimate the benefits of improved safety in the worksplace because of the implicit underlying assumption that safety is a choice variable to workers and not an exogenous attribute imposed on them. If instead, workers are constrained from reaching equilibrium, then this type of evaluation makes sense. In the presence of such constraints, using the slope of the hedonic wage function in the usual fashion to estimate workers' risk valuations will be inappropriate, however.

Before demonstrating the connection between welfare measures and the hedonic wage function, we define compensating and equivalent variation in this expected utility framework where risk is treated as exogenous so that changes in risk can be valued. It is common in this literature to consider the compensating variation of a change in the level of risk to be the amount of money, *with certainty*, that we would need to give to, or take away from, an individual after the change, such that his expected utilities before and after the change are equated.[9] The expression 'with certainty' implies that the same amount (equal to the CV) would need to be added to or taken away from income, irrespective of the uncertain outcome–that is, irrespective of whether the injury/death occurs. More precisely, starting from the expression in (7.4), the CV of a change in ρ is defined implicitly by:

$$
\begin{aligned}
& (1-\rho^0)\cdot u_0(\bar{y}+w)+\rho^0\cdot u_1(\bar{y}+\tilde{w}) && (7.9)\\
= \; & (1-\rho^1)\cdot u_0(\bar{y}+w-CV)+\rho^1\cdot u_1(\bar{y}+\tilde{w}-CV), && (7.10)
\end{aligned}
$$

where ρ^0 is the initial probability of an injury/death outcome and ρ^1 is the subsequent probability, after the policy change has taken place.

Just as in the hedonic housing analysis, the hedonic wage function can tell us nothing about the 'pure willingness to pay' for a discrete change in a workplace attribute. Unless we have a way of identifying the underlying indifference curves, we only know that the change along the hedonic wage function will overestimate the gains from a reduction in risk and underestimate the losses from an increase. The problem of identifying underlying preferences (indifference curves) is as difficult here as it is in the housing hedonic literature. With the exception of papers such as Biddle and Zarkin (1988) and Kahn and Lang (1988), few attempts to identify underlying preferences in the wage hedonic setting can be found. The studies that attempt to estimate structural models have resulted in even less robust estimates than the simple wage hedonic models (Viscusi, 1993). Given the tenuous standing of the literature on uncovering preferences, we limit our discussion to efforts to estimate marginal values.

At least in the US, environmental policy aimed at reducing mortality often involves quite small changes in risk, largely because baseline mortality risks of this sort are very small to begin with. Consequently, most researchers have felt justified in assuming that information about the slope of the wage hedonic is sufficient for valuing reductions in risk. If marginal changes in risk are being evaluated, then the marginal valuation a worker has for safety will be given by the slope of the hedonic wage function evaluated at the worker's location on that function. In theory, this is true only if workers perceive risks

---

[9] Equivalent variation would, of course, require the adjustment in income to be made in the initial state thus holding constant the ultimate expected utility level.

accurately and labor markets operate approximately competitively. In practice, the requirements for revealing this marginal valuation using available data and estimation methods are even more demanding.

7.3 Estimating the 'Value of a Statistical Life'

Governments face numerous opportunities of reducing people's risk exposure through regulation. Possible regulations vary in their ability to reduce the likelihood of accident or disease and in the costs they impose on society. In a paper that has attracted much attention from the legal profession, Morrall (1986) reported estimates of the cost-effectiveness of a number of federal regulations aimed at risk reduction–estimates ranging from a cost of \$100 per life saved to a cost of \$72 million.[10] The costs of some risk reductions are clearly so great as to make nonsensical any consideration of undertaking them. But at what point does a potential risk reduction become too expensive? Economists have argued that we should apply criteria based on the value individuals themselves put on safety, as revealed by the decisions they make in the market place or work place.[11]

Based on the idea of revealed preferences for risk-income trade-offs, economists have used results from the hedonic wage estimation for the valuation of reduced mortality risk, specifically by providing estimates of the 'value of a statistical life' (VSL). The VSL is a term used by US federal agencies to denote the benefit measure to be placed on every 'statistical life' saved when undertaking a cost-benefit analysis of a new regulation. A policy saves one statistical life if it diminishes the likelihood of mortality by $1/N$ where N is the size of the affected population. The VSL is defined as that population's aggregate willingness to pay for an increase in one expected life saved.

In the absence of impediments to equilibrium, the slope of the hedonic wage has the interpretation of the value of a marginal change in workplace risk for the individual found at that baseline level of risk. In Figure 7.6, whoever chooses to be at (w_2, ρ_2) reveals his marginal trade-off between money and risk as the slope of the hedonic wage function at that point. These wage-risk trade-offs, it is argued, could be converted to VSL estimates.

To make the logic of the VSL clear, suppose that a group of similar indi-

---

[10] A more recent paper (Morrall, 2003) responds to critics and adds several new regulations to the list.

[11] A slightly different stream of literature has considered risk-risk trade-offs. Given that increases in real income have been shown to lead to better health status, reductions in real income to workers and consumers from increased regulatory costs would lead to increased mortality. This literature argues that no regulation should be considered for which the cost-induced mortality would exceed direct mortality reductions.

viduals in a labor market are all found at a given point (a given 'baseline' risk) on a common hedonic wage function. At this point, suppose the slope of the wage hedonic equals 0.4, where annual wages are measured in thousands of dollars and risk is measured in deaths per 10,000. This implies a trade-off of \$400 in annual wages to reduce the risk of mortality by $1/10,000$. If the policy reduces the risk of mortality by this amount (by 1 in $10,000$) for $60,000$ identical people subject to the same baseline risk, then the aggregate willingness to pay of these $60,000$ people for this risk reduction would be calculated as $\$400 \times 60,000 = \$24$ *million* and the expected reduction in mortality is $(1/10,000) \times 60,000 = 6$. The VSL is calculated as aggregate willingness to pay divided by lives saved. Put another way, it is the implied change in income along the hedonic for a one unit change in risk. Given the estimated slope of 0.4, the $VSL = 0.4 \times \$1000/(1/10000) = \$4$ *million*.

An obvious problem arises in practice. If all individuals were found at the same baseline risk then a hedonic wage *function* would not exist. On the other hand, if a hedonic wage function does exist, then individual workers must be at different points along it, and unless the hedonic wage function is linear, different workers will be at different marginal risk valuations. There will be no common slope and no simple way to calculate a single VSL measure. In practice, researchers typically estimate a hedonic wage function using data on a sample of workers and then calculate the mean marginal wage-risk trade-off to be used to calculate a VSL estimate.

VSL estimates have played an important role in regulatory analysis. Policies that result in the curtailment of pollution bring about reductions in mortality risk. Using an estimate of the reduction in risk, the size of the exposed population, and a VSL estimate, agencies regularly calculate the benefits of policies aimed at risk reduction. Risk reduction valuation is central to two recent efforts to assess environmental policy benefits: the prospective and retrospective analyses of benefits from the Clean Air Act Amendments (CAAA).[12] The prospective study estimates the benefits accruing between 1990 and 2010 due to the CAAA as \$110 billion (in 1990\$, mean level of benefits).[13] Of the \$110 billion in benefits, 90% are attributable to reductions in risk of mortality.

Not all estimates of the VSL depend on hedonic wages analysis; stated pref-

---

[12]The prospective study is *The Benefits and Costs of the Clean Air Act 1990 to 2010: EPA Report to Congress.* November 1999. Environmental Protection Agency, Office of Policy and Office of Air and Radiation. Report # EPA-410-R-99-001. 1 v. The retrospective analysis is entitled *The Benefits and Costs of the Clean Air Act, 1970 to 1990. October 1997.* Environmental Protection Agency, Office of Policy, Planning and Evaluation and Office of Air and Radiation. Report # EPA 410-R-97-002.

[13]The total benefits estimate for 2010 can be found in Appendix H (page H-30) and the total cost estimate for 2010 on page iii of the Executive summary, both in *The Benefits and Costs of the Clean Air Act, 1990 to 2010.*

erences methods are also used. But hedonic wage analysis provides the only behavioral evidence, and so it is not surprising that the approach has come under considerable scrutiny–all the more so because of the variation in estimates that can be found in the empirical literature. At last count there were more than 40 published papers analyzing the wage hedonic function for the purposes of estimating the value of a statistical life. In a meta-analysis, Mrozek and Taylor (2002) review 33 studies and over 200 specifications of the hedonic wage equation, with statistically significant VSL estimates ranging from \$16,000 to over \$30 million (in US 1998\$).[14] Viscusi and Aldy (2003) summarize estimates of the VSL from labor market studies, finding the estimates range from \$0.5 million to \$20.8 million (US 2000\$).[15]

In subsequent sections we outline the basic structure and the differing specifications used in U.S. hedonic wage applications and, because hedonic wage analysis exhibits such substantial influence on policy evaluation, we also review attempts to determine the sources of variation in these estimates.

7.3.1 Data Sources for Wage and Risk Variables

Of all valuation methods, only hedonic wage analysis of risks relies on national markets and a common pool of datasets. Part of the challenge in estimating the VSL arises in the use of these datasets to construct risk estimates and corresponding wage rates. Given the importance of the VSL for policy evaluation and the well-known variability of estimates from hedonic wage analysis, we discuss the data construction process.

All hedonic wage analysis employs individual data for a sample of workers. Occasionally these data are obtained directly through a researcher-implemented survey, but more commonly the sample and data are drawn from an existing source, such as the Current Population Survey, the Panel Study of Income Dynamics, the Quality of Employment Survey or the National Longitudinal Study of Youth. These sources include wage rate; worker characteristics such as age, education, experience, gender, race, marital status, and union membership; and a few job-related attributes such as industry, occupation, blue vs white collar job, and whether supervisory duties or strenuous physical activity are involved.

The two key variables in the analysis are the wage and risk variables. The

[14]This is the range of positive and significant estimates, but at least 16 of the 33 studies report some specifications with insignificant or negative estimated wage-risk trade-offs.

[15]The very low estimates come from studies using Society of Actuaries data, which is typically an order of magnitude higher because it includes all causes of fatality, not just on the job risk, and hence would like find wages less responsive to differences in this risk measure.

worker survey supplies the wage rate. This is usually a before-tax wage rate because of lack of tax information, even though we might prefer to measure the after-tax trade-off between income and risk. If we assume that workers make their employment choices on the basis of after-tax wages, then using pre-tax wages has two implications. One concerns the errors that are systematic across states due to different state tax systems, though these could possibly be accounted for by state dummy variables. The second arises because income taxes are roughly proportional so that using before-tax wages leads to a steeper wage-risk function than would be obtained if an after-tax wage was used. This will not be a problem, however, if the relationship is estimated in semi-log form, since the percentage change in wage is all that matters. In their meta-analysis, Mrozek and Taylor did not find a significant effect on VSL estimates due to before-tax vs after-tax differences, but this could have been due to the small number of studies in which the latter were available.

All VSL studies must incorporate a value for mortality risk for each worker in the sample, but no such measure can be found in worker surveys.[16] Wage hedonic studies must match individual workers to risk estimates available from other sources, linking the two by industry or occupation of the worker. Although a few studies in the 1970's and 1980's used fatality risk data from the Society of Actuaries (SOA), most studies over the last 25 years have depended on either the Bureau of Labor Statistics (BLS) estimates based on their Survey of Working Conditions (available from the late 1960's to about 1990), the Bureau of Labor Statistics (BLS) estimates based on their Census of Fatal Occupational Injury (beginning in 1995), or the National Institute of Occupational Safety and Health (NIOSH) estimates from their National Traumatic Occupation Fatality Survey (which are reported for 5-year averages beginning in the 1981-1985 period). The SOA data reports fatalities for 37 occupation classes, for all causes of death whether job-related or not. For that reason and because their data is skewed towards higher risk jobs, the fatality rates in the SOA data are an order of magnitude larger than those in the BLS or NIOSH data. The latter include risks for on-the-job fatalities only. Although alike in this regard, the BLS and NIOSH risk data differ substantively in ways we will discuss more thoroughly in a later section.

A compensating wage differential for risk can be expected only if there is a mechanism by which workers (or their representatives) can perceive not only *differences* in workplace risk across jobs, firms, and/or industries but also *absolute* risk levels. If their perceptions do not match actuarial risks, then it is presumably the perceived not the actuarial risk that should be used to esti-

[16] This is not quite true as the Quality of Employment Survey asks workers whether they face job hazards or not.

mate the worker's value for safety. A few researchers have attempted to include 'self-reported' risk. The Quality of Employment Survey of workers supplies information on whether the worker believes he faces hazards on his job. Some researchers have included this information in their wage hedonic, most notably by reducing the risk variable to zero for any worker who does not perceive himself to be in danger. For example, as a measure of risk, Moore and Viscusi (1988) use the industry risk level when the worker regards the work as hazardous, and zero if the worker does not perceive the work as hazardous (p 377). More quantitative levels of self-reported risk were used by Gegax, Gerking and Schulze (1988) who asked workers to place their job on a risk ladder in which different steps were related to different numbers of fatalities per 4000 workers. Several examples of jobs with these risk levels were supplied as examples. The authors found positive and significant effects for unionized and blue collar workers and not for others. The implied VSL measures were lower than generally found in the literature, but this is completely consistent with the fact that the absolute risk levels used by the authors to construct the risk ladder were based on SOA estimates which have since been identified as over-estimating risk levels.

As we do not have accurate actuarial risk measures for individual workers, it is difficult to test whether workers' perceived risks are accurate. In addition, it remains a question whether workers have sufficient information about the small risks and differences in those risks that exist in the labor market to make well informed trade-offs.

7.3.2 Variability in Specifications

Before returning to the risk data issues introduced in the last section, we consider other types of variation in model specification that can affect estimates. Here we turn to some of the conceptual issues raised earlier in this chapter and consider how researchers have handled these issues empirically. In discussing these variations, we rely heavily on Mrozek and Taylor (2002) who provide a careful review of the bulk of this literature.

Functional Form

Given that the ultimate aim of the literature is to recover the slope of the hedonic wage function, the functional relationship specified between wage and risk is clearly important. The functional form will also affect how one obtains a single VSL estimate from a sample of workers and related risk levels. For example, some studies estimate a linear relationship between wage and risk, and interpret the coefficient on risk as a sort of average effect of risk on wages. However, more than three-quarters of the 200-plus specifications included in

Mrozek and Taylor's meta-analysis involved a log transformation of wages. One of the implications of a semi-log form is that if wages are to be increasing in risk, they must be increasing at an increasing rate. Smith (1979) suggests that this functional form may often be chosen under the misguided notion that the shape of the hedonic wage function should be the same as the shape of an individual's indifference curve. But there is nothing in the theory that prevents the locus of equilibrium points in the labor market from taking some other shape. On the one hand, if workers have identical or very similar preferences then the long-run hedonic wage function should be increasing and convex. However, as we mentioned earlier, Viscusi (2003), Herzog and Schlottman, Smith (1979) and Rosen (1974) maintained that the hedonic wage function will be concave from below if marginal costs of supplying safety are increasing across firms; and Kniesner and Leeth repeatedly produced a concave hedonic wage function in their simulations.

At least four papers have included risk as a quadratic in their wage hedonic function and found both risk and risk-squared to be significant (Leigh and Folsum, 1984; Olson, 1981; Dorsey and Walzer, 1983; Scotton and Taylor, 2003).[17] All of the results imply a concave wage-risk specification. Black, Galdo and Liu (2003) conclude that there is a highly non-linear relationship between wages and risk levels. A related phenomenon appears at the meta-analysis level. Mrozek and Taylor found that VSL estimates increased at a decreasing rate in the average baseline risk of the study and turned downward at risk levels well within the range of risk in the data. This was true for the entire sample, even in specifications that included a dummy variable for high risk samples, and continued to a lesser degree in the analysis that omitted studies with particularly high risk samples or studies that used the SOA risk data. Ultimately the shape of the hedonic wage function is an empirical issue, and so it makes good sense to let the data determine the shape with a flexible function.

Controlling for Differences in Skill Levels and Risk Preferences

Raw correlations between wages and risk levels tend to reveal no association. That is, high risk levels are found in conjunction with high and low wages. The absence of raw correlation between wages and risk means that any empirical finding of a wage premium for risk depends heavily on econometric specification. Only by controlling for other factors will such a premium, if it exists, be revealed.

If workers have different productivity in jobs then they will face different hedonic wage functions. This is the motivation for including education and/or

[17] These include models in which wages (the dependent variable) are included in linear form and models in which wages have been transformed into logs.

experience in the wage hedonic. It is also the motivation for including distinctions between blue and white collar jobs. Where samples include both these types of workers, researchers regularly include a dummy variable for this difference.[18] This distinction together with whether the job is supervisory or not are the only additional variables describing job requirements that are typically available for inclusion in the model. Smith (1979) has argued that where market segmentation is expected, the functional form should explicitly allow for different wage-risk trade-offs at any given level of baseline risks. The alternative more often chosen is to select as homogeneous a sample as possible–most commonly male, blue collar workers between the ages of 30 and 50, as these are workers more likely to face perceptible risk in their jobs.

Researchers regularly include an array of socio-demographic variables in their wage hedonic models, such as gender (where relevant), race, marital status, and age, but less frequently provide a theoretical explanation for their inclusion. If these socio-demographic differences are not proxies for productivity differences, then one possible explanation for their inclusion in a hedonic wage model is that these personal characteristics are correlated with different preferences for risk. It is widely believed, for example, that women are typically more risk averse than men and that married individuals are more risk averse than those who are single.

Worker age is particularly interesting to contemplate in this light. Analyses based on life cycle models (e.g. Shepard and Zeckhauser 1984) find that in a world in which consumption is spread evenly over all years of life (i.e. one in which a riskless rate of interest exists equal to the subjective rate of time preference), then WTP for risk reduction should decline monotonically with age. However, if the individual cannot be a net borrower and consumption is constrained by net income, then it is possible to find WTP increasing up to a point and then falling. These results are based on a model in which the utility of living is a function only of consumption and not on length of life itself (Freeman, 1993b).

The conceptual models have motivated empirical research on the effect of age on the VSL, but without providing structure for estimating wage equations. Viscusi (1993, 2003) maintains that older workers should have lower VSL's because of a shorter expected remaining lifetime. Moore and Viscusi (1988) refined this idea by including the discounted loss in life expectancy in their hedonic wage model, taking the form $\rho(1-e^{-rT})/r$ where r is the discount rate, T is the expected remaining years of life and ρ is the annual risk of mortality. Using this model they estimated a discount rate of 10-12% and an implicit

[18]The evidence for a compensating wage premium for risk among blue collar workers is regularly found to be much stronger than among white collar workers, presumably because the latter do not face perceptible on-the-job risks.

value per discounted expected life year lost of $175,000. The basic tenet that older people have a lower implicit value of life (even if they have the same or higher value for a discounted expected life year) has been challenged by Smith, Evans, Kim and Taylor (2004) who found that the 'near elderly' may possess even higher marginal valuations for risk reduction than do younger workers.

Socio-demographic variables may be correlated with preferences in another way. Because safety is a normal good, individuals with more wealth are also generally expected to have stronger preferences for safety. To the extent that wealth (or the wealth an individual experienced while growing up) is correlated with socio-demographic variables, the latter could be systematically correlated with varying risk preferences.

A less often considered explanation for including socio-demographic variables in a hedonic wage model is that some individuals may face different hedonic wage functions due to discrimination. Analyses of this question generally exploit the uneven distribution of individual types over risky occupations or risky industries to attempt to test alternative hypotheses (see Leeth and Ruser, 2003; Viscusi, 2003). For example Viscusi finds evidence that blacks are discriminated against and actually face different offer curves than whites in blue collar jobs. But because risk data are only available by industry or occupation, and not by gender, age, or race, alternative hypotheses are difficult to test. For example, an alternative hypothesis includes the possibility of self-sorting into jobs with differing but unmeasured risk. Specifically, consider what might happen when using industry risk measures. Suppose women are indeed more risk averse and tend to sort themselves into lower risk jobs in any industry. Given limitations in data, the same risk levels are generally attributable to all types of jobs within a given industry. Women in industries with higher average risk may be in jobs involving no more risk than other industries, but it will appear as though women are being paid less than men to accept high risk. Lower apparent compensation for risk for women could be due to discrimination but could also be due to this self-sorting phenomenon.

One final control variable needs to be discussed because it has been found to have such an important empirical effect. A large and increasing number of wage-hedonic studies control for union status. Viscusi and Aldy (2003) cite ten papers that specifically test for differences in the wage-risk trade-off over union status, and all but one of these finds a significantly larger wage response to risk for unionized jobs. Of the nine studies that find a significant difference, five fail to find a positive significant effect of risk on wages in non-unionized jobs and three find significant effects for both but the magnitude of the implied VSL from the unionized job sample is at least twice as great (see for example Olson, 1981; Dillingham and Smith, 1983; Dorman and Hagstrom, 1998; Viscusi, 1980). Some researchers have argued that unions, because they possess an institutional memory and act on behalf of workers, can process and disseminate risk information better. If unions are instrumental in conveying accurate risk information to their members or if they understand the risk preferences of their members and bargain accordingly, then this might support the notion that compensating wage differentials actually capture individual worker's trade-offs between money and risk for union workers. However, some have argued that the wage premia merely reflect union bargaining power and do not reveal worker willingness to pay for safety. In any event, the results cast doubt on whether non-union workers possess either the knowledge or bargaining power to capture the compensating wage differential.

Controlling for Different Forms of Compensation

Job packages typically include other forms of compensation besides wages. Also, there will potentially be other disamenities associated with the job besides risk. At best, omission of these will increase the amount of variation in wages left unexplained and reduce the precision of parameter estimates. But more likely, some of these factors will be correlated with risk, biasing its estimated coefficient. If firms that supply more safety also tend to supply more of other desirable workplace amenities, then the VSL estimates will tend to be biased upwards when the other amenities are omitted. This direction of bias will hold, as well, if jobs that are dangerous are also inherently unpleasant. Alternatively if firms use non-wage incentives (such as better benefits packages) to attract workers to risky jobs, then omission of these non-wage job amenities will bias the VSL estimates downward.[19] Unfortunately, few variables are available in most worker data sets to reflect job and work place characteristics. Mrozek and Taylor report that more than 40% of the hedonic wage specifications they analyze in their meta-analysis include no job characteristics.

Certainly the risk of injury is a disamenity of a job that will be correlated with risk of death and should, in theory, be included in the wage hedonic. As such, omitting the risk of injury can be expected to bias upward the coefficient on the risk of fatality. But the correlation between the two is so great that multicollinearity often poses an obstacle to obtaining precise enough coefficient estimates for fatality risk. Several papers have omitted injury risk for this reason (e.g. Dillingham and Smith, 1984; Kniesner and Leeth, 1991).

An important element in the wage package of a worker facing a risky job is the compensation he would receive should he be injured on the job or the death benefits his beneficiaries would receive should he be killed. Omitting information about workers' compensation payments may bias the results because

[19]It is possible that some firms (or perhaps more correctly, some industries) offer higher wages and better amenities, irrespective of risk levels, to attract workers with better (but unobserved) characteristics.

this compensation is part of the wage package and may vary with risk. Dorsey and Walzer (1983) and Viscusi and Moore (1987) have argued that workers may accept a combination of higher wages and better worker compensation packages in exchange for more risk. Arnould and Nichols (1983) were among the first to include worker compensation measures in a wage hedonic function. Specifically they included the income replacement rate available through the particular worker's state mandated compensation scheme. Viscusi and Moore (1987) included a worker specific measure of the expected value of compensation received–that is, the probability of accident times the benefit that would be paid. In general, inclusion of workers compensation measures has led to increases in the estimated wage-risk trade-off and therefore increases in the implied VSL estimate. This is to be expected, especially if the compensation measure is included in expected value form, because the compensation measure will be correlated with the risk of injury.

7.3.3 Fragility of Estimates of the Wage-Risk Trade-off

Wage hedonic studies have produced a wide range of VSL estimates and this has left many troubled. Some of this variation can be expected on theoretical grounds, while some is due to the sorts of variations in model specification outlined above. Arguably the greatest systematic variation in VSL estimates arises from a different source, however. Researchers are confronted by two interrelated problems in hedonic wage analysis: the measurement error implicit in available risk data and the persistence of inter-industry wage differentials. Differences in the treatment of these related problems have led to vastly different wage-hedonic results and implied VSL estimates.

Measurement Error in Risk Data

Since the first empirical research on the VSL by Thaler and Rosen (1976), researchers have been aware of the problems of matching workers and their wage rates with the risks that they face. For example the use of risk that varies only by the worker's industry would result in assigning the same risk of fatality to a coal miner and a secretary in the coal mining industry; using risk by occupation would attach the same risk to night-shift clerks in convenience stores as clerks in department stores. Leigh (1995), Dorman and Hagstrom (1998), and Black and Kniesner (2003), as well as a recent report to EPA by Black, Galdo, and Liu (2003) have argued that VSL measures are highly sensitive to the risk data source and, because of the nature of these risk data, also sensitive to the occupation and industry controls included in the hedonic wage specification. Recently, these problems have been reduced to some extent, but far from eliminated, by the collection of occupational fatality data by industry and occupation in the Census of Fatal Occupational Injury.

As we indicated earlier, most hedonic wage studies depend on job-related risk data from one of two agencies–the Bureau of Labor Statistics (BLS) or the National Institute of Occupational Safety and Health (NIOSH). BLS data has changed over time. Procedural changes altered the way in which BLS data were collected and reported when OSHA was established, so that data before and after 1972 are not comparable. Most studies have used the post-1972 data from the Survey of Current Workers (BLS-SCW) which, like its predecessor, depends on surveys of workers and thus suffers from sampling and self-reporting bias. In 1992 the BLS began the Census of Fatal Occupational Injuries (BLS-CFOI), which, in keeping with its name, is far more complete and accurate than the earlier survey counts. Based as it is on at least two independent source documents (e.g. death certificates, workers compensation reports, etc.), it is also considered more reliable in this respect than NIOSH data which record fatalities from death certificates alone.[20] The BLS-CFOI data are available at the 2 or 3-digit SIC industry and occupation level but do not vary regionally, while NIOSH data are available at only the 1-digit occupation or industry level but are available by state.

Until the construction of the CFOI in 1992, risk data from both sources were available only by occupation or industry. When working with either occupational or industry risk data, researchers face the irreducible problem of assigning the same risk to jobs of very different natures. A second source of error arises in that workers may not be accurately assigned to occupation or industry, with assignment errors particularly prevalent for occupational groups. A third error in measurement problem arises because absolute numbers of fatalities are reported and must be translated into risk rates with the use of independent estimates of total employment. Risk cannot be directly observed, even under the best of circumstances, introducing yet another problem. All we ever observe are random draws from a distribution, and therefore any realization will be an imperfect measure of the 'true' risk level. Hence, most studies use mean death rates over multiple years, smoothing the annual variability but not eliminating errors of observation and assignment. The sampling variation can be particularly large for occupation or industry groups with small numbers of deaths, which is the reason that both BLS and NIOSH suppress risk data for cells with less than 5 deaths. Small numbers of deaths may occur because an industry or occupation has low risk or because it has relatively small total employment. If risk levels and employment totals are negatively correlated, as some have argued, the necessity of dropping cells with few deaths can have

[20]Several researchers report that it is often difficult to judge from death certificates issued in some jurisdictions whether the fatality is job-related or not.

consequences for the range of risk levels present in the data.

Finally, where it is possible to compare BLS and NIOSH risk estimates, they are discouraging in their inconsistency. Dorman and Hagstrom have argued in favor of the BLS data, at least in part because less aggregation across industry or occupation levels is to be preferred over regional variation. The choice among data sources is not an academic question. Because the risk levels are different and not highly correlated, the same specification will generate different implied $VSL's$ depending on which data set is used. One of the few robust results from the Mrozek and Taylor meta-analysis is that NIOSH data produce significantly larger VSL estimates than do BLS data.[21] All of the above casts doubt on the researcher's ability to produce accurate measures of one of the two most important variables in the model.[22] Black and Kniesner compare risk levels implied by four different risk data sources for a given set of workers drawn from the 1995 Outgoing Rotational Group of the Current Population Survey (ORG-CPS). The four measures are BLS-SCW risk by industry, BLS-SCW risk by occupation, NIOSH risk by industry and NIOSH risk by occupation. For the sample of 51,140 workers drawn from the ORG-CPS, Black and Kniesner find pair-wise correlations among the four alternative risk variables that range from 0.3 to 0.53.

In an extensive investigation using three sources of worker information–the March CPS, ORG-CPS, and the National Longitudinal Study of Youth (NLSY)–in conjunction with risk data from BLS-SCW and NIOSH, Black, Galdo and Liu (2003) illustrate the instability of the resulting valuations of marginal risk reduction. When the same model is estimated using a) different sources of data on workers, b) different sources of risk data, and c) different controls–the inclusion or elimination of state of residence, occupation, and/or industry dummies, the authors find wide variation in the wage-risk trade-off.[23]

[21]Of the specifications investigated by Mrozek and Taylor that used either BLS or NIOSH data, BLS data were employed 88% of the time.

[22]But even if accurate measures were possible, would individual workers perceive and appreciate these same measures? This remains an underlying problem and one that has not yet been adequately addressed. Clearly we need evidence that workers are exposed to the same risk estimates as the researchers are able to calculate.

[23]Each model is estimated using pairwise combinations of the three worker surveys and the two risk data sources. Both male and female worker subsamples are employed, with individuals ranging in age from 25 to 60. The workers from each survey are linked to BLS risk data on the basis of their 3-digit occupation or industry SIC classification and to the NIOSH risk data on the basis of their state of residence and their 1-digit occupation or industry classification. In each case, a hedonic wage model is estimated and takes the usual semi-log form including a standard set of controls including age, education, union status, marital status, race and ethnicity. Six dummy variables for firm size are also included in the March CPS and ORG-CPS estimations but only a dichotomous firm-size dummy variable is available for the NLSY estimations. However the latter includes additional information about the individual,

The Black *et al.* results reveal a basic conundrum embedded in wage hedonic analysis using either NIOSH or BLS-SCW data. Clearly much unobserved heterogeneity will exist across types of jobs and across types of industries. Inclusion of industry dummy variables in particular is regular procedure in wage studies in the labor economics literature to correct for this. But attempting to control for the unobserved heterogeneity in the wage-risk model using industry (or occupational) dummy variables (the only tools available, given the data) introduces serious collinearity because it is only at aggregate industry or occupation levels that risk measures are available. Thus, inclusion of these dummies robs the risk variables of at least some of their explanatory power. On the other hand, omitting these controls introduces potential bias and attributes much of inter-industry differentials to risk.

In their study, Black *et al.* find that the occupational measures of risk from BLS-SCW data produce estimates that are almost uniformly contrary to expectations. And BLS-SCW industry risk data, used in the majority of existing studies, produces highly unstable results with significant negative coefficients on risk arising in about three-quarters of the specifications. NIOSH industry risk data provide the most plausible results, but an independent rationale for using these data is difficult to come by. In any event, adding industry dummy variables in all specifications where industry risk is used reduces the size of the estimated wage-risk coefficient by more than 50%. As further indication of the relative fragility of wage-risk trade-off estimates, Black *et al.* find the effect on wages of other variables, such as education, to be very robust as compared with the instability of estimated risk coefficients. Black *et al.* find

> 'compelling evidence that there is a great deal of measurement error in the various measures of job risk....In addition, there appears to be a systematic bias that is correlated with many of the covariates that labor economists often include in wage equations' (Black, Galdo, and Liu, 2003, p 36).

Their results cast doubt on the reliability of VSL measures based on wage-risk trade-offs estimated from these data sources. While more detailed data sources are likely to become availability, their investigation of the impact of measurement error in estimating the hedonic wage equation provides a guide for future research.

---

including scores from the Armed Forces Qualification Test and length of tenure in current job. Each regression is then re-estimated including state dummies to account for differences in cost of living, as well as differences in state tax rates and compensation structures. The third estimation adds 1-digit industry (occupation) dummies if the risk measurement is by occupation (industry). The fourth estimation adds the occupation (industry) dummies.

Controlling Inter-industry Wage Differentials

One especially important finding of the Black *et al.* study is that the results vary dramatically depending on whether industry/occupation controls are included in the regression. This result is supported by the Mrozek and Taylor meta-analysis, which shows that the inclusion of multiple industry dummies significantly lowers VSL estimates.[24] Directly related to this finding, papers by Leigh (1995) and Dorman and Hagstrom (1998) argue that significant relationships between wage and risk found in the empirical literature are largely spurious and are due, not to risk differences, but to systematic inter-industry wage differentials of the sort discussed in an earlier section. They argue that the empirical evidence of inter-industry wage differentials should cause some skepticism over the ability of wages to reflect accurately the impact of small differences in risks that vary only at highly aggregate levels. This argument draws special strength from the fact that accounting for inter-industry wage differentials is routine practice in labor economics.

Dorman and Hagstrom provide the same type of comparative analysis as Black *et al.* using all four measures of fatal risk (BLS-SCW and NIOSH, by occupation and industry), except that they use the Panel Study of Income Dynamics (PSID) for their worker survey and include a larger array of explanatory variables. They compare results using a basic control model with one using industry dummy variables and a third using an array of industry specific variables that have a record of providing explanatory power in the extensive inter-industry wage differential literature. These industry-specific variables include capital-labor ratio, density of female workers, union density, average establishment size, value added per worker, and unemployment. As with the Black *et al.* results, the wage-risk parameter is highly unstable–sensitive to both the measurement of risk and the inclusion of industry controls. Except in the regressions using the NIOSH industry risk variable, the wage-risk coefficient is either insignificant or negative. Only the NIOSH industry risk variable, when crossed with union membership, appears to perform well, but the authors argue that the NIOSH risk measure 'possesses the least plausibility' because it varies only over state and 1-digit industry and because a further inquiry into correlation patterns suggests that it does not exhibit the sort of relationship with variables such as education that we would expect. Even if one accepts the accuracy of the NIOSH data, positive significant results can be found only for union workers.

---

[24]The authors include two different measures reflecting the inclusion of industry dummies. One equals the number of such industry dummies included and the other equals 1 if at least 4 industry dummies are included and 0 otherwise. Each of these measures is signifcant and negative in the meta-analysis explaining the variation in VSL estimates across studies.

The BLS-CFOI data have two advantages over the data discussed so far: these data provide the number of fatalities by industry *and* occupation and each fatality is authenticated by several sources of information. With these characteristics, the BLS-CFOI data would appear to eclipse other data sources for risk. Increasingly studies are using this new source, but not necessarily to its full potential. Leeth and Ruser (2003) use risk data only by occupation, while Black and Kniesner (2003) use risk data only by industry. Scotton and Taylor (2003) and Viscusi (2004) make full use of the industry/occupation risk data but using different specifications and generating vastly different VSL estimates. Viscusi includes measures of injury compensation that are omitted from Scotton and Taylor, but the latter include a full set of industry dummy variables not included in Viscusi's specification.

A more complete treatment of the problem along the lines of the Black *et al.* study but using BLS-CFOI data would be useful. Nonetheless, analysis of the measurement errors in the occupation-only risk data and the industry-only risk data by Black *et al.* and others makes past estimates of the wage-risk trade-offs of questionable value as benchmarks. It makes little sense to look back at these estimates to see whether current efforts are in the right ball park. In any event, a good deal of convincing evidence exists that the inter-industry wage differential must be addressed for VSL estimates to have credibility.

7.3.4 The Challenge of Transferring VSL Estimates

One can readily understand the enormous appeal that hedonic wage analysis has for economists and government agencies seeking a means to estimate the benefits from regulations aimed at saving lives. Valuing lives saved is a difficult and contentious issue, and the labor market provides a ubiquitous setting in which individuals reveal real, rather than hypothetical, trade-offs between money and risk. Such behavior seems a good source of evidence on people's preferences for risk-reduction. Whether it is appropriate to extrapolate value of life estimates thus obtained to settings outside the labor market is the question we take up in this section.

In most efforts to value changes in environmental amenities, the application is locally specific. Data needed to generate welfare estimates involves behavior related to the specific resource affected by the policy, project or event to be valued. But VSL estimates are applied more broadly–for example, a likely use would be to evaluate a policy designed to reduce air pollution throughout the US. Because of the importance of hedonic wage analysis in constructing VSL estimates and because of the far-ranging use made of these estimates, this type of revealed preference analysis will necessarily be subject to particular scrutiny.

Consider the conditions required to make the average slope of an estimated hedonic wage function a good measure of the marginal value of reducing the risk of fatality from environmental and other publicly determined hazards. For one thing, the estimate of the slope of the hedonic wage function with respect to risk should be unbiased. Yet, as we have seen, the nature of the data available to calculate the risk variable leads to measurement error and to correlation between risk measures and unobserved heterogeneity, making VSL estimates very sensitive to model specification.

A second condition requires that the slope of the hedonic wage function at an individual worker's wage-risk location actually reflect his marginal valuation of safety. This requires that the worker perceive risk accurately and that his wage-risk location be optimal–i.e. there should be no serious impediments to equilibrium. As we have seen, the literature contains several challenges to the assumption that the labor market is in equilibrium. Herzog and Schlottman argue that imperfect information, ineffective bargaining, and transactions costs prevent labor markets from being in a state of wage-risk equilibrium. They provide empirical evidence that for many workers the marginal value of risk reduction exceeds the labor market price for marginal changes in safety. Likewise, Dorman and Hagstrom argue that inter-industry wage differentials support the notion of imperfect competition in the labor market: '[T]he role of rent-sharing or other forms of strategic bargaining behavior...and the gender distribution of both wage and risk demonstrate that noncompetitive elements in US labor markets are sufficiently strong to overcome the competitive tendency toward equalizing differentials' (p 133-134). Finally, their result that only for union workers (and only using the NIOSH risk data) can significant positive compensating wage differentials be found empirically leads the authors to further question the interpretation of the slope of the hedonic wage function as a marginal willingness to pay for risk, since only those workers 'most insulated' from labor market competition appear to receive a wage premium for accepting additional risk.

From the perspective of public policy that seeks a unique VSL estimate to be applied to regulatory cost-benefit analysis, perhaps the most troubling aspect of the VSL literature is the wide range of VSL estimates reported. Much of this variability has been attributed to errors in measuring risk, and especially to the connection between measurement error and inter-industry wage differentials. Less thought has been given to other, legitimate sources of variation in the wage-risk trade-off. We have seen that there are many valid reasons why wage-risk trade-offs in the workplace should vary, and these reasons will not disappear with better empirics.

Of these, the most commonly considered source of variation is in workers' preferences for risk. Different preferences, represented graphically by workers with different indifference curves, can be induced by different tastes or different socio-demographic variables. To the extent permitted by data, researchers have included variables that influence tastes in wage hedonics. For example, when Black *et al.* use the NLSY they include marital status, age and race, variables that might influence or be correlated with tastes. Viscusi (2004) includes race, nationality, gender, and marital status. The inclusion of measures such as education, which could control for both productivity and tastes, is also common.[25]

Even if differences in preferences supply the primary source of variation in observed wage-risk trade-offs, a number of challenges remain. For one thing, any hope of finding a unique answer to the VSL question is eliminated. And, even if we could explain all the variation in preferences with variation in socio-demographic variables, we still need to reduce the range of estimates to one statistic since, as a society, we are unlikely to be willing to apply different values of life to different sub-populations.

In principle, if the sample of workers were representative of the US population, then a sample mean of the slope of the wage hedonic would be an estimate of the mean marginal value of the population. When the sample used in hedonic wage analysis is not representative of the population to which the VSL is to be transferred, we are on shakier ground. The population most affected by risk reductions from a proposed environmental policy will not necessarily have the same risk preference as those in the workforce sample included in a particular analysis. This latter point has been raised by many. The male blue collar worker in his 30's or 40's is the dominant profile in hedonic wage studies. The major beneficiaries of environmental risk reduction, however, are often children and the elderly, because their health status is thought to be more sensitive to environmental stressors. Yet, neither group is represented in the workforce, let alone in the special samples chosen for hedonic wage analysis.[26] To the extent that differing risk preferences are really due to differences in income or wealth, the problem is further complicated. Because safety is a normal good, we would expect to find higher implied VSL's among the well-to-do, suggesting that the typical wage hedonic underestimates these individuals' values for safety.

There is a sense in at least some papers that while there may be differences across workers, any worker has one marginal valuation for risk–a marginal valuation that is somehow independent of the market terms of trade. That is, the slope of the wage-hedonic is often unconditionally linked to that individual's willingness to pay for risk reduction, as if the latter were a constant. If not,

---

[25] Workers with identical preferences may face different offer curves because of discrimination as suggested by Viscusi's work (2003). Those discriminated against may be forced into circumstances in which their wage compensation for risk is much lower than less disadvantaged workers.

[26] We do not advocate attempting to measure the risk preferences of children, but rather their parents' preferences for their safety. Yet parents' preferences for their children's safey will likely diverge considerably from their preferences for their own workplace safety.

why would we be so ready to extrapolate these money-risk trade-offs to other settings? Yet, an observed wage-risk trade-off, at best, reflects a worker's marginal willingness to pay for a reduction in risk, *given a particular level of risk and a particular wage*. A unique pair of wage-risk values places the individual on a unique indifference curve at a particular point. But the individual's marginal valuation of risk in the workplace can change, if the opportunities he faces in the labor market change. His marginal valuation for safety will be given by the location in wage/risk space he finds himself at and that location is a function not only of his risk preference but also of the opportunity set (i.e. offer curves) that he faces. This fact seems largely to be ignored in the VSL literature.

The literature in hedonic housing markets recognizes the idea that a sufficiently large change in circumstances (such as a sufficiently wide-spread improvement in an environmental amenity) will result in a shift in the hedonic locus. The reason is that the equilibrium points we observe in a market are the outcomes of the interaction of exogenous distributions of buyer preferences and housing attributes. If the distribution of housing attributes changes substantively, then a new set of equilibrium points emerges. The wage hedonic literature largely ignores the fact that something like a technology change that makes safety cheaper to supply would shift the hedonic wage locus. With cheaper provision of safety, workers could now be on higher indifference functions, and this could easily change each worker's, as well as the average, observed wage-risk trade-off. To see this, assume two types of workers as indicated by their indifference curves (v_A^1 and v_B^1) depicted in Figure 7.9. They face the initial opportunity locus described by O_1, where we use the term 'opportunity locus' to denote the envelope of offer curves of firms in the market. Now imagine that technology reduces the cost of safety, so that firms now have lower costs in general and their marginal costs of increasing safety at any level of risk are also lower. The new opportunity locus might look like O_2. In this new circumstance, both groups of workers find themselves on higher indifference curves with optima at lower levels of risk, even though risk preferences have not changed for either group. What's more, there is no reason to believe that the marginal value of risk that each group reveals remains constant over the technological change.

The importance of the firm side of the problem in the solution cannot be ignored. Consider the common explanation of the lower $VSL's$ estimated in developing countries–that these people have lower incomes and therefore lower willingness to pay for safety. This explanation is, no doubt, partially true, but it is also likely that safety is more expensive to provide in these countries. Ignoring the latter leads logically to the rather dangerous conclusion that it is less valuable to save a life in a market environment where safety technology is relatively expensive to produce. Yet why should the value of a life saved by

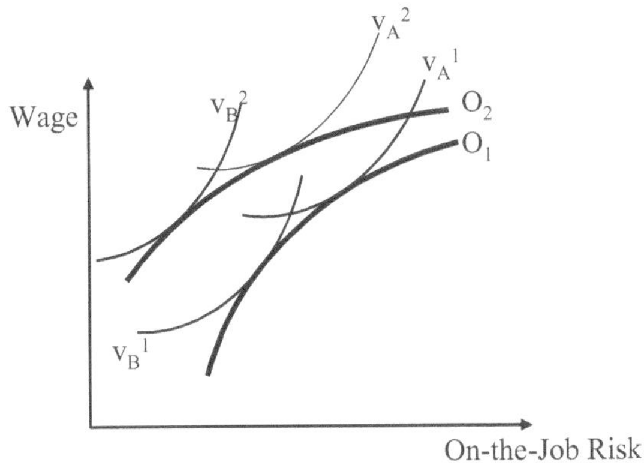

FIGURE 7.9. The Impact of a Decrease in the Cost of Providing Safety

reductions in air pollution or increases in drinking water standards depend on technology in the workplace?

This aspect of hedonic wage analysis has as much do to with marginal valuations of safety in the US as in developing countries. Over the last eighty years, workplace risk in the US has decreased by as much as an order of magnitude.[27] Whether this change is due to regulations that constrain firms or to technological change that makes safety cheaper to produce, the result is a changing opportunity locus facing any individual worker and therefore a changing tangency between any individual's highest indifference curve and his relevant offer function. Given that exactly the same people with exactly the same preferences can be observed at different wage-risk trade-offs, depending on technologies of producing workplace safety brings into question the appropriateness of using workplace wage-risk trade-offs to calculate a unique VSL to be transferred to a completely different setting. This is an important point because, unlike benefit transfer for activities like recreation, there is no individual arbitrage over time and space that would allow individuals to equate risk levels from different sources.

A final difficulty in transferring VSL estimates concerns the *types* of fatality risks. Psychologists have studied the way people perceive different risks. Slovic (1987) argues that risk means more than expected number of fatalities. He shows the importance of various characteristics of risk. One conclusion that accords with intuition is that risks over which individuals exercise some control

[27] See Costa and Kahn (2004) for the period 1940 to 1980 and Bailer, Stayner, Stout, Reed and Gilbert (1998) for the decade 1983 to 1992. For the period since 1992, see the charts at http://stats.bls.gov/iif/oshwc/cfoi/.

appear less threatening than risks which are imposed involuntarily. A common manifestation of this tendency is that despite the greater riskiness of automobile trips, many travelers have greater fear of air travel. Occupational fatalities are likely different in nature from the fatalities avoided by environmental regulation. This is particularly true given that the fatality risk measured in hedonic wage studies is limited to on-the-job accidents and ignores illness-related occupational fatality. Yet the VSL estimates thus derived are applied to reductions in such hazards as cancer risk.[28]

7.4 Wage Hedonics and Locational Amenities

We have investigated two uses of hedonic models for environmental valuation. Economists employ wage hedonic analyses for the estimation of individual willingness to pay for risk reduction, calculated as compensating differentials in wages paid to workers who assume riskier jobs. These analyses typically use cross-sectional data from a national sample of workers. The housing hedonic analysis explored in Chapter 6 sought to estimate willingness to pay for environmental improvements, where the environmental dimension of interest varied within a metropolitan area. Examples of the latter include air quality and distance to hazardous waste sites, each varying within a metropolitan housing market.

Except in cities such as Los Angeles, with its substantial air quality gradients, it is usually necessary to look across metropolitan areas rather than within them to find sufficient variation to make valuation of air quality improvements viable. More often than not, many environmental amenities vary most dramatically over broad geographical areas and less significantly within cities. Climate provides an especially compelling example. Using hedonic methods to value changes in climate dimensions requires relying on inter-city variations in climate.

Economists have long argued that inter-city compensating differentials should exist when amenities differ. However, the early literature is divided on the issue of whether these compensating differentials should show up in wages (e.g. Henderson, 1982) or land rents (e.g. Graves, 1983). Rosen's (1979) original argument, later expanded upon by Roback (1982), posited that individuals make locational choices among bundles of city attributes including wages, housing prices, and amenities, and that both wages and housing prices (or land rents) adjust to establish equilibrium.

---

[28]Evidence on response to different sources of risk is quite limited. Scotton and Taylor (2004), who disaggregate the fatalities by cause in the BLS data, have initiated a study of the implicit value of risk reduction from different sources of risk in the workplace.

7.4.1 The Roback Model

Roback (1982) provides a complete statement of equilibrium with locational amenities. She shows that the theory of compensating differentials does not necessarily imply that wages are lower and rents higher in locations with better amenities, as one might think. Roback's simplest model illustrates her key result. The model assumes that all households are alike and supply a fixed amount of homogeneous labor. Firms are alike, employing land, labor and capital and producing a composite commodity whose price is set in international markets.

Households maximize utility subject to a budget constraint:

$$\max_{z,\mathbf{q}} u = u(z, \mathbf{q}) + \lambda(w - z - r) \tag{7.11}$$

where z is the consumption of a composite commodity by the household and $\mathbf{q}$ is a vector of amenities. Annual wages are designated as w and land rents as r.[29] Equation (7.11) implies the indirect utility function

$$v(w - r, \mathbf{q}) = k \tag{7.12}$$

where k is a common level of utility. In equilibrium, the identical households must be able to achieve the same utility, independent of location, or they would have the incentive to move.

A major contribution of Roback's work is to include the producer's situation in the analysis. Firms are assumed identical and their production subject to constant returns to scale. The unit cost function is given by

$$c(w, r, \mathbf{q}) = 1, \tag{7.13}$$

which equals the fixed price of the composite commodity, z. This condition ensures a spatial equilibrium for firms. Firms with higher costs will need to move their capital to lower cost locations or go out of business.

Roback includes $\mathbf{q}$ in the unit cost function directly to allow for the possibility that amenities could affect production costs. For example, average winter temperatures might be expected to affect costs, because lower temperatures raise heating bills for the firm. As another example, better air quality might lower costs by reducing deleterious effects on worker productivity through health impacts. In both these examples, the marginal cost of an increase in the environmental variable, c_q, is negative because an increase in winter temperatures

[29] Variations of the model include a conversion between land rents and housing prices based on the amount of land consumed by households, a quantity that may be endogenous and may vary if intracity spatial location is considered.

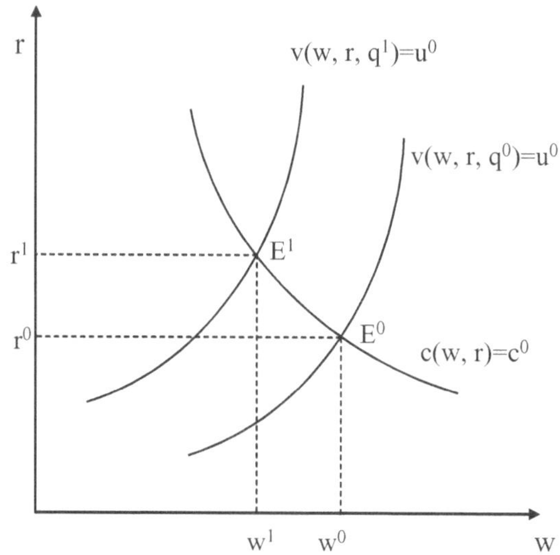

FIGURE 7.10. Change in Equilibrium Due to a Change in q when q Does Not Affect Firm Costs Directly

or in air quality lowers costs. The term c_q could be positive if, for example, increasing environmental quality or safety is achieved through costly regulations imposed on firms. In this case increases in q negatively affect firms, but positively affect workers.

Roback's key result can best be seen in graphical form, where we restrict the $\mathbf{q}$ vector to one element, q. First, in the case in which $c_q = 0$, the amenity has no direct effect on firms' costs. An initial equilibrium could be described as in Figure 7.10 by point E^0. This is the intersection of a common indifference curve, that must be upward sloping when drawn in (r, w) space, with an isocost function also drawn in (r, w) space. Increasing land rents must be paired with increasing wages to hold utility constant, given any level of the amenity. In contrast, firms' unit cost functions require declining wages to offset rising land rents. If utility is to be held constant, an increase in q from q^0 to q^1 leads to a new equilibrium at point E^1 at which land rent is higher and wages are lower. The individual's change in real income that compensates for an increase in q comes in the form of higher land prices and lower wages. Both are necessary since if only wages drop, firms in this area would be better off than firms in other areas. Likewise if only land rents increase, the firms in this area would be worse off. The two prices need to adjust to keep firms, as well as individuals, in equilibrium.

Now suppose q is a productivity-*increasing* locational attribute to firms as well as a locational amenity to individuals. Then an increase in q induces a new

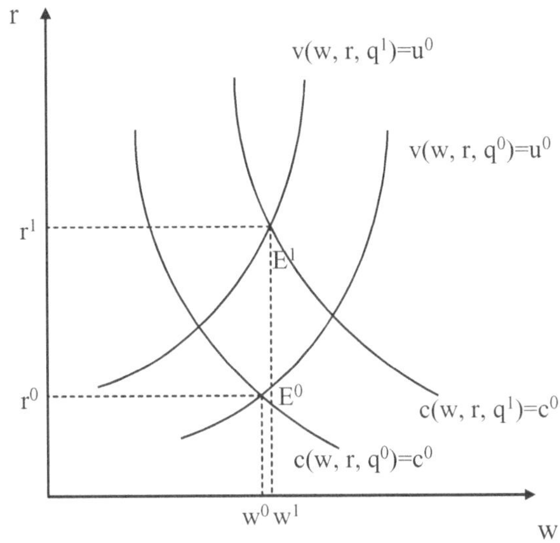

FIGURE 7.11. Change in Equilibrium when Increases in q Lower Firm Costs

equilibrium such as E^1 in Figure 7.11. Land prices unambiguously increase. We show wages also increasing slightly but the direction of change in wages is, *a priori*, indeterminate. Alternatively, consider the case in which the individual's locational amenity is a productivity-*decreasing* locational attribute to firms. Then, as shown in Figure 7.12, the new equilibrium will be at lower wages. In this case land prices are indeterminate, although shown in the diagram as decreasing slightly.

From the diagrams it is clear that the levels of exogenous amenities in an area affect wages and rents. The underlying story is presumably one in which differences in amenities cause migration of households and firms, and this migration puts upward or downward pressure on wages and rents. Equilibrium wages and rents just compensate for differences in amenities so that no incentive exists for further moves by households or firms.

Equilibrium levels of both r and w can now be expressed as functions of the amenity levels because both r and w are assumed to adjust as migration occurs among regions in response to changes in amenity levels. In equilibrium, any change in q must be compensated by a change in wages and/or rents to hold utility constant across locations, so that

$$v_w \frac{dw}{dq} + v_r \frac{dr}{dq} + v_q = 0. \tag{7.14}$$

Likewise, for firms to be in equilibrium, it must be true along an isocost function

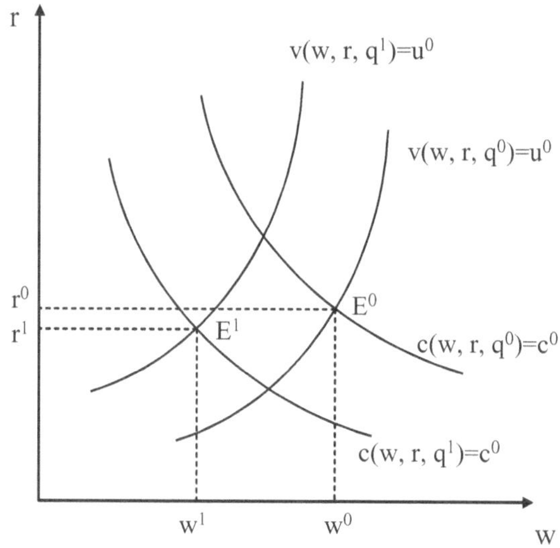

FIGURE 7.12. Change in Equilibrium when Increases in q Raise Firm Costs

that

$$c_w \frac{dw}{dq} + c_r \frac{dr}{dq} + c_q = 0 \tag{7.15}$$

where c_q could be of either sign or could be zero.

Comparative statics results for the simple model can be written as the joint solution of equations (7.14) and (7.15), which yields

$$\begin{aligned} \frac{dw}{dq} &= \frac{v_r c_q - c_r v_q}{v_w c_r - v_r c_w} \\ \frac{dr}{dq} &= \frac{c_w v_q - v_w c_q}{v_w c_r - v_r c_w}. \end{aligned} \tag{7.16}$$

The partial derivatives are signed as: $v_r < 0$, $v_w > 0$, $c_r > 0$, $c_w > 0$, and the denominator is positive. Because q is defined as a public *good*, $v_q > 0$ in all cases. The term c_q will be signed according to whether the locational attribute is productivity enhancing ($c_{q_a} < 0$), productivity decreasing ($c_q > 0$), or neither ($c_q = 0$). Changes in wages will be negative or ambiguous and changes in land prices will be positive or ambiguous, depending on the sign of c_q.

Roback's paper is important because it shows that when looking across regions, land prices need not necessarily be increasing in amenities, nor wages necessarily decreasing in amenities. Of course, this model is quite simple. In reality individuals are not identical, nor are firms. In a simple extension to two groups of workers with different skills and different preferences, Roback (1988) shows that heterogeneity further increases the chances that the 'expected' pattern of signs on dw/dq and dr/dq will fail to hold.

Hoehn, Berger and Blomquist (1987) develop a model much like Roback's but with two differences.[30] First, although their paper explicitly includes unit cost functions for firms, land is not a factor of production. All firms are assumed to be located at the central business district (CBD). Although ignoring this aspect of Roback's model, they add richness by introducing city size as an endogenous variable. While the price of land at the city's edge is assumed to be set by agricultural rents, the radius of the city and its population are endogenous to the problem. In addition, by including city size, Hoehn *et al.* permit agglomeration effects associated with city size. These are assumed to increase productivity for small city sizes and diminish productivity for large cities. The system of equations takes the form

$$\begin{aligned} v(w - tm - r_m, q) &= k \\ v(w - tm^* - r^*, q) &= k \\ c(w, g(N, q)) &= 1 \\ \int_0^{m^*} (2\pi m / L_m) dm &= N \end{aligned}$$

where m is the household's distance from the CBD, m^* is the radius of the city, r_m is land rent at distance m, r^* is exogenous agricultural rent that sets the land price at the city edge and t is cost per mile of commuting. L_m is the amount of land consumed by an individual residing at distance m from the CBD and is equal to $-v_r/v_w$. N, the population of the urban area, is defined by the density gradient and city radius. Firm costs are a function of wages, the public good and city size.

Ambiguities in the signs of dr_m/dq and dw/dq also emerge from this model, in which wages, land rents, and city size are all endogenously determined. Depending on assumptions about the effects of the amenity on firm costs (c_q) and of population size (agglomeration effects) on firm costs (c_N), a variety of outcomes are possible for dr_m/dq and dw/dq. The results support the Roback contention that when looking across cities, we need not find land rents necessarily increasing, nor wages necessarily decreasing, in amenity levels.

7.4.2 Migration and Disequilibrium

The locational equilibrium model employed in the papers discussed above is one of instantaneous adjustment. The equilibrium assumption creates something

[30] Blomquist, Berger, and Hoehn (1988) adopt a very similar model to that of Hoehn *et al.*, but allow cities to be comprised of two parts with varying intracity amenities.

of a contradiction. Migration of population and firms motivates the underlying story of wage and rent adjustment, although these movements are not explicitly modeled. Instead the system is assumed always to be observed in equilibrium. The equilibrium assumption allows researchers to estimate reduced form hedonic wage and property models as functions of amenities. Not all of the literature that deals with interregional differences in wages and rents assumes equilibrium, but the assumption of equilibrium is especially prevalent and important in papers that use the model for environmental valuation, as we will see. Thus, the question of whether the system is observed in (approximate) equilibrium has become the subject of considerable discussion and debate. Graves and Mueser (1993) and Graves and Knapp (1988), in particular, contend that equilibrium is an acceptable 'approximate' assumption, but others using both conceptual reasoning (e.g. Evans, 1990) and empirical evidence (e.g. Topel, 1986; Goldfarb and Yezer, 1987) have argued otherwise.

Returning to the original Roback and Hoehn, Berger, and Blomquist models, we observe that the behavior that causes the system to return continually to equilibrium is migration. Both firms and households are assumed to be freely mobile across regions. The persistence of equilibrium requires that no shocks to the system have occurred for a long time or that households and firms adjust instantaneously. In the presence of relatively recent shocks, sluggish migration of households and firms reflects disequilibrium. The observation of persistent migration (Evans, 1990; Rappaport, 2004) contradicts the notion of no shocks and reason denies the assumption of instantaneous adjustment. It is unreasonable to expect that households can sell their houses, find new jobs, and buy newly constructed houses in these expanding areas in immediate response to changing signals; or that firms can transfer capital between locations just as quickly.

Researchers have provided several explanations for the observed persistent internal migration in the U.S. One source is the continually changing circumstances in the production sector–changing technologies, instability in world prices for commodities such as oil, etc. Another is the life cycle argument. Individuals' relative preferences for amenities change over their lifetimes as they move from the status of single worker, to worker with family, to retirement status. If amenity preferences change over these life stages, persistent migration will be observed. This will cause interregional disequilibrium unless the age structure of the population remains constant over time. Increases in real incomes can also cause shifts in demands for amenities over time, and this can contribute to migration and disequilibrium, as will technology change that affects households. For example, preferences for certain aspects of climate may have shifted over time with the universal availability of air conditioning (or the invention of the snow mobile).

The persistence in migration has spawned several attempts to recast the

Roback model in a disequilibrium framework. These are models of migration in which households (and sometimes firms) are induced to move when the equilibrium conditions in (7.12) and (7.13) are violated. Consistent with the actual process, households and firms respond with a lag. This approach requires the explicit introduction of population size and number of firms at each location. An example is the model of Mathur and Stein (1991,1993) which takes the general form

$$\begin{aligned} p_j &= \alpha[v(w_j - r_j, \mathbf{q}_j) - k] \\ n_j &= \beta[1 - c(w_j, r_j, \mathbf{q}_j)] \\ w_j &= w(P_j, N_j, \mathbf{q}_j) \\ r_j &= r(P_j, N_j, \mathbf{q}_j) \end{aligned}$$

where p_j is the change in population (P_j) in region j, and n_j is the change in number of firms (N_j) in region j. The parameters α and β reflect speeds of adjustment by households and firms. These authors also extend the model to include endogenously determined amenities such as congestion. Mathur and Stein's model is conceptual only. Mueser and Graves (1995), Greenwood, Hunt, Rickman and Treyz (1991) and Treyz, Rickman, Hunt and Greenwood (1993) estimate simpler variations of this model. For example, the Treyz *et al.* paper estimates the rate of change in net economic migration into region j as a function of differentials between region j and the rest of the U.S. in income and amenities and obtain estimates of the speed of adjustment. Greenwood *et al.* estimate compensating differentials in the context of this sort of model.

7.4.3 Welfare Interpretations

The importance of the locational amenity literature for environmental valuation began with Rosen's (1979) insight that amenity values should be capitalized into wages and rents. To see this, return to the constrained maximization problem in Roback's model (equation (7.11)). From first order conditions, the household's optimum locational choice occurs where

$$\frac{\partial u/\partial q_k}{\partial u/\partial z} = \frac{\partial r}{\partial q_k} - \frac{\partial w}{\partial q_k}, \tag{7.17}$$

for all $k = 1, ..., K$ amenities. This expression represents (minus) the slope of the indifference curve between the amenity and the composite commodity. Just as in previous hedonic discussions, the expression $(\partial u/\partial q_a)/(\partial u/\partial z)$ can be interpreted as the marginal value of the amenity because $\partial u/\partial z$ is equal to the marginal utility of income. The expression in (7.17) suggests the need for information on both housing prices and wages to estimate the value of a marginal change in a locational amenity. Only a slight modification need be made in the Hoehn *et al.* model, where the marginal value of an amenity is given by

$$\frac{\partial u/\partial q_k}{\partial u/\partial z} = L_m \frac{dr_m}{dq_k} - \frac{dw}{dq_k}, \tag{7.18}$$

(L_m is the amount of land consumed at distance m from the CBD). Even though firms are assumed not to consume land in this model, the value of the amenity is capitalized into both rents and wages because agglomeration externalities are considered. This expression is almost identical to Roback's except that land consumption and land rents vary with the distance to the CBD. The importance of the equilibrium model is that given the slopes of a wage hedonic and rent hedonic with respect to some amenity, it is possible to calculate the willingness to pay for a marginal change in the amenity evaluated at a given level of the amenity.

Roback (1982, 1988) estimates reduced form functions for both wages and rents, so that mean marginal willingness to pay estimates can be obtained. In concept this is possible because one can solve the two equilibrium conditions, (7.12) and (7.13), for the two endogenous variables, wages and rents, as functions of amenities. Although the Hoehn *et al.* model is more complex, it too can be expressed in similar reduced form.[31] In practice estimation of the wage and rent hedonic functions is typically based on different data sources and estimated separately; and amenity levels are linked to individual observations on the basis of geographic area. Once the rent and wage hedonic equations are estimated, marginal amenity values are calculated as in (7.17) or (7.18) where the slopes are evaluated at the means of r and w, respectively, because the hedonic functions are non-linear in their arguments.[32]

The empirical applications of Roback, Hoehn *et al.* and Blomquist *et al.* support the two important contentions of the Roback model: a) that amenities are capitalized in both land prices and wages, and b) that wages are not necessarily falling, nor land prices rising, in amenities, although the marginal amenity as represented in (7.17) or (7.18) should be positive. Little in the way of consistent

---

[31] Earlier locational equilibrium hedonic models were estimated by Hoch and Drake (1974) and Cropper and Arriaga-Salinas (1980) but were limited to wage hedonics. These early studies controlled for cost-of-living differences, but since these indices include a large housing price component and since housing prices were later shown to absorb some of the compensating differential, the interpretation of results from these models is confounded (Roback, 1988).

[32] Some researchers (e.g. Maddison and Bigano, 2003) estimate one function in which the dependent variable is annual wage minus housing costs. There is no theoretical development in this paper so one is left to assume that firms play no part in establishing equilibrium. In their study of the marginal value of climate amenities in Italy, Maddison and Bigano use average data over 95 provinces.

and pervasive empirical results for any specific amenities (including measures of climate) emerges from this literature. Results are understandably sensitive to specification which is guided more by availability of data than anything else. For example, Roback includes a limited list of amenities: crime rate, air quality, population growth (or density), and one climate variable (either heating degree days, cloudy days, clear days or snowfall, depending on the model), while the Hoehn *et al.* and Blomquist *et al.* papers use a longer list of amenities including teacher-pupil ratios, pollution discharge, landfill waste, Superfund sites, precipitation, humidity, windspeed, etc.

There is, of course, a near-infinite list of city characteristics that could matter to people and many will be correlated. The potential for serious multicollinearity or omitted variable bias plagues hedonic analysis of all sorts, but these problems are arguably even more acute in locational equilibrium models where observations are typically over metropolitan statistical areas. For example, Roback's empirical result that the average household would pay about $150 (converted to 2005 dollars) to experience one fewer cloudy day per year seems unreasonable and is five times the marginal value obtained by Blomquist *et al.* who include a much longer list of climate amenities. Many climate variables are correlated because they are physically interrelated–variables such as heating degree days, cooling degree days, snowfall, cloud cover, precipitation, etc. And these climate attributes are also physically related to topographical characteristics such as mountainous terrain or coastal access that have their own amenity value.

Additional interrelationships are likely to arise due to endogeneity. In-migration increases the size and density of a city, which in turn can produce disamenities such as congestion and air pollution. Higher and denser population can also generate amenities in the form of cultural activities, sports teams, etc. Areas with low wages are often those found to accept more hazardous waste or other noxious facilities in an attempt to attract jobs or tax revenues to the locality. These complex interrelationships, the subject of analysis in other literatures, are largely ignored in the locational equilibrium literature. All of these factors make establishing evidence of causation rather than just correlation a difficult enterprise. It brings into question the accuracy of estimates of the marginal contribution of a single amenity.

Several locational equilibrium papers have estimated marginal values of amenities using one or both terms of the marginal value expression in (7.17) or (7.18). Some (e.g. Roback and Blomquist *et al.*) have constructed quality of life indices from these marginal values. The implicit marginal value of amenity q_i, designated as φ_{q_i}, is calculated as the marginal value expression evaluated at the mean, and the quality of life index for each city is calculated as the product of the amount of each amenity times its marginal value, summed over all amenities:

$$QOLI = \sum_i \varphi_{q_i} q_i.$$

The welfare significance of this expression is open to question. Calculating an index by attributing a marginal value for an amenity, estimated from a non-linear function, to all units of the amenity is, of course, incorrect. It would be incorrect even if the non-linear hedonic function had welfare meaning, but of course it does not. As discussed in the previous chapter, the hedonic price (or wage) function is a locus of equilibrium points. Only at the margin does this locus tell us anything about preferences. Even without this serious problem, the notion of a quality of life index estimated from the locational equilibrium model is puzzling. The premise of the model is that wages and housing prices adjust so that household utility is constant across cities. Since income net of housing expenses generates utility and since the model suggests that this utility just compensates for the locational amenities, then no city can be better than another. Clearly, the term 'quality of life index' refers only to the non-priced amenities.[33] These papers, while having popular appeal, provide no policy implications.

Abstracting from the empirical difficulties of estimating marginal values, we can see quite clearly why the interest in environmental valuation requires the assumption of an interregional system in equilibrium. Only when amenity values are fully capitalized into wages and rents will the marginal value functions in (7.17) or (7.18) hold and thus only in equilibrium will the slopes of the hedonic wage and rent functions yield marginal welfare information. However, as Evans (1990) and Greenwood *et al.* (1991) argue, if regional markets do not tend to clear quickly, assuming equilibrium will lead to biased estimates of amenity values.[34] Greenwood *et al.* attempt to demonstrate this bias empirically, and find evidence of interregional disequilibrium. Yet they find that the errors generated by assuming equilibrium appear relatively minor.

In all of this literature it is worth asking: what is the welfare question? Presumably this line of research seeks to contribute to the debate: are the benefits to society from preventing further climate change or from reducing pollution less than the regulatory costs? In the context of a hedonic model we must always ask: what benefits are we measuring? Are these benefits equal to

---

[33] However, the results cast doubt on the credibility of even that interpretation. Any ranking that lists Norfolk, VA as the second most amenity-rich city in the U.S. but places the Marin County portion of San Franscisco at 145th is suspect (see Blomquist, *et al.*).

[34] Some have argued that since climate has been quite stable over the past century, the system should be in equilibrium with respect to this amenity dimension at least, but this argument seems flawed. If interregional equilibrium does not exist for any reason, even if it is not instigated by climate change, then differences in wages and rents cannot generally be assumed to reflect compensating differentials for any amenity differences.

individuals' 'pure willingness-to-pay' in the sense we used this term in Chapter 6? Or are they the monetized change in utility that *actually* occurs as a result of all the market adjustments that the proposed change would set in motion? If we are interested in only marginal changes, then there will be no difference and the marginal willingness to pay will equal the implicit amenity price in equilibrium.

Non-marginal changes in amenities complicate welfare estimation as usual. First, 'pure willingness to pay' for such a change cannot be measured along estimated hedonic functions, since the latter are loci of equilibrium points and not preference functions. Furthermore, the non-marginal change should set in motion forces that will cause firms and individuals to relocate resulting in equalization of utility and costs across regions. Thus the long run welfare effect would seem to be zero. Given the existence of transactions and moving costs, however, this answer cannot be quite right; firms and households will move only if the stimulus is great enough to overcome their transaction costs.

A more fundamental question arises when considering large changes in amenities in the location model. The literature universally assumes that shocks ultimately return households to the same utility level, k in equation (7.12). This result is based on a premise that a 'small' region is affected by the shock and that its wages and rents adjust to equate utility to the common level reached elsewhere in the country. Presumably a shock as far-reaching as climate change will have broad geographical implications and can be expected to change the common level of utility across regions. None of the literature considers this possibility. Yet it is at the heart of the broader question: what are the welfare consequences of policies that alter the future path of climate change.

7.4.4 Locational Amenities in a Discrete Choice Framework

In what remains we review three recent attempts to value locational amenities in discrete choice models. A discrete choice model offers an alternative to the continuous hedonic model, with the advantage that welfare measures no longer depend on slopes of hedonic functions presumed to represent equilibrium in labor and land markets. Whether this approach is a panacea for the ills of the more conventional models remains to be determined.

Cragg and Kahn (1997) use data on location decisions by migrants to estimate willingness to pay measures for climate variables. Migrants are chosen as the sample because they are assumed to be a sufficiently small group such that their actions do not affect prices. An additional argument can be made, along the same lines as that made for using only sales data in housing hedonics. There may be obstacles to moving which keep many individuals from being at equilibrium. Migrants have overcome these transactions costs and, once the decision to move is made, can be expected to optimize. Although appealing from that respect, the use of only migrants introduces sample selection problems that are difficult to avoid.

The underlying theoretical structure of the problem is not developed explicitly. Yet, we observe migrants changing locations. In order to estimate the discrete choice model, wages and rents must first be imputed for all alternatives (the 48 contiguous states) for each observation in the data set. This is done using conventional hedonic regressions for wages and rents, much like those used in the papers reviewed earlier. Each regression is estimated as a function of amenities, which suggests an implicit assumption that wages and rents are in equilibrium. Willingness to pay measures for changes in climate variables are then interpolated from the results of the discrete choice analysis. This is accomplished by interpreting the discrete choice model as a random utility model and investigating how individuals trade wages minus housing prices for amenities.

The discrete choice approach does not provide an escape from multicollinearity problems. Cragg and Kahn's treatment of entire states as alternative migration destinations introduces errors in measurement in the climate variables and prevents controlling for all those locational amenities and disamenities that vary within large states. As such the empirical results probably suffer even more from omitted variable bias than would conventional hedonic analyses.

Although Cragg and Kahn's model appears to assume interregional equilibrium, is this necessary in discrete choice models? There are really two components to equilibrium. One is that the observed decisions of the agents included in the model reflect optimality, given existing wages and rents. The second is that wages and rents adjust rapidly to those decisions. Discrete choice models do not require the latter, but they do require the former. It must be true that we observe individuals in their optimal locations. If there are obstacles to mobility or lags in adjustments then simple discrete choice models will produce biased results.

Clark, Herrin, Knapp and White (2003) adopt a binary discrete choice approach to modeling migration behavior in a paper that explicitly attempts to address disequilibrium. Deviations from 'complete compensation' are calculated by first estimating a fixed effects model of wage variation over metropolitan statistical areas and then regressing these fixed effects on locational amenities. In a third stage the binary migration decision of whether to migrate is modeled as a function of individual characteristics, locational characteristics at both origin and destination locations, and the calculated 'incomplete compensation' measures at origin and destination. For those who do not migrate, origin and destination represent the same location. The sample includes only households who moved, but those who moved within a metropolitan area are considered non-migrants. This three stage analysis is *ad hoc* and suffers from endogeneity problems, but represents one of the few attempts to model the behavior that ultimately establishes equilibrium.

Whether the discrete choice framework offers a way around problems that are all too obvious in conventional hedonics remains to be determined. In an unpublished paper, Bayer, Keohane, and Timmins (2005) offer a careful econometric treatment of a discrete decision model in which at least one type of impediment to mobility is introduced. This is the tendency (for whatever reason) of individuals to remain in the region where they were born. This is not a trivial modification to the discrete choice model, because this feature of a location varies over individuals and thus introduces important interpersonal variation. This modification does not slow migration; instead it changes the optimal pattern. Nonetheless, it is an advance over the simplistic assumptions made in much of the literature. Their preliminary results–that benefits from reductions in particulate matter are understated by about 75% by conventional hedonic analysis that does not take into account mobility constraints–give pause for thought.

7.5 Conclusions

Efforts to amass evidence on household preferences for natural amenities from labor markets and labor and housing markets jointly remains an open field for economists. The reliance on hedonic wage models for estimates of the VSL creates substantial incentives to strengthen the empirical evidence for risk-wage trade-offs. Two research directions promise rewards for hedonic wage models. Researchers know little about how workers perceive risks in different industries and occupations. Given the enormous leverage of small differences in risks, it makes sense to investigate how perceptions of risk are formed. A second line of investigation concerns econometric specifications. There is substantial research under way concerning individual issues in the specification of hedonic wage models. For example, topics under separate investigation include the effect of age, unions, measure error in risk, functional form of hedonic models (especially quadratic modeling of risk), inter-industry effects, worker compensation and the risk of non-fatal accident. Insights into the robustness of wage-risk trade-offs are limited when each of these topics is investigated separately. The more important issue concerns the joint investigation of these issues. For example, the impact of unions on the wage-risk trade-off may depend strongly on the inclusion of a set of inter-industry fixed effects. Advancement in understanding hedonic models will be greatly enhanced when researchers provide evidence of the impact of a range of specifications on the wage-risk trade-off, much like in the Black, Galdo and Liu (2003) report.

In the analysis of migration and joint housing-wage hedonics, the assumption of equilibrium plays an especially strong role. Recent research, especially in the context of discrete choice models, adds insight and richness to these models and will likely provide the basis for future research.

Chapter 8

Public Goods in Household Production

8.1 Introduction

In this chapter we continue the analysis of the household production models introduced in Chapter 3. Now we investigate models in which the non-market or public good is used as an input together with privately purchased goods to produce a commodity that the household values. At first the problem is framed in terms of a public 'good'. For example, a household might combine fishing effort with public fish stocks to produce recreational catch. In our subsequent discussion of defensive expenditures we adopt the more usual practice of framing the problem in terms of a public 'bad'. Households defend themselves against the consequences of environmental degradation by making a variety of decisions, such as purchasing bottled water or pesticides, spending extra time in food preparation, installing water or air filters, etc. In either case, the public input has the effect of altering the costs of 'production' of some household produced commodity rather than affecting utility directly.

This chapter begins with a generic treatment of the problem but, as is usually true in non-market valuation, the general case offers little promise for welfare measurement. This is followed by a consideration of the alternative restrictions that allow exact welfare measurement. Applications based on averting behavior or defensive expenditure models are more prevalent and motivate much of the remainder of the chapter. These commonly used approaches are applicable in more general circumstances but, as we will see, provide only bounds on welfare measures. Finally, we discuss the often used 'cost of illness' model in which increases in pollution reduce the health state and lead to monetary consequences. We postpone discussion of a related problem–the welfare effects of environmental change on the production of a good bound for the market–until Chapter 9.

Averting and defensive behavior models have received far less attention than the recreational demand and hedonic models of earlier chapters, and their use in welfare measurement is not always straightforward. Yet they offer considerable promise because defensive and averting actions are applicable to such a wide variety of settings.

8.2 The Structure of the Problem

The story begins with an environmental or public good, q, that can be viewed as an input into some sort of household production process. The utility function used here has a simple specification. One of its arguments is a composite commodity purchased on the market and denoted z_2. The other is the commodity, z_1, which the household produces with a combination of purchased goods, $\mathbf{x}$, and the public good, q. Household time is likely to be an input as well, but will be subsumed in the vector $\mathbf{x}$ for simplicity. Doing so presumes a well-defined price for time, yet we know from Chapter 4 that the value of time is difficult to measure and often non-parametric. All the issues that arose in that chapter remain problems here. We will not repeat that discussion here, but will implicitly assume a 'price' for time. Where this is a clear violation of reality, the arguments of Chapter 4 will be relevant.

Unlike our earlier treatments of the environmental or public good, q is now an non-priced input in production rather than an argument of the utility function. The preference function is given by $u(z_1, z_2)$ and the production function for z_1 by $z_1 = f(\mathbf{x}, q)$. The only way in which q affects the household is as an input into the production of z_1. We expect $f(\mathbf{x}, q)$ to be increasing in the purchased inputs and the public input. To fix ideas, we could think of z_1 as drinkable water and q as a measure of the water quality of the public drinking water supply. In this context, one of the $\mathbf{x}'s$ might be water filtering services. An alternative story might view z_1 as health, q as public programs for reducing pest populations that cause disease, and one of the $\mathbf{x}'s$ as privately purchased pesticides. We could also frame a recreational demand problem in these terms, so that z_1 could be recreational fish caught and q a measure of stock abundance. An x of importance would be the purchased input: trips to the fishing site. In each case we have mentioned one purchased input of importance and will, for the time being, suppress all other inputs.

The household buys a composite commodity, z_2, at price p, and x at price r, maximizing utility subject to the budget constraint and production technology:

$$\max_{\mathbf{z}_2,x}\{u(z_1,z_2)|y \geq pz_2 + rx,\ z_1 = f(x,q)\}. \tag{8.1}$$

Alternatively, the utility function could be written in terms of x, q,and z_2. By substitution,

$$u(z_1,z_2) = u(f(x,q),z_2) = \tilde{u}(x,q,z_2),$$

so that the household's maximization problem becomes

$$\max_{z_2,x}\{\tilde{u}(x,q,z_2)|y \geq pz_2 + rx\}. \tag{8.2}$$

This looks just like a standard utility maximization problem, with parametric prices. However, stating the problem in this form obscures the behavioral and technical assumptions that are important to understanding the role of the public input and that are needed to establish the restrictions necessary for welfare measurement.

The expenditure function, which can be derived from equation (8.1) or (8.2), is

$$m(p,r,q,u^0) = \min_{x,z_2}\{rx + pz_2|u(f(x,q),z_2) = u^0\}, \tag{8.3}$$

and equals the minimum income needed to achieve utility level u^0 with prices p and r and public good level q. The compensating variation of a change in q is the change in expenditures necessary to achieve the original utility level after the change and is given by the change in the expenditure function:

$$CV = m(p,r,q^0,u^0) - m(p,r,q^1,u^0). \tag{8.4}$$

In (8.4) q^0 is the initial level of the public good and q^1 is its final level. By construction, the welfare measure will be positive for increases in a desirable public good, defined as one with a positive marginal product in the production of the commodity. It is a simple matter to write the welfare measure as in equation (8.4). The challenge is to find behavior that will reveal this welfare measure or some approximation of it.

8.2.1 A Simple Result for Constant Marginal Costs

An obvious approach would be to use information on how the demand for the household produced commodity, z_1, changes as q changes, much like the route taken in Chapter 3. Indeed, this can give us useful information if the demand for z_1 is well defined and can be estimated. Suppose, for example, that the technology for producing z_1 with x and q is such that the cost function is linear in z_1. We denote the general cost function as $c(z_1,r,q)$, but in this special case

$c(z_1, r, q) = \tilde{c}(q, r)z_1$. Marginal costs are constant in z_1 and decreasing in q. This reduces nicely to a conceptually simple welfare problem, because (8.3) can be written as

$$m(p, r, q, u^0) = \min_{z_1, z_2}\{\tilde{c}(q, r)z_1 + pz_2 | u(z_1, z_2) = u^0\}.$$

Changes in q affect the individual only by raising or lowering $\tilde{c}$, the constant marginal cost of z_1. Now $\partial m/\partial \tilde{c} = z_1^h$, so that

$$CV_{\Delta q} = -\int_{\tilde{c}(q^0, r)}^{\tilde{c}(q^1, r)} z_1(\tilde{c}(q, r), u)dq, \tag{8.5}$$

where $\tilde{c}(q, r)$ is the parametric marginal cost of z_1 and, as the notation indicates, $z_1(\tilde{c}, u)$ is the utility constant demand curve. The problem becomes one of measuring the welfare of a price rather than quality change. All the results of Chapter 2 now apply, including those that relate areas behind Hicksian demands (as employed in equation 8.5) to their observable Marshallian counterparts.

The simplicity of this case and its similarity to previous discussions is appealing, but its lack of generality is not. One *can* construct cases in which marginal cost is constant with respect to z_1 and varying with q, but they are very special cases and ones that are awkward to explain. Take for example a situation in which the public drinking water supply is compromised. In its cleanest state, individuals purchase all their water uses from the public water supply at a price per unit, p_p, but when the source becomes contaminated they are forced to buy bottled water at some higher price per unit. This is an oversimplification of reality, though. There are often many technologies for dealing with the drinking water contamination problem and the least cost method may depend on the volume of water consumed. Also, some technologies, such as filtering systems, have large fixed costs, making the cost per unit dependent on quantity.

In their paper on household production technology, Pollak and Wachter (1975) argued that the commodity demand function will often be ill-defined in a household production model. Either non-constant returns to scale or jointness in production will cause problems. The former is evident in many of the previous examples, especially those with large fixed costs. The latter occurs often when household time is an input into production because it may also generate utility or disutility directly. Examples include time used for recreational trips that is also a measure of utility from the recreational experience. The consequence of either non-constant returns or jointness is that the marginal cost z_1 becomes a function of z_1, and the analysis must account for the non-linear nature of the budget constraint. As we have shown in Chapter 2, one cannot use the standard approaches for measuring welfare when the budget constraint is non-linear. Although there is no easy way to use information about z_1 to obtain empirical welfare measures, there are other means of extracting welfare information.

8.3 Restrictions on the Demand for an Input

In this section we determine in what sense information about the demand for a purchased input can help us value a change in q. In all cases the object of interest is a household production function that is a function of the public good, q, and at least one purchased input, x. As introduction, we consider examples of the different ways q can relate to a purchased input, x.

First, recall one of the examples we used to motivate the chapter: the commodity (z_1) is sport-caught fish, the public good (q) is fish stock abundance, and the purchased input (x) is trips to the fishing site. The production function for this problem might take a form such as

$$\begin{aligned} z_1 &= g(\textit{stock abundance}) \times \textit{trips} \\ &= g(q) \cdot x \equiv f(q, x). \end{aligned} \tag{8.6}$$

Given that $g'(q)$ is positive, by definition $f_{qx} > 0$. An increase in q has the effect of increasing the marginal productivity of fishing trips. More fish are caught per trip when stock abundance is higher.

A second example offers a different picture. Here the story is more naturally told in terms of a public 'bad' (which we will denote b), rather than the public good. Suppose the commodity is health status, the public bad is the level of malaria carrying mosquitoes, and the privately purchased input is some form of pesticide. A simple way of representing this problem might be to denote α as a rate of disease transmission that maps an appropriately measured variable indicating the level of disease carrying mosquitoes into the probability of illness. The production function might be written as $z_1 = 1 - \alpha \cdot b$, where z_1 is the probability of remaining well, b is the stock of mosquitoes, and α is a function of the use of pesticides by the household, so that

$$\begin{aligned} z_1 &= 1 - \alpha(\textit{pesticides}) \cdot b \\ &= 1 - \alpha(x) \cdot b. \end{aligned} \tag{8.7}$$

More pesticide use lowers the transmission rate of disease from any initial stock of malaria-carrying mosquitoes, implying that $\alpha'(x) < 0$.

With the problem framed as a plausible story, we rewrite the production function as a function of inputs with positive marginal products. The public *good*, q, related to b might be described as the reduction in malaria carrying mosquitoes due to public programs and could be expressed implicitly as $b = M - q$, where M is the exogenous initial level of malaria-carrying mosquitoes and b is the level remaining after the public programs have exerted their influence on the pest populations. Now, the production function can be written as:

$$z_1 = f(x, q) = 1 - \alpha(x) \cdot [M - q]. \tag{8.8}$$

Written this way, $f_q > 0$, $f_x > 0$, and now $f_{xq} = \alpha'(x) < 0$. Pesticides have a lower marginal product the lower the levels of malaria-carrying mosquitoes.

In both the recreational demand and the health production examples, x and q are substitutes in the sense that they have positive elasticities of substitution. This must be true where there are only two inputs, as long as each input exhibits a positive marginal productivity. In each case, the isoquant between x and q is downward sloping and convex to the origin. What is different is the effect an increase in q has on the marginal productivity of the purchased input.

There is a third case–the one in which $f_{xq} = 0$. Here the production function is additively separable in the publicly supplied good and a privately supplied input. It is always possible in such a case to completely mitigate the loss of either x or q by augmenting the other. In the next sections we consider three strategies for obtaining exact welfare measures, each of which aligns itself with one of the above relationships between the public input and a purchased input. In each case, however, additional restrictions will be necessary to obtain the measure we need.

8.3.1 The Case of a Separable Production Relationship

One of the most surprising results arises when $f_{xq} = 0$ and the production function is linear in x. We will refer to this case as one of perfect substitution between the purchased input, x, and some function of q, such as in the following:

$$z_1 = f(x, q) = \delta x + \psi(q). \tag{8.9}$$

Increases in the purchased input increase z_1 at the rate δ, a positive constant parameter in the production function, while increases in the public good increase the commodity at the rate $\psi'(q)$. With this form of the production function, exact welfare measurement requires only a knowledge of technology and the price of the purchased input.

We demonstrate this in two stages, first defining the minimum cost function for achieving a given amount of z_1. The cost function implied by this production function is

$$c(z_1, r, b) = \min_x \{rx | z_1 = \delta x + \psi(q)\} = \frac{r}{\delta}(z_1 - \psi(q)). \tag{8.10}$$

Requiring that $f(x,q)$ be additively separable in x and q and linear in x ensures that $c(z_1, r, b)$ will be additively separable in z_1 and q and linear in z_1. The maximization problem is given by

$$\max_{z_1,z_2}\{u(z_1,z_2)|y = p_2 z_2 + \frac{r}{\delta}(z_1 - \psi(q)\} \tag{8.11}$$

which can be rewritten as

$$\max_{z_1,z_2}\{u(z_1,z_2)|y + \frac{r}{\delta}\psi(q) = \frac{r}{\delta}z_1 + p_2 z_2\}$$

implying an indirect utility function of the form

$$v(\frac{r}{\delta}, p_2, y + \frac{r}{\delta}\psi(q)). \tag{8.12}$$

The exact form of the indirect utility function depends on the form of $u(z_1, z_2)$, but the way in which the public good enters the indirect utility function is a consequence of the simple form of the household production function. The public good has no effect on the marginal cost of z_1 which is constant.[1] Instead it acts to augment income. One can think of income as given by $y + \frac{r}{\delta}\psi(q)$. The only qualification is that $\frac{r}{\delta}\psi(q)$ can only be 'spent' on z_1 and not on z_2. But as long as $\psi(q)$ is less than z_1 at the optimal level of z_1, thinking of this term as an addition to income makes sense.

The form of the indirect utility function, as given in (8.12), implies that the compensating variation for a change in q is defined implicitly by the expression

$$v(\frac{r}{\delta}, p_2, y + \frac{r}{\delta}\psi(q^1) - CV) = v(\frac{r}{\delta}, p_2, y + \frac{r}{\delta}\psi(q^0)). \tag{8.13}$$

Using this equation the compensating variation must be

$$CV = \frac{r}{\delta}[\psi(q^1) - \psi(q^0)]. \tag{8.14}$$

Given that the compensating variation of an income change is simply that income change, the compensating variation of a change in this augmentation to income is simply the change in this income augmentation. The welfare measure holds for any utility function with the usual properties and depends on the appropriateness of the technology restriction.

The CV measure is illustrated in Figure 8.1. In this figure, the budget constraints play the key role. In the initial circumstance, the individual faces a

[1]This is different from the constant marginal cost assumption of the earlier section. In that case the constant marginal cost was a function of q and changes in q acted to change the implicit 'price' of z_1.

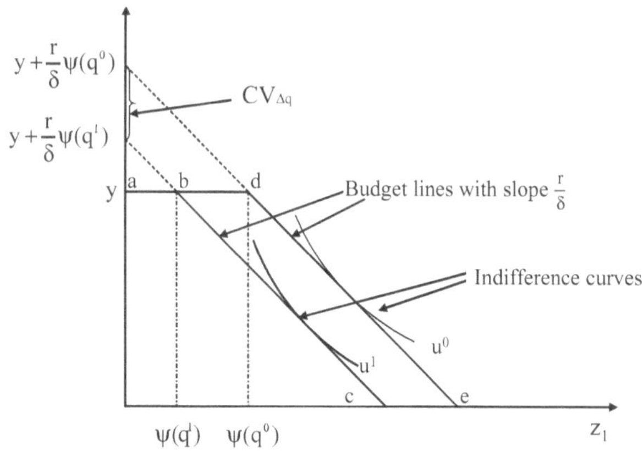

FIGURE 8.1. Compensating Variation when x and q Are Perfect Substitutes

budget constraint defined by the kinked curve *ade*. Given a level of the public good equal to q^0, $\psi(q^0)$ units of z_1 are effectively available to the individual for free. This is the *ad* portion of the budget constraint. After the point at which $z_1 = \psi(q^0)$, all further amounts of z_1 must be produced by the household at a cost per unit of r/δ and described by the portion of the budget constraint marked *de*. Solution occurs where the indifference curve marked u^0 is tangent to this budget constraint. With a decline in q from q^0 to q^1, the budget constraint becomes the kinked line *abc* because less of the good z_1 (specifically $\psi(q^1)$ of it) can be obtained with no direct expenditures. Now the individual must settle for a lower indifference curve, the one labelled u^1. The change in q implies a change in real income denoted by the difference between the implicit intercepts of the budget constraints along the 'y' axis. By definition, this difference equals $\frac{r}{\delta}[\psi(q^1) - \psi(q^0)]$ and in this case is negative.

Note that the extensions of the budget constraints (the dotted lines between d and the 'y' axis and between b and the 'y' axis) are not feasible solutions. The 'real income' $\frac{r}{\delta}\psi(q)$ can only be used to 'purchase' z_1 and not the composite commodity. Circumstances that would otherwise lead to a solution along one of these dotted-line segments will instead result in corner solutions, and corner solutions will typically cause violations of the above result. Consider an individual who initially optimizes at a corner solution at point d on indifference curve, u^0, in Figure 8.2 when $q = q^0$. At this point, the public good supplies him with sufficient z_1 given his preferences, and he chooses to spend all his income on z_2. Now consider a decline in the public good to q^1 and a new optimum at the interior solution marked by point f. The minimum amount of money necessary to return him to his initial utility (u^0) before the decline in q will be something less than $\frac{r}{\delta}[\psi(q^0) - \psi(q^1)]$ and so his loss from the decline in

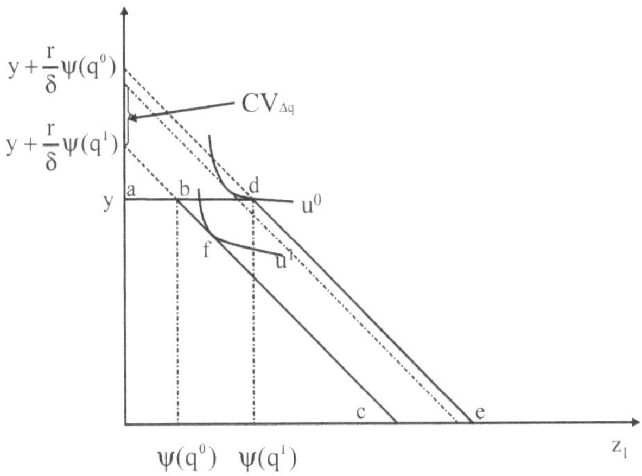

FIGURE 8.2. A Corner Solution

q will be a smaller negative number than $\frac{r}{\delta}[\psi(q^1) - \psi(q^0)]$ marked in the figure as $CV_{\Delta q}$.

To give an example, consider a household in a developing country. Suppose that part of the household's water supply comes from an underground stream that provides q^0 units of water per unit of time, free for the household's use. The remainder of the household's water needs must be filled by purchasing water (x) at price r from traveling vendors who procure water from elsewhere. The production function is a simple one: $z_1 = q + x$. Now suppose the diversion of water elsewhere causes stream flows to decrease to q^1. The loss to the household will equal $r[q^1 - q^0]$, but this will only be true if the household was found at an interior solution before and after the loss.

8.3.2 Demand for Essential Inputs

The above restrictions make welfare measurement feasible with information about technology and prices only. When these restrictions can not reasonably be assumed, other more demanding methods may still be available. Here we set out the restrictions that allow the welfare effect of a change in the environmental input, q, to be measured as the area above price and between two (Hicksian) demand functions for a privately purchased input, one demand function conditioned on the initial level and one on the subsequent level of q. This will look very much like the results in Chapter 3 that depend on a different restriction–that of weak complementarity. While the restrictions themselves *seem* quite different, the same problem can often be cast in either framework.

The restrictions needed to make the current story viable are the following. First, as we made clear earlier, q is only useful to the household as an input into the production of z_1. This means that if z_1 is not produced for whatever reason, then changes in q are of no consequence to the household. Second, the *purchased* input, x, must be 'essential' in the production of z_1. In their treatment of welfare measures for the firm, Just, Hueth, and Schmitz define an essential input as one that is necessary in the production of the firm's output. If a firm ceases to purchase some input that is essential, by definition the firm shuts down. The same concept has meaning for household production; if an essential input in the production of z_1 is not purchased, then the household can neither produce nor consume z_1.

Putting our two restrictions together, we have the result that if the essential good is not purchased, then no z_1 is produced or consumed and changes in q do not matter to the household. This set of restrictions is easiest to motivate and most likely to be met when $f_{qx} > 0$, i.e. when increases in the public good increase the marginal product of the private good. Returning to our sportfishing example, the output (sport-caught fish) can be increased by more trips or higher catch rates per trip (as indicated by stock abundance). However, trips are an essential input because if no trips are taken, no level of stock abundance will be high enough to produce any sport caught fish. Put another way, if the cost of travel rises sufficiently so that the individual chooses to take no trips, then changes in the stock of fish at the site do not matter to him, as he produces and consumes no sport-caught fish.

To see why these restrictions are useful in a technical sense, note that the Hicksian demand for the input x, which will be assumed essential, is

$$x(p, r, q, u^0) = \partial m(p, r, q, u^0)/\partial r.$$

Let r^0 be the current price (which, in our example, includes money and time costs of travelling to the site) and let $r^*(q^0)$ be $x's$ choke price when the public good is at level q^0. The area behind the Hicksian demand for x conditioned on the initial value of q is given by

$$\int_{r^0}^{r^*(q^0)} x(p, r, q^0, u^0)dr. \tag{8.15}$$

This is the amount of income that would compensate the household if access to the fishing experience were denied, given that the public good is at level q^0. Likewise, we can define such an area conditioned on the public good being at some new (increased) level, q^1, as

$$\int_{r^0}^{r^*(q^1)} x(p, r, q^1, u^0)dr. \tag{8.16}$$

The area between the two demands and above input price is

$$\int_{r^0}^{r^*(q^1)} x(p,r,q^1,u^0)dr - \int_{r^0}^{r^*(q^0)} x(p,r,q^0,u^0)dr \tag{8.17}$$
$$= m(p,r^*(q^1),q^1,u^0) - m(p,r^0,q^1,u^0) \tag{8.18}$$
$$-[m(p,r^*(q^0),q^0,u^0) - m(p,r^0,q^0,u^0)].$$

This area, illustrated in Figure 8.3 for an increase in q, is composed of four expenditure function terms and will equal the compensating variation measure, as expressed in (8.4), if and only if

$$m(p,r^*(q^1),q^1,u^0) = m(p,r^*(q^0),q^0,u^0). \tag{8.19}$$

Equation (8.19) holds if x is essential in the production of z_1. At any r^*, x is zero and z_1 must also be zero, so differences in q do not matter.[2] Given (8.19), the compensating variation for the public good will be

$$\int_{r^0}^{r^*(q^1)} x(p,r,q^1,u^0)dr - \int_{r^0}^{r^*(q^0)} x(p,r,q^0,u^0)dr \tag{8.20}$$
$$= m(p,r^0,q^0,u^0) - m(p,r^0,q^1,u^0)$$

which is the CV definition in equation (8.4) and area A of Figure 8.3.

The restriction in equation (8.19) looks like weak complementarity, but it is a technical relationship, not one that involves tastes and preferences. The resemblance to the weak complementarity case is not without basis, though. Whether the weak complementarity or the essentiality restriction fits best is sometimes a matter of the way the story is told and not of fundamental differences in the underlying problem. Returning to our earlier example, z_1 could be relabelled as recreational fishing trips, in which case stock abundance could be viewed as a quality characteristic of trips, exhibiting weak complementarity with z_1. One way of telling the story casts the public good as a characteristic of a commodity that enters the utility function; the other way portrays it as an input into production. The two ways of telling the story result in the same function estimated and the same areas measured. In both cases the proper area will be the area above trip costs and between the demand functions for trips conditioned on different levels of the public good.

It is not surprising that the same sorts of problems that arose in Chapter 3 arise here. The compensating variation is the area between two Hicksian

---

[2]Essentiality is not the only technical relationship we could invoke to ensure that (8.19) holds. Equally useful would be the slightly more general restriction that the marginal product of the public good equals zero when the input is zero: $f(0,q) = f(0,q+\Delta q)$.

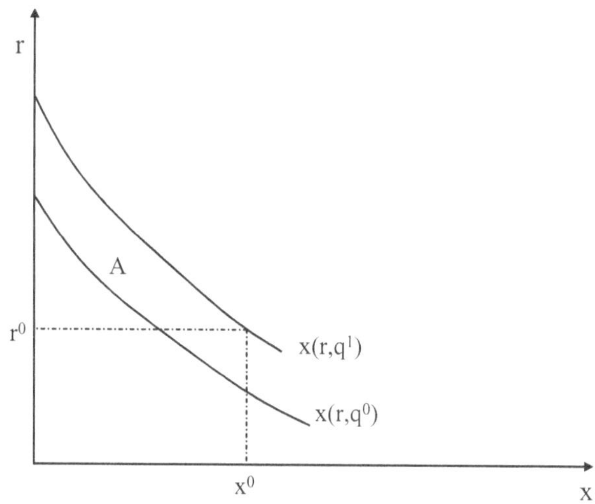

FIGURE 8.3. Compensating Variation with an Essential Input

demands, but it is Marshallian demands that we are likely to estimate. The resolution depends on the Willig condition holding for the demand for x. A second, more practical problem arises in estimating the Marshallian demand for x. It must be estimated as a function of both r and q, which means one needs to observe systematic variation in the price of x and in the public good, and concurrent observations on the consumption of x. The systematic variation in prices might come in the form of cross-sectional differences in travel costs, but variation in q may only be observed over time. The econometric requirements for estimating such a demand curve are not difficult but data requirements may be, as these are not data that are usually collected. It is especially important to obtain variation in q that is independent of individual stochastic influences. In the recreational fisheries case, for example, it would not work to use observed levels of q that depend on individuals' catch rates, because catch rates differ in part due to unobserved heterogeneity in individuals.

In addition we can encounter unusual relationships between x and q. In Figure 8.3 we depicted the demand for x shifting unambiguously outward with an increase in q, but this need not always be true. Recall the pure repackaging case from Chapter 4. In that example we considered juice from oranges, but the same problem can arise in the sport-caught fish case. Suppose the only reason for going sportfishing is to catch fish for food. Then an increase in stock abundance (and as a result, an increase in catch per trip) may actually cause the individual to take fewer trips. The increase in stock abundance has two effects. It lowers the 'price' of sport-caught fish, increasing the demand for this commodity, but it makes sport-fishing trips more productive, reducing the number of trips needed to achieve a given level of utility. The two effects work in opposite directions. Analogous to what we saw in Chapter 4, the Hicksian demand for trips may actually end up shifting backwards rather than outwards–at the current price–with an increase in q (stock abundance). Even so, the area between the two Hicksian demands for trips, conditioned on the two levels of q, will equal the correct welfare measure because trips are an essential input. This requires that the two demands cross at some price greater than the current price.

There appear to be few empirical examples in the literature that explicitly exploit the concept of an essential input in the context of environmental goods and household production. This may be because any case that meets the restrictions is better told as a weak complementarity story, and that other cases (outside of recreation, for example) rarely fit the required circumstances. Dickie and Gerking (1991) provide one of the few applications. They are interested in measuring the willingness to pay for air quality and ultimately ozone control. Their model introduces the public good in two ways. It is an input into the production of health and it enters the utility function directly because individuals enjoy air quality through increases in visibility. Usually in such a case, welfare measurement would be confounded, but the authors make two intriguing assumptions. The first is that there is a privately purchased essential input into the production of health. The private good is medical care, and the assumption is that in the absence of medical care, health status drops to 'poor'.[3] The implicit assumption is that at this level of health status changes in air quality do not alter health. The second assumption is that air quality and health status are weakly complementary, such that if health status is 'poor', the individual does not care about visibility and other aesthetic services of air quality. This combination of assumptions is an interesting way to induce correct theoretical measures of welfare. But one may find it difficult to accept the idea that a medical visit during the time frame of the study is an essential input in the production of health or that changes in air quality have no effect on the health of someone already in poor health.

Agee and Crocker (1996) also attempt to use the essentiality condition but ultimately question whether it applies in their case. Their study, which attempts to measure the willingness to pay for changes in 'lead-burden' imposed on children, defines the purchased input as chelation therapy. But the authors admit that the essentiality condition–changes in lead-burden do not matter if chelation therapy is not purchased–does not seem to fit. Ultimately they use

[3]In the model formulation, this essentiality is effective only if an individual's previously acquired stock of health knowledge is inadequate, although the distinction disappears in the application.

the expenditures on chelation therapy as a lower bound on losses–an argument better supported by the defensive expenditure literature reviewed in the next section. These papers help illustrate the difficulty in finding plausible cases of essential inputs outside the recreational demand domain that can aid welfare measurement. To add to the problem, we generally must use intuition rather than econometric tests to establish the necessary restrictions. As is true in the Agee and Crocker example, the defensive expenditure story is often a more useful way to use information from the household production function to reveal welfare effects.

8.3.3 Weak Substitutability

In an early work on measuring the benefits from environmental improvement, Feenberg and Mills (1980) suggested an additional restriction somewhat parallel to weak complementarity. Here we explore this restriction–which the authors labeled 'weak substitutability'–in the context of household production. Suppose the relationship between x and q is as described in the malaria-carrying mosquitoes story. Specifically, suppose as in that example that x is used to mitigate declines in q. This fits most naturally with a production technology in which $f_{xq} < 0$. Now add the restriction that it is possible to completely mitigate the effects of a change in q by sufficient increases in the use of x, and that at some level of x (call it $\bar{x}$) further declines in q do not matter. If these conditions hold, then it is possible to use the area between the two Hicksian demands for x (conditioned on the two levels of q) and *below* price. Graphically, this unusual measure is portrayed as area A in Figure 8.4.

Why might this area have welfare significance? Mathematically the area equals

$$\begin{aligned} Area\ A &= \int_{r(\bar{x})}^{r^0} x(r,p,q^0,u^0)dr - \int_{r(\bar{x})}^{r^0} x(r,p,q^1,u^0)dr \\ &= m(r^0,q^0,u^0) - m(r(\bar{x}),q^0,u^0) - m(r^0,q^1,u^0) + m(r(\bar{x}),q^1,u^0) \\ &= [m(r^0,q^0,u^0) - m(r^0,q^1,u^0)] + [m(r(\bar{x}),q^1,u^0) - m(r(\bar{x}),q^0,u^0)] \end{aligned}$$

where $r(\bar{x})$ is defined as the price at which $x(q^0)$ and $x(q^1)$ converge, since at $\bar{x}$ differences in q do not affect x. Area A in the graph will be equal to the compensating variation measure we seek (the terms in the first set of brackets), because the terms in the last set of brackets equal zero. By definition, at levels of $x \geq \bar{x}$ changes in q do not matter.

Unfortunately, cases in which such a threshold makes sense are not easily identified. It might seem reasonable to assume that at $r = 0$, the condition would hold automatically, because at that point as much x as is needed can be acquired costlessly. And as long as increases in x can completely mitigate

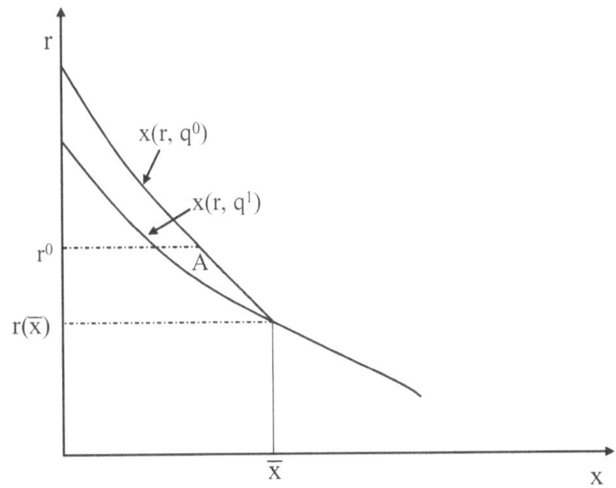

FIGURE 8.4. Compensating Variation and Weak Substitutability

any changes in q, then at $r = 0$, changes in q would not matter. However, at $r = 0$ the household may more than mitigate the decline in q. For example, suppose the initial public health program (before it was cut back) reduced the mosquito population only partially, but the household could completely eliminate the pest problem by using enough x.[4] At $r = 0$, the household would choose to eliminate pests completely, making it better off than it was under the original public health program. The area between the Hicksian curves would overstate the loss in the public programs. Feenberg and Mills (pp. 80-81) used the example of measuring the benefits of increases in the quality of public education by observing the demand for private education. If the price of private education falls to zero, they argue that the quality of public education becomes irrelevant. In that case, assuming we had Hicksian demand curves, we could measure the value of improved public education by an area such as A in Figure 8.4, where $\bar{x}$ is such that $r(\bar{x}) = 0$.[5]

Another example suggests itself, but it too falls short of meeting the require-

[4] In the pesticide example, we implicitly assume that the household recognizes no adverse effects from using pesticides. Introducing these adverse effects would require introducing pesticides directly into the utility function as well as in the household production function, or introducing them into a second household production function for different health consequences. This would obviously complicate the problem and prevent the measure described from being an accurate welfare measure of the change in the public program.

[5] This example appears to take as given that the quality of public education is always at least as great as that of private education.

ments of Feenberg and Mills. Suppose agricultural chemicals are found to be leaching into a household's well-water. If the problem gets sufficiently bad, the household can install a filtering system that removes any amount of the agricultural chemicals that are present. Once having purchased this system, the level of q (the well water quality) no longer matters, so that further increases in chemical leaching will have no effect on the household. In this example, the purchased input is a discrete technology. There is no continuous demand associated with it, as pictured in Figure 8.4, and therefore an area such as A can not be measured. Although these stories do not quite fit Feenberg and Mills' conception of weak substitutability, we will see in the next section that information about mitigating behavior can tell us something about welfare effects of degradations in environmental quality.

8.4 Bounds Using Defensive Expenditures

Each of the strategies reviewed so far for measuring welfare effects of changes in a public good requires multiple technical restrictions applied to the production function. There may exist situations in which these strategies will succeed, but to date surprisingly few applications can be found in the literature, and those that do exist are often strained. By far the more popular approach is to use expenditures on defensive actions as some bound on the welfare measures we seek. When individuals face increasing pollution, intuition suggests that *expenditures* made to defend the household against these increases ought to have some relationship to damages. After all, a household should not be willing to spend \$50 to protect itself from contaminated water if the damage from contamination is perceived to be less than \$50. Inherent in this statement is the notion that the *chosen* level of defensive expenditures is viewed as a *signal* about the household's willingness to pay to avoid the consequences of increased pollution. For example Bresnahan and Dickie (1995, p 378) write 'Inferences about WTP for improved health, safety or environmental quality are often derived from individuals' *choice* of protection action.' This is somewhat different from the alternative view that defensive expenditures are a *cost* imposed on individuals by rising pollution (see for example, Dasgupta, 2004, p 83).

A seminal paper by Bartik (1988a) sets out the important theory, but the two measures he uses–savings in defensive expenditures holding output constant (which we will label DS) and actual savings in defensive expenditures chosen by a utility maximizing individual (which we will label ADS)–are still often confused in the literature. In this section we will present what results are available, being careful to distinguish between these two defensive expenditure measures and paying particular attention to the common errors in the applied literature.

8.4.1 Framing the Problem and Finding Marginal Values

It was arguably Courant and Porter (1981) who first set environmental economists on the track of defensive expenditures as a possible basis of welfare measurement. This paper appears to have been the first to state the marginal conditions that provide a link between defensive expenditures and the marginal compensating variation of a change in environmental quality. We restate these results using our notation to provide the starting point for this analysis.

Because of its emphasis on defending against degradation, most of this literature frames the problem in terms of a public bad, as we did in equation (8.7) above. For ease of discussion we will often refer to b as pollution, but all our arguments are equally relevant for any public bad. The marginal product of b in the production of z_1 will be negative, so that according to our convention increases in b will have negative compensating variation. Although the household cannot alter b, it can mitigate the effects of increases in b by undertaking defensive or averting actions that might involve one or more inputs. Items such as air filters, sun screen, bottled water, and medical treatment have all been used in the literature as examples of purchased inputs that can mitigate the effects of pollution.[6] Focusing as it does on the implied cost function for producing z_1 and not on demands for individual inputs, this approach is really quite general and can be made to work for all forms of technology, irrespective of the sign of f_{xq}. The logic of the approach depends on purchased inputs being able to compensate for increases in the public bad. Because individuals could take more trips to make up for declines in fish stocks due to increases in pollution, the logic is as applicable to the recreational demand case specified in (8.6) as to the more obvious pesticide protection example specified in (8.7).

The cost function for producing z_1 represents the expenditures on $\mathbf{x}$ necessary to maintain z_1 when the public bad is at some level of b. The cost function is derived as

$$c(z_1, \mathbf{r}, b) = \min_{\mathbf{x}}\{\mathbf{rx} | z_1 = f(\mathbf{x}, b)\}, \tag{8.21}$$

where $\mathbf{r}$ is a vector of prices of the privately purchased inputs and $f(\mathbf{x}, b)$ is the household production function stated in terms of the public 'bad' rather than the public good, as before. The cost function, $c(z_1, \mathbf{r}, b)$, specifies the level of defensive expenditures necessary to achieve a level of the household produced commodity, given the level of b, and as such is often referred to in the literature

[6] Dasgupta (2004) suggests that defensive activities may involve expending household time (e.g boiling water) rather than purchasing inputs. As long as time can be valued at a fixed opportunity cost, nothing in the model changes. If not, then the model becomes more complex. As has been true throughout this book, the valuation of time–especially in developing countries–poses interesting challenges for the researcher and is an area where good research will have high payoffs.

as the 'defensive expenditure function'.

The explicit statement of the defensive expenditure function in (8.21) reveals its reliance on technology and not preferences. It is distinct from the expenditure function, $m(p, \mathbf{r}, b, u)$, which equals the minimum expenditures necessary for the household to achieve a given level of utility and *does* depend on preferences. The expenditure function is defined as

$$m(p_2, \mathbf{r}, b, u) = \min_{z_1, z_2} \{p_2 z_2 + c(z_1, \mathbf{r}, b) | u(z_1, z_2) \geq u^0\}. \tag{8.22}$$

We use both of these constructs, so it is important to watch for the distinction between the expenditure function and the *defensive* expenditure function (or cost function). From here on, we will suppress the input prices, $\mathbf{r}$, as they never change in the subsequent story. We also normalize relative to the composite commodity price, so that p_2 is set equal to 1 and disappears from our notation. For reference we write the CV for a change in b as:

$$CV_{\Delta b} = m(b^0, u) - m(b^1, u). \tag{8.23}$$

This conforms with our signing convention because an increase in pollution $(b^1 > b^0)$ implies negative CV.

By the envelope theorem, the compensating variation of a marginal change in b is

$$-m_b(b, u) = -c_b(z_1, b). \tag{8.24}$$

The term on the left is the *marginal* compensating variation and the term on the right is the savings in the costs of achieving a given level of z_1 with a marginal increase in the pollutant. Signing can be confusing here, so it is worth taking time to be sure this is correct. The term $-m_b(b, u)$ will always be the marginal compensating (or equivalent) variation measure and, as b is a 'bad', $m_b > 0$ so $-m_b < 0$. Likewise as b increases, the defensive expenditures necessary to maintain z_1 at its original level increase, so that $c_b > 0$ and $-c_b < 0$.

The simple result in (8.24) initially led researchers to conclude that all one needed to know to measure the welfare effects of a change in pollution was the technology of the averting or mitigating behavior.[7] With knowledge of the costs of mitigating an increase in pollution, it was believed one could obtain the desired compensating variation measure. One need only figure out how much it

---

[7] Although alluding to this theoretical result as one which does not require information on preferences, researchers have sometimes further confused things by empirically estimating the $c(z_1, b)$ function using *observable* behavior. For example Gerking and Stanley (1986) estimate a model of medical expenditures as a function of air pollution, whether the individual has a chronic illness, and number of years the individual has suffered from this illness. They call this a health production function. They then interpret the change in the dependent variable with a 30% change in air pollution as a willingness to pay measure.

would cost, using available household 'technology', to reverse the consequences of the pollution. No knowledge of preferences seemed necessary as $c_b(z_1, b)$ is the slope of an isoquant and includes no information on behavior or preferences. It turns out that this is exactly correct for marginal changes. It is also correct for non-marginal changes but only under the special assumption about technology discussed in the earlier section on separable production functions–the case when an x is a perfect substitute to q. But it is non-marginal changes that we are most often asked to evaluate, and often the linear and separable restrictions on technology that are necessary in non-marginal cases cannot be assumed to hold.

Courant and Porter developed the marginal conditions set out in equation (8.24). However, their analysis leaves room for confusion. Having developed conditions analogous to (8.24), they appear to interpret these as meaning that the savings in defensive expenditures *holding output constant* that result from a discrete change in b equals the appropriate compensating variation. They then go on to show that this is different from the actual savings in defensive expenditures.[8] The latter is correct, but the first statement is correct only for marginal changes, as is their statement that willingness to pay depends only on technology and not on tastes. Some researchers have incorrectly assumed that these results hold for discrete changes as well.

To see that this is not so, we can illustrate using Courant and Porter's simplest example, but rewritten in our notation. In this example, the individual maximizes utility, which is a function of a composite commodity, z_2, and the household produced commodity, z_1. The production technology is such that z_1 is linear in the purchased input and the marginal product of this input decreases with b. The result is a cost function for z_1 of the form $c(z_1, b) = z_1\tilde{c}(b)$, where the marginal cost of z_1 is constant and a function of the level of the public bad. This is exactly the model developed in an earlier section (see equation 8.5). The only difference is that we write the welfare measure in terms of a public bad instead of a public good:

$$CV = -\int_{\tilde{c}(b^0)}^{\tilde{c}(b^1)} z_1^h(\tilde{c}(b), u)db.$$

For this simple problem, defining the CV of a non-marginal change in the

[8]Courant and Porter's 'actual change in defensive expenditures' is a utility held constant measure, rather than the usual income held constant measure we will be discussing later in this chapter. Because they tell their averting behavior story in the context of open cities and migration, incomes fall whenever environmental quality rises, as utilities must remain equated across space. This distinction is irrelevant to the problem alluded to above which involves equating the savings in defensive expenditures holding output constant with compensating variation.

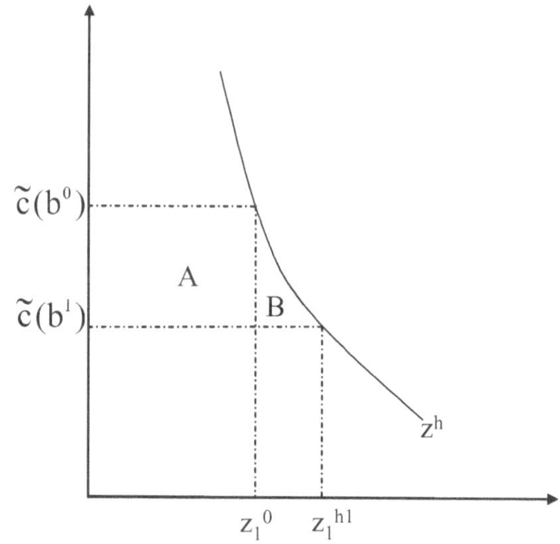

FIGURE 8.5. A Constant Marginal Cost Case

public input is straightforward because the change in b maps monotonically into a change in the 'price' of z_1. Figure 8.5 tells the story. The function $\tilde{c}(b^i)$ is the constant marginal cost of producing z_1 conditioned on the public bad being at level b^i. The CV of a decrease in b is given by area $A + B$, while the savings in defensive expenditures holding output constant is area A. The distinction between these two measures holds true for more complex cost functions as well, but is most easily illustrated in this simple example.

8.4.2 Bartik's Principal Results

An important paper by Bartik (1988a) set out some practically useful welfare measurement results and went a long way to resolving some of the confusion that had arisen in the literature. In this paper he demonstrated that for non-marginal changes the savings in defensive expenditures necessary to keep output (z_1) at its initial level after a change in the public good provides a *bound* for the exact welfare measure. Again, this 'savings in defensive expenditures' measure of Bartik does not refer to actual changes in expenditures but rather to the changes that would be necessary to return the output of the household produced commodity (e.g. quality of the home environment, health) to its initial level after the change in b. The Bartik result–that the benefits of non-marginal reductions in pollution can be bounded by this 'savings in defensive expenditures'–is widely cited, although the distinction between savings in defensive expenditures holding output constant and *actual* savings in defensive expenditures is sometimes lost.

Savings in Defensive Expenditures as a Bound on *CV*

Bartik's development implicitly relies on two assumptions. First, the privately purchased inputs do not contribute directly to utility but only indirectly through the household produced commodity. They do not generate utility in their own right nor through any other household production. An example of a violation is an air conditioning system which, although purchased to enhance health by filtering particulates from the air, also provides cooling, which also is valued separately from health. An alternative example arises when use of some purchased input to produce z_1 has some negative side effects, such as potential health risks from using pesticides to control disease-carrying insects. Second, the damage done by increasing levels of the public 'bad' can be completely mitigated by private defensive expenditures. As an example of a violation of this assumption, medical attention may lessen the consequences of a respiratory illness, but may not prevent all the discomfort of the disease.

Consistent with our early definitions, Bartik defines the 'savings in defensive expenditures' (holding output constant) as the reduction in expenditures on purchased inputs that would return the household to its initial level of z_1 after environmental quality changes. This information gives a lower bound on the true compensating variation of the pollution change. In our notation, and ensuring that the signs are consistent with our convention, the result is simply

$$CV_{\Delta b} \geq c(z_1^0, b^0) - c(z_1^0, b^1) = DS(z_1^0), \tag{8.25}$$

where DS is the 'savings in defensive expenditures', b^0 and b^1 are the initial and terminal levels of the public bad, and z_1^0 is the initial level of the household produced commodity that is being held constant. Keeping track of the signs of the changes in defensive expenditures is not always easy. Given the way we have written it, the savings in defensive expenditures (right hand side of (8.25)) is a positive quantity when b falls and a negative quantity when b increases.

To prove the result in (8.25), denote the initial level of utility by $u^0 = u(z_1^0, z_2^0)$, and define a restricted expenditure function in the following way:

$$\tilde{m}(b, u^0|z_1^0) = \min_{z_2} \; z_2 + c(z_1, b) \text{ s.t. } u^0 = u(z_1, z_2) \text{ and } z_1 = z_1^0. \tag{8.26}$$

This is the amount of income needed to reach utility level u^0 when the household produced commodity is fixed at z_1^0. For any particular level of z_1^0, the restricted expenditure function is given by $\tilde{m}(b, u|z_1^0) = c(z_1^0, b) + z_2(b, u|z_1^0)$.

Now consider the difference between the expenditure function at the initial level of b and the restricted expenditure function at the subsequent level of b:

$$\begin{aligned} & m(b^0, u^0) - \tilde{m}(b^1, u^0|z_1^0) \\ = \; & [c(z_1^0, b^0) + z_2(b^0, u^0)] - [c(z_1^0, b^1) + z_2(b^1, u^0|z_1^0)]. \end{aligned}$$

In analyzing the second line of this expression, the key issue is the behavior of $z_2(b^1, u^0|z_1^0)$. What amount of the composite commodity is needed to achieve utility level u^0 when $z_1 = z_1^0$? The answer is independent of the level of b, because b does not enter the utility function. By definition, $z_2(b^1, u^0|z_1^0)$ must equal z_2^0, regardless of the cost of achieving z_1^0, because z_2^0 is the only level of z_2 that will achieve utility level $u^0 = u(z_1^0, z_2^0)$ when $z_1 = z_1^0$.[9] Hence we can write

$$\begin{aligned} & m(b^0, u^0) - \tilde{m}(b^1, u^0|z_1^0) \\ = \; & [c(z_1^0, b^0) + z_2^0] - [c(z_1^0, b^1) + z_2^0] = c(z_1^0, b^0) - c(z_1^0, b^1). \end{aligned} \tag{8.27}$$

Expression (8.27) is not compensating variation, because the second term of the first line is a restricted expenditure function. By the nature of optimizing behavior (and more specifically by Le Chatelier's Principle), we know that the minimum restricted expenditure function cannot be smaller than the minimum unrestricted expenditure function. Hence when households are restricted to consume their initial environmental quality, the cost of achieving utility level u^0 must be at least as great as when the household produced commodity can be adjusted. That is,

$$m(b^1, u^0) \leq \tilde{m}(b^1, u^0|z_1^0).$$

This result leads logically to Bartik's bound:

$$CV_{\Delta b} = m(b^0, u^0) - m(b^1, u^0) \tag{8.28}$$

$$\geq m(b^0, u^0) - \tilde{m}(b^1, u^0|z_1^0), \tag{8.29}$$

which implies

$$CV_{\Delta b} \geq c(z_1^0, b^0) - c(z_1^0, b^1) = DS(z_1^0). \tag{8.30}$$

From equation (8.21), we can see that $c(z_1^0, b^0) - c(z_1^0, b^1) = \mathbf{r}[\mathbf{x}(z_1^0, b^0) - \mathbf{x}(z_1^0, b^1)]$, so that CV is bounded by the change in expenditures on purchases

[9]It might appear that this result is an artifact of the composite commodity construct and that if we were to introduce two or more additional commodities having varying relationships with z_1, these simple results would no longer hold. This is not the case, however. The results do hold for any number of goods for the simple reason that b does not enter utility, z_1 is forced to remain constant, and no price ratios change. Because of this, there is no reason for the optimal consumption levels of other commodities to change when b changes, as long as utility is being held constant.

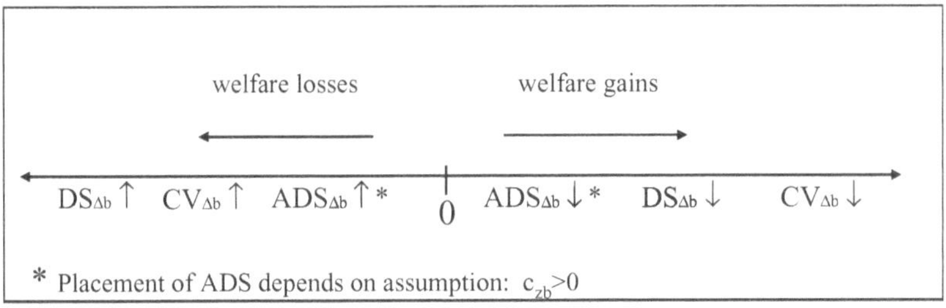

FIGURE 8.6. Ordering of Savings in Defensive Expenditures (DS), Actual Savings in Defensive Expenditures (ADS), and Compensating Variation (CV)

of defensive inputs, $\mathbf{x}$, necessary to achieve the original level of the household produced commodity, z_1, after the change in the undesirable public input.

We have not needed to be specific about the direction of pollution change. The change in defensive expenditures (holding output constant) forms a lower bound on compensating variation *along the real number line*, irrespective of the direction of the change in pollution. When pollution decreases, CV in (8.28) is positive and (8.29) represents the smaller positive savings in defensive expenditures. If pollution increases, so that CV is negative, the savings in defensive expenditures will be more negative than the compensating variation. To summarize, for reductions in pollution, DS will be a lower bound for CV, and since both are positive, DS understates the gains. For increases in pollution, DS will be negative because more expenditures will be required to maintain the household produced commodity at z_1^0 and this savings will be greater in absolute value than CV. Thus DS will overstate losses. This ordering is summarized in Figure 8.6, where the term ADS should be ignored for the time being.

Perhaps the most technically detailed study that uses the Bartik bound is that of Murdoch and Thayer (1990). They estimate an upper bound on the losses that would occur from the expected increase in the incidence of nonmelanoma skin cancers should ozone layer depletion continue.[10] The authors develop a complex technological function for 'effective UV-b radiation' exposure. The task is further complicated by the need to play out this scenario over 50 years. The upper bound on the losses equals the present value of the added expenditures on sun protection products that would be necessary, holding the likelihood of skin cancer constant, should the ozone layer continue to deplete as projected over the next 50 years.

We can cast the Murdoch-Thayer analysis in the current framework for clar-

[10]Strictly speaking, the authors interpret the result as an upper bound on the benefits of a policy aimed at preventing expected ozone depletion. This awkward definition was required given that, in the absence of policy, a change in b will automatically take place over time.

ity. Let the probability of remaining free of non-melanoma cancer be represented by z_1 and let its production function be $f(x, b)$ where b represents deterioration in the ozone layer. Increases in b reduce the probability of remaining free of non-melanoma cancer. Consider the expenditures that could be made to control non-melanoma cancer. These are the expenditures on sunscreen products, given by $c(z_1, b)$. Bartik's savings in defensive expenditures is the reduction in costs due to a change in the ozone layer, holding the risk of skin cancer constant. With greater declines in the ozone layer, the 'savings in defensive expenditures' $(c(z_1^0, b^0) - c(z_1^0, b^1))$ will negative. Of course, CV will be negative as well. The Bartik bound given in equation (8.30) is

$$CV_{\Delta b} \geq DS(z_1^0),$$

so that the savings in defensive expenditure is a larger negative number than CV. The authors provide extensive calculations to obtain an estimate of DS which they maintain is an upper bound on the losses from failure to address the deterioration in the ozone layer or, equivalently, an upper bound on the benefits from a policy that prevents the ozone layer from further deterioration.

Bartik's bounds are useful because they depend on knowing about household technology only and not about preferences. If pollution (or some other public bad) increases and we know how much it would cost to return the individual to the initial level of the household produced commodity, then we have an upper bound on the losses incurred. Likewise if it falls, knowing how much the individual can save in reduced defensive expenditures and still maintain the old level of the household produced commodity gives a lower bound on gains. In neither case do we need to know anything about preferences for the household produced commodity and so behavioral functions need not be estimated. This approach is potentially most useful for pollution abatement. Lower bounds in absolute value terms are usually far more useful pieces of information. A DS measure associated with a pollution increase rarely has policy or legal relevance, as the true loss may be any smaller number (possibly indistinguishable from zero), but it may be very helpful to know that the benefits from a policy are at least as great as some estimated DS measure.

Savings in Defensive Expenditures as a Bound on EV

Using the same general approach set out in equations (8.25) to (8.29), Bartik provides an upper bound on the equivalent variation of a change. If one were to know the chosen level of the commodity after the change in b (which we will label z_1^1), then

$$EV \leq c(z_1^1, b^0) - c(z_1^1, b^1) = DS(z_1^1).$$

The two results that $DS(z_1^0) \leq CV_{\Delta b}$ and $EV_{\Delta b} \leq DS(z_1^1)$ are intriguing. If we could assume that $CV_{\Delta b} \leq EV_{\Delta b}$ then the two defensive expenditure measures would bound CV and EV. But this is exactly what we can not assume. Unlike in the price change case, it is not necessarily true that $CV_{\Delta b} \leq EV_{\Delta b}$. The ordering depends on the sign of m_{ub}, with $m_{ub} > 0$ implying $CV_{\Delta b} \leq EV_{\Delta b}$.

To determine $sign(m_{ub})$ we begin with the envelope (marginal) result that

$$m_b(b,u) = c_b(z_1, b).$$

Differentiating both sides with respect to u gives

$$m_{bu} = c_{bz_1} \cdot \frac{\partial z_1^h}{\partial u},$$

where $\partial z_1^h / \partial u$ is directly related to the income effect:

$$\frac{\partial z_1^h}{\partial u} = \frac{\partial z_1^m}{\partial m} \frac{\partial m}{\partial u}.$$

Hence we have

$$sign\{m_{bu}\} = sign\{c_{bz_1} \cdot \frac{\partial z_1^m}{\partial m} \frac{\partial m}{\partial u}\}.$$

The sign depends on behavior, through the sign of the income effect, and on household technology, through the properties of the defensive expenditure function. If the income effect for z_1 is positive, then the condition $c_{bz_1} \geq 0$ will ensure that $m_{bu} \geq 0$ and $CV_{\Delta b} \leq EV_{\Delta b}$. Intuitively we might expect c_{bz_1} to be positive (or at least non-negative)–the marginal cost of producing z_1 would usually be expected to rise as pollution increases.[11] For example, as a public water supply becomes more contaminated, bottled water must be substituted in more and more household uses. But, as Bartik points out, there can be some unusual cases in which c_{bz_1} is negative–when higher levels of pollution make x significantly more productive than at lower levels. Because the sign of c_{bz_1} is important for several results in this section, we show its derivation and explain the unusual case in the appendix to this chapter.

If $c_{bz_1} \geq 0$, then the two 'savings in defensive expenditures' measures will bound CV and EV, such that

$$DS(z_1^0) \leq CV_{\Delta b} \leq EV_{\Delta b} \leq DS(z_1^1). \tag{8.31}$$

This set of results is certainly appealing, although in some circumstances the two bounds may be very inaccurate. In any event, expression (8.31) is useful only if we can anticipate the level of actual adjustment in z_1, that is, if we have a way of knowing or guessing at z_1^1. In many cases this will be difficult; in all cases it will involve information on preferences and not just technology.

---

[11] Note that c_{bz_1} will be non-negative for the technology examples presented in equations (8.6) and (8.9). It will also be true for (8.7) given most choices for the function $\alpha(x)$, given that the range of x satisfies $0 \leq \alpha(x) \leq 1$. For example if $\alpha(x) = \beta x^n$ where $n < 0$ then $c_{bz_1} > 0$.

8.4.3 Actual Savings in Defensive Expenditures

Bartik's bounds, depending as they do on savings in defensive expenditures holding output constant, have nothing to do with choices or decisions and so cannot reflect anything about preferences. They use technology to determine the costs of defending against increased pollution or the savings made possible by reductions in pollution. Therefore, they must (except in special circumstances) be bounds and not exact measures. This is most obvious for the measurement of losses due to degradation. It may be extremely costly to mitigate completely a decrease in visibility brought about by increases in particulate matter in the air. However, the payment necessary to return the individual to his initial utility level may be quite small, if the change in visibility has little value to the individual. So it would *seem* that information on behavior–that is, *actual* savings in defensive expenditures–should provide a more accurate approximation to the desired CV measure, reflecting as it does information on preferences. But, as Bartik showed, this is not necessarily the case.

Actual savings in defensive expenditures is defined as the savings that occur after b has changed and the *utility maximizing* household has adjusted its consumption of z_1. Intuitively, if public programs to treat malaria-carrying mosquitoes are cut back, then the utility maximizing household may increase its defensive expenditures on pesticides to mitigate the resulting increase in pest populations. But its exposure to malaria is likely to rise, because the 'price' (or more correctly, the marginal cost) of protection from malaria has increased.

The actual savings in defensive expenditures measure is given as:

$$ADS = c(z_1^m(b^0, y), b^0) - c(z_1^m(b^1, y), b^1) = c(z_1^0, b^0) - c(z_1^1, b^1). \tag{8.32}$$

Compare this to the savings in defensive expenditures holding output constant used in the Bartik bounds:

$$DS = c(z_1^0, b^0) - c(z_1^0, b^1). \tag{8.33}$$

The difference,

$$ADS - DS = c(z_1^0, b^1) - c(z_1^1, b^1),$$

depends on the sign and size of the Marshallian response to a change in b. If $z_1^1 \leq z_1^0$ then $ADS \geq DS$. Intuitively we expect the demand for z_1 to fall with increases in b, but in fact a definitive proof depends once again on the signing of c_{bz_1}. The Marshallian response can be expressed as:

$$\partial z_1^m / \partial b = \partial z_1^h / \partial b - (\partial z_1^m / \partial y) \cdot m_b, \tag{8.34}$$

where $(\partial z_1^m / \partial y) m_b$ will be positive if z_1 is a normal good. In the appendix we show that $\partial z_1^h / \partial b$ (the Hicksian response to a change in b) will have the opposite sign from c_{bz_1}. Thus, if the marginal cost of producing z_1 rises with increases in pollution, as we expect in most cases, then $\partial z_1^h/\partial b < 0$. This will be true as long as large levels of pollution do not make the purchased input *significantly* more productive. Thus $c_{bz_1} > 0$, together with the assumption that z_1 is a normal good, ensures that the Marshallian response to changes in b will be negative.

Given this result, a comparison of (8.32) and (8.33) implies that for degradations in the environment (i.e. increases in b) $ADS \geq DS$ and for improvements in the environment (i.e. declines in b) $ADS \leq DS$. Put another way, the absolute value of the savings in actual defensive expenditures will always be smaller than the absolute value of savings in defensive expenditures holding output constant. The former will measure smaller gains and smaller losses than the latter. This ordering is depicted in Figure 8.6.

This together with the fact that $DS \leq CV$ for both improvements and degradations leads to the unexpected result that ADS is a worse lower bound for CV than is DS when the environment improves. This is most easily illustrated using the simple case that we have referred to so often–the case when changes in b simply shift the marginal cost of z_1. Though not general, this case helps with intuition. In Figure 8.7 environmental improvement implies a decrease in b which causes a decrease in the 'price' of z_1 from $\tilde{c}(b^0)$ to $\tilde{c}(b^1)$. The area behind the Hicksian demand function for z_1 is the exact CV measure and is denoted by $(A+B)$. In contrast the savings in defensive expenditures holding output constant (DS) is area (A) which is clearly smaller than CV. The actual savings in defensive expenditures is given by $\tilde{c}(b^0)\, z(b^0, y) - \tilde{c}(b^1)\, z(b^1, y)$ which equals $(A+E) - (E+F+G) = (A-F-G)$ in Figure 8.7. And this area is smaller than DS.

So, for improvements we find that $ADS \leq DS \leq CV$, but for degradations all we know is that $|ADS| \leq |DS|$. Bartik's results do not supply enough information to establish the relationship between ADS and CV. In fact Bartik gives no information on this relationship. A number of papers have assumed, however, that when environmental degradation occurs, the increase in actual defensive expenditures will be a lower bound on the true welfare losses. That is, they have assumed that $|CV| \geq |ADS|$ (see for example Abdalla, Roach, and Epp (1992) and Harrington, Krupnick, and Spofford (1989)).

Once again assuming that $c_{bz_1} > 0$ and that z_1 is a normal good seems to give us what we need. To show this, compare the following statements of CV and ADS:

$$\begin{aligned} CV &= m(b^0, u^0) - m(b^1, u^0) = c(z_1^0, b^0) + z_2^0 - c(z_1^{h1}, b^1) - z_2^{h1} \\ &\quad \textit{and} \\ ADS &= c(z_1^0, b^0) - c(z_1^{m1}, b^1) \end{aligned} \tag{8.35}$$

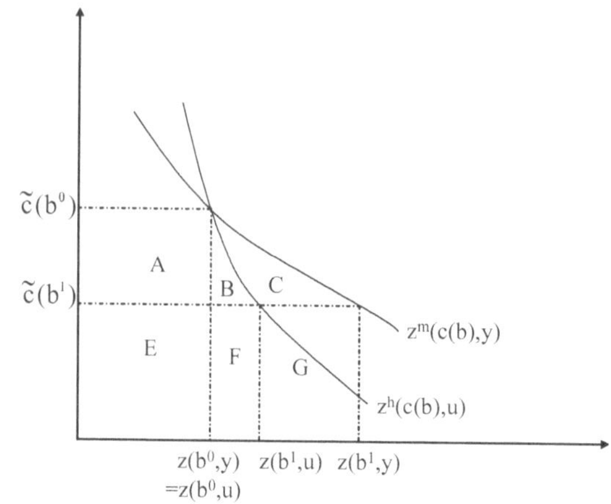

FIGURE 8.7. Bounds for CV, DS and ADS

where the superscript $h1$ indicates the Hicksian levels of the choice variables after the increase in pollution and $m1$ indicates the Marshallian levels of the same choice variables. CV is negative because it measures the welfare effect of an environmental degradation. The difference in these two measures is

$$\begin{aligned} & CV - ADS \qquad (8.36) \\ = \; & c(z_1^0, b^0) + z_2^0 - c(z_1^{h1}, b^1) - z_2^{h1} - [c(z_1^0, b^0) - c(z_1^{m1}, b^1)] \\ = \; & c(z_1^{m1}, b^1) - c(z_1^{h1}, b^1) + (z_2^0 - z_2^{h1}). \end{aligned}$$

By equation (8.34) the Marshallian response to the increase in b will be more negative than the Hicksian one, so that $z_1^{m1} < z_1^{h1}$, as long as $c_{bz_1} > 0$ and z_1 is a normal good. This ensures that $c(z_1^{m1}, b^1) - c(z_1^{h1}, b^1) < 0$. The last term must also be negative, since z_2^{h1} must exceed z_2^0. This is necessary because when b increases, the Hicksian demand for z_1 decreases. The only way to maintain utility at u^0 is to increase consumption of the composite commodity. The two results together imply that $CV - ADS < 0$. For degradation, the entire ordering becomes $DS < CV < ADS$ where all are negative. When the above assumptions hold, Figure 8.6 portrays the relationships among CV and the two defensive expenditure measures.

The literature contains more examples of bounding using ADS than DS. For one thing, ADS can be observed, but DS must be reasoned from evaluating available household technologies. For another, DS is an upper bound on losses–not a very useful bound for damage assessment, but ADS is a lower bound. Laughland, Musser, Shortle, and Musser (1996) use ADS as a lower bound on the losses from a Giardia contamination incident in the water supply of a small Pennsylvania town. The averting behavior included boiling and hauling water and buying bottled water. Similar averting behaviors were considered by Harrington, Krupnick and Spofford (1989) in an analysis of the losses from a Giardia outbreak in another Pennsylvania town, and by Abdalla, Roach and Epp (1992) in analyzing a trichloroethylene groundwater contamination incident in southeast Pennsylvania.

8.4.4 Lumpy Technologies and Discrete Choices

One can imagine cases in which the household production technology would be approximately continuous. Households may buy more and more bottled water to substitute for increasing deterioration in public drinking water supplies, using the bottled water at first for drinking, then for cooking, then for washing clothes, etc. Individuals may use more sun screen to defend against ozone depletion. They may make more medical visits and incur more medical testing to mitigate the consequences of increased carcinogenic exposure or use more insecticide with the increase of malaria carrying mosquitoes. But in many circumstances, the most obvious defensive activities are ones that involve discrete investments. Some individuals install water filtering devices to defend against drinking water contamination, they buy air conditioning systems to protect home air quality, or they install basement venting systems to reduce radon gas accumulation. To what extent does the possibility of discrete technologies complicate the analysis?

One of the advantages of Bartik's DS bounds is that they require neither a well-defined demand function for the household produced commodity nor a continuous production technology. However, if few alternatives are available and technology is lumpy, the inaccuracies in the bounds may be exacerbated. Consider defending against a decline in indoor air quality brought about by the construction of a nearby incinerator. There may be no technology that would return indoor air quality to exactly its initial level. The only recourse the home owner may have is an air filtering system that costs c^* and actually improves air quality to levels much higher than initial levels. To use Bartik's DS bounds we would like to know the least cost way of achieving z_1^0, the original level of the household produced commodity, but the least cost way of achieving *at least* z_1^0 may be to purchase this air filtering system. This measure remains an upper bound on losses (or a lower bound on CV along the real number line), but the discrete nature of this technology compounds inaccuracies. Discrete or lumpy technologies need not necessarily lead to very different results from continuous technologies, though, if there are enough alternative technologies offering different outcomes at different prices. In such a case, the story will not differ substantially from the continuous one told in the previous sections.

Where defensive activities are investments, a rather interesting asymmetry may arise, as suggested by Bartik. The absolute value of the costs of taking a defensive action may not equal the savings in defensive expenditures of the reverse action. To make this clear, first consider the continuous defensive activity–the purchase of more and more bottled water as a response to increasing water pollution. If water quality is subsequently improved, the defensive expenditure savings equals the amount that the household would save on bottled water. However, if the technology chosen is a water filtration system, subsequent reductions in pollution could not save the full cost of that system–only future operating and ultimate replacement costs.

Can *actual* changes in defensive expenditures be used to approximate welfare measures when technology is lumpy? The implications are best illustrated through example. Consider a case in which the individual faces pollution at level b^0, chooses to undertake no defensive activity, and produces a commodity (perhaps level of soiling) at level z_1^0. Now suppose air pollution increases to b^1 and there is only one technology to combat it. That technology involves a fixed investment cost of $\widehat{c}$ with no variable costs. The individual faces a discrete choice between $u(z_1(b^1,0),y)$ and $u(z_1(b^1,\widehat{x}),y-\widehat{c})$ where $z_1(b^1,0)$ is the new level of soiling if no defensive action is taken and $z_1(b^1,\widehat{x})$ is the level if the defensive technology is adopted.

Analyzing this discrete choice will tell us something about the demand for z_1, especially if there is enough variation over circumstances in a sample of individuals. This will be revealed by people choosing between two discrete levels of z_1 and accompanying discrete levels of net income (y or $y-\widehat{c}$). But we really do not learn anything directly about the valuation of changes in b unless the technology just happens to produce the original level of z_1. If so, then $z_1(b^1,\widehat{x}) = z_1(b^0,0)$, so that the discrete choice we witness is equivalent to a discrete choice between $u(z_1(b^1,0),y)$ and $u(z_1(b^0,0),y-\widehat{c})$. If the individual chooses to make the defensive investment, then we know that

$$u(z_1(b^0,0),y-\widehat{c}) > u(z_1(b^1,0),y)$$

which implies that losses from the change in b are at least equal to $\widehat{c}$, and if he does not, then the losses are less than $\widehat{c}$. In this case we can apply all the results that were developed in Chapter 5 on welfare measurement using discrete choice models. However, if no available defensive option returns the individual to the original level of the commodity, then no trade-off allows a direct comparison of levels of b.

What the discrete choice problem does yield is information about the valuation of discrete packages of z_1 and $y-\widehat{c}$, and sometimes this is useful in its own right. Akerman, Johnson and Bergman (1991) define z_1 as cancer risk and study individuals' discrete responses to knowledge of radon infiltration in their homes. In their study, the 'pollutant' (b) is not under the control of the public sector and no policy question about the value of its reduction is at issue.

However, the authors argue that observing households' choices among discrete defensive technologies for reducing radon infiltration reveals information about their valuation of changes in cancer risk.

8.5 Cost of Illness and Defensive Expenditure

One of the most commonly treated problems in the literature is the case in which b is a contaminant and z_1 is health status.[12] In early applications health professionals began measuring the gains from reduced pollution in terms of the reduced costs of illness, defined largely as medical expenses and opportunity costs in terms of lost wages. Dose-response functions, relating pollution levels to days of illness, formed the basis of their analysis. These 'predicted days of illness' were then multiplied by monetary costs of illness (medical expenses plus lost wages) to come up with the social costs of higher pollution–or the gains from reducing pollution. Although some economists have argued against their relevance, cost of illness (COI) measures are still found in agency guidelines as ways of obtaining the benefits of reduced pollution, and applications can be found in published literature in which COI is taken as a welfare measure.

The connection between defensive expenditures for valuing changes in non-market goods and the cost of illness approach has been the source of considerable confusion. Some of the confusion stems from the fact that the theory is often developed in marginal terms but applied to empirical problems in which discrete changes take place. In addition the inappropriate inclusion of defensive expenditures has continued to plague applications. Much of the literature is further confounded by differences between Marshallian and Hicksian adjustments and ambiguities about what can be mitigated. And, once again, a common source of confusion comes from the distinction between savings in defensive expenditures holding the household produced commodity (in this case, health status) constant (DS) and actual (i.e. observed) defensive expenditure changes (ADS).

A further complication relates to the fundamental distinction between ex-

[12] Introducing health risk suggests that we must reformulate the problem in terms of expected utility. Some authors (e.g. Shogren and Crocker, 1991) have suggested that the basic result–that the willingness to pay to avoid a marginal increase in b equals the marginal cost of mitigating that change in b–fails to hold when the household produced commodity is health risk. Quiggin (1992) produces far more positive results under the plausible assumption of decreasing absolute risk aversion. The papers on this issue deal almost exclusively with marginal changes and require developing theories of behavior under risk and uncertainty. As such they are beyond the scope of this book. From here on, then, we assume that z_1 is health status and that individuals have preferences over levels of health status as well as access to technologies that can mitigate the hazards that can impair health.

penses that can be classified as costs of illness and those that are defensive expenditures. In some cases the distinction is clear. An individual may use sunscreen to protect against the possibility of skin cancer, but will then incur costs of illness (e.g. medical expenses and lost work days) if he nonetheless succumbs to the illness. In papers such as these, medical expenses are viewed as an exogenous consequence of illness and not a choice variable. The choice variable relates to preventive defensive actions. Some researchers, however, have viewed the costs of medical treatment as defensive expenditures in their own right. Viewed in this way, medical treatment relates to curative actions which defend against the *consequences* of illness, but then these costs must be treated as a choice variable rather than an exogenous consequence, becoming part of the defensive expenditure function. The model we set out in the next section implicitly allows for both types of illness costs–those which occur as a consequence of illness and those that are endogenously chosen defensive actions.

We first introduce the cost of illness model and restate the welfare problem in the cost of illness case, demonstrating the inherent difficulty in obtaining exact welfare measures. We then show why defensive expenditures and other endogenous expenditures should not be included in a cost of illness welfare approximation. Given the bounding results for defensive expenditures, it is helpful to examine the cost of illness measure by recasting it in Bartik's framework. This allows us to address the problem specifically as a non-marginal change and to consider whether any of Bartik's results are altered. Finally we examine the implications of calculating costs of illness as a welfare measure.

8.5.1 The Model with Cost of Illness

In this section we trace out the results presented by the most cited early paper on this topic, Harrington and Portney (1987), using our own notation, and we make the model a bit more general. To accommodate the idea of costs of illness in the household production model, we define the household produced commodity as health status, so that

$$z_1 = H(x,b) = \bar{H} - S(x,b) \tag{8.37}$$

where we have simply substituted $H(x,b)$ for $f(x,b)$ of the previous section. $H(x,b)$ is health status and $S(x,b)$ is a function that shows the deleterious effects of pollution on health as well as the ability of behavior to avert or defend against changes in pollution. Hence we write utility as $u(H(x,b),z_2)$, where increases in health obviously increase utility. For simplicity we assume one purchased input, x, that is used to defend against pollution, so that Harrington and Portney's 'defensive expenditure' variable equals our rx. The results hold for more inputs, but are easiest to see in this form. In much of the literature, S is measured as number of sick days, so that $S(x,0) = 0$ and $\bar{H}$ is health status in the absence of illness. One expects $S(x, b)$ to be decreasing in x and increasing in b. Defensive inputs reduce sick days and increase health status. Increases in pollution cause increases in sick days and reductions in health status.

The individual maximizes utility which is a function of a composite good and health status, subject to both time and income constraints. The two constraints catalogue the types of time and money costs sickness might induce. The model typically specified in the literature includes medical costs charged at a fixed price, g, per visit M, where visits are exogenously determined by the days of sickness. The income constraint is

$$\bar{y} + wT_w = z_2 + rx + gM(S(x, b))$$

where $\bar{y}$ is exogenous income, w is the wage rate, T_w is time spent working, and rx are defensive expenditures written in terms of the defensive input x. In the literature, the time constraint generally reflects a loss of work time, and therefore a loss in wage earning capability, equal to sick days, as well as a loss of time for leisure (L):

$$T_w = T - L - S(x, b)$$

where T is total available time.[13] These constraints could be written in a number of ways, and all sorts of costs could be included, but we use this simple model much like that presented in Harrington and Portney, for illustration. Note that these 'costs of illness' are exogenously imposed in this model. Once sick days are incurred, medical expenses and lost wages follow automatically. They are not endogenous variables like defensive expenditures.

If the implicit time constraint can be collapsed into the money constraint, the model can be rewritten as

$$\begin{aligned} &\max_{x, L, z_2} \; u(\bar{H} - S(x, b), L, z_2) + \\ &\lambda[\bar{y} + wT - wL - wS(x, b) - gM(S(x, b)) - z_2 - rx]. \end{aligned} \tag{8.38}$$

Restating in expenditure minimization form yields

$$\begin{aligned} m(b, w, r, g, T, u) \;=\; & \\ & \min_{x, L, z_2} \; z_2 + rx + wL + wS(x, b) + gM(S(x, b)) - wT \\ & + \mu[u - u(\bar{H} - S(x, b), L, z_2)]. \end{aligned} \tag{8.39}$$

The first order condition for x, which will be useful later, can be written

$$r + w\frac{\partial S}{\partial x} + g\frac{\partial M}{\partial S}\frac{\partial S}{\partial x} - \mu\frac{\partial u}{\partial S}\frac{\partial S}{\partial x} = 0. \tag{8.40}$$

---

[13] Workers with sick leave benefits do not incur the full costs of lost work days, but instead share these costs with employers. Some papers have attempted to take this into consideration in their models.

In structure, $m(b, w, r, g, T, u)$ looks like the expenditure functions that we have employed earlier in the chapter. However, the problem now includes the impact of sickness variables–lost work and medical expenses.

Exact Welfare with Medical Expenses and Lost Work Days.

Using the expenditure function in equation (8.39), the exact welfare measure with the additional cost of illness components is

$$CV = m(b^0, w, r, g, T, u^0) - m(b^1, w, r, g, T, u^0). \tag{8.41}$$

In the context of this problem, the input x appears to be the only choice variable that could possibly help reveal this welfare measurement. The variable x is an input to health such as preventive medical care, bottled water, sunscreen, etc. As before we have assumed that the input conveys utility only because it contributes to health.

Let us look briefly at the potential for welfare measurement through observations on input demand. When we account for the explicit exogenous effects of illness–lost work time and curative medical expenses–the input demand function will have a different structure. Nevertheless, Shepherd's lemma gives us

$$\partial m(b, w, r, g, T, u^0)/\partial r = x(b, w, r, g, T, u^0).$$

Pursuing the approach used in equations (8.17) and (8.18), the change in the area under the demand for x is:

$$\begin{aligned} &\int_{r^0}^{r^*(b^1)} x(b^1, w, r, g, T, u^0)dr - \int_{r^0}^{r^*(b^0)} x(b^0, w, r, g, T, u^0)dr \quad (8.42) \\ = \; & m(b^1, w, r^*(b^1), g, T, u^0) - m(b^1, w, r^0, g, T, u^0) \\ & -[m(b^0, w, r^*(b^0), g, T, u^0) - m(b^0, w, r^0, g, T, u^0)]. \end{aligned}$$

We know that the differences in areas under the Hicksian demand curve will equal the exact CV in equation (8.41) if

$$m(b^1, w, r^*(b^1), g, T, u^0) = m(b^0, w, r^*(b^0), g, T, u^0).$$

The likelihood that this condition will be fruitful for welfare measurement is small. For this condition to hold, the input could be an essential good–i.e., without sufficient inputs there is no health produced, and hence no impact from the environmental pollution. But a market good that is an essential good in health seems farfetched. A second interpretation, useful in other contexts, posits that the pollutant is embodied in the input as a quality characteristic. In the health case this requirement seems unlikely to prevail. Hence we pursue approximations and bounds on welfare measures.

Marginal Values

The basic model with cost of illness and lost sick days, as well as other elaborations, comes from Harrington and Portney. Recognizing the difficulty in estimating demand functions, they offered an approximation based on an expression of the marginal value of pollution. They argued that the marginal willingness to pay to avoid an increase in b is

$$\frac{\partial wtp}{\partial b} = w\frac{\partial S}{\partial b} + g\frac{\partial M}{\partial S}\frac{\partial S}{\partial b} - \mu\frac{\partial u}{\partial S}\frac{\partial S}{\partial b} + r\frac{\partial x}{\partial b}. \tag{8.43}$$

This expression has since been interpreted to mean that willingness to pay for a non-marginal change in b equals the sum of medical costs, lost wages, monetized lost utility and defensive expenditures. Much like the confusion generated by Courant and Porter's results, this expression may not have been meant to apply to non-marginal changes, but has repeatedly been employed in this way. In addition, the term $r\,\partial x/\partial b$ does not belong in this expression because of the envelope theorem.[14]

Starting with the expenditure function in (8.39), the correct statement of the *marginal* compensating variation of due to a change in b can be shown to be

$$m_b = (w + g\frac{\partial M}{\partial S} - \mu\frac{\partial u}{\partial S})\frac{\partial S}{\partial b}. \tag{8.44}$$

Note that this marginal willingness to pay for a reduction in pollution does not include anything about defensive expenditures. The results of optimally choosing the defensive actions are implicit in the expression above because the optimal level of defensive action dictates the actual level of sickness experienced. At the margin the choice of defensive expenditures is made so that the benefits of reducing sickness in this way just equal the costs of doing so. Substituting the first order condition for the optimal level of x yields

$$(w + g\frac{\partial M}{\partial S} + \mu\frac{\partial u}{\partial S})\frac{\partial S}{\partial x} = -r,$$

a result that can be found in Harrington and Portney and is the equivalent of the result in (8.24):

$$m_b = -\frac{rS_b}{S_x}, \tag{8.45}$$

[14]Note that if $r\,\partial x/\partial b$ is included, we must also include all the other terms involving x. Doing so yields the following terms:

$$[r + w\frac{\partial S}{\partial x} + g\frac{\partial M}{\partial S}\frac{\partial S}{\partial x} + \mu\frac{\partial u}{\partial S}\frac{\partial S}{\partial x}]\frac{\partial x}{\partial b},$$

but the expression in brackets equals zero from the first order condition, equation (8.40). The incorrect marginal value expressed in equation (8.43) continues to be found in some averting behavior papers (e.g. Alberini and Krupnick, 2004).

where S_b and S_x are shorthand for $\partial S/\partial b$ and $\partial S/\partial x$.

Equation (8.45) shows that the compensating variation associated with a *marginal* change in b is equal to (minus) the increased cost necessary to hold S at a fixed level as pollution increases. Once again, simply knowing the technology of the health production function tells us what we need to know about *marginal* CV. All that is needed in expression (8.45) is knowledge of the least cost way of achieving the same level of health status (or the same number of sick days) using privately purchased inputs. Nothing about the various costs generated by illness enters the problem. The reason is simply that b does not affect medical costs or lost wages directly, but only through its impact on sick days and its impact on sick days can be mitigated by averting behavior. At the optimum, averting behavior is adjusted such that the marginal costs just equal the marginal benefits.

8.5.2 Cost of Illness and Bartik's Results

It is clear that we cannot use the expression in (8.43) to develop welfare measures for non-marginal changes, so it is best to return to the Bartik framework and see what we can learn. Our first point is that none of the Bartik results change with the inclusion of costs of illness. In fact many of the applied papers that use Bartik's bounds as developed in the last section do so in the context of a model that is expanded to incorporate these exogenous costs of illness (e.g. Harrington, Krupnick, and Spofford. 1989; Laughland, Musser, Shortle, and Musser, 1996).

To make the model both more general (to include more inputs) and more analogous to the one developed earlier, we rewrite the utility maximization and the expenditure minimization problems in terms of the defensive expenditure function. To do so it is advantageous to change the decision variable from the input into health production to the number of sick days, S. It is consistent with the model, though somewhat counter-intuitive, to explain choices in terms of sick days, but doing so makes the connection between the Bartik bounds and the cost of illness transparent. Define the indirect utility function and the expenditure function as

$$\begin{aligned} v(b,w,g,T,\bar{y}) &= \max_{S,L,z_2} u(\bar{H}-S,L,z_2) \\ &\quad +\lambda[\bar{y}+wT-z_2-wL-wS-gM(S)-c(S,b)], \end{aligned}$$

and

$$\begin{aligned} m(b,w,g,T,u) &= \min_{S,L,z_2} z_2+wL+wS+gM(S)+c(S,b)-wT \quad (8.46) \\ &\quad +\mu[u-u(\bar{H}-S,L,z_2)], \end{aligned}$$

where $c(S, b)$ is the cost function and stands for $c(\bar{H} - S, b)$ but $\bar{H}$ does not change and so is omitted for simplicity. However, we need to remember that $c_S < 0$ while $c_H > 0$, where $H = \bar{H} - S$ is equivalent to z_1 from the previous section. Despite the complexity introduced by including money and time costs of illness into the optimization problem, the savings in defensive expenditures (holding output constant) will still be an underestimate of CV. The logic remains the same as that proposed by Bartik. The restricted expenditure function, holding S at its initial level, can still be defined as

$$\tilde{m}(b, w, g, T, u|S^0) = \min_{L, z_2} z_2 + wL + wS + gM(S) + c(S, b) - wT \qquad (8.47)$$

$$\text{subject to } u^0 = u(\bar{H} - S, L, z_2) \text{ and } S = S^0.$$

This is the amount of income needed to reach utility level u^0, given the other arguments, with sick days fixed at S^0. Even though we now have two additional variables in the utility function, Bartik's results still hold. The optimal way to hold utility at u^0 when $S = S^0$ and prices have not changed is to set $z_2 = z_2^0$ and $L = L^0$. So we have

$$\begin{aligned} & m(b^0, w, g, T, u^0) - \tilde{m}(b^1, w, g, T, u^0|S^0) \\ = \; & [c(S^0, b^0) + z_2^0 + wL^0] - [c(S^0, b^1) + z_2^0 + wL^0] \\ = \; & c(S^0, b^0) - c(S^0, b^1). \end{aligned}$$

The last line is Bartik's change in defensive expenditures (holding sick days constant). Invoking the Le Chatelier Principle yields Bartik's results because

$$\begin{aligned} CV &= m(b^0, w, g, T, u^0) - m(b^1, w, g, T, u^0) \\ &\geq m(b^0, w, g, T, u^0) - \tilde{m}(b^1, w, g, T, u^0|S^0), \end{aligned} \qquad (8.48)$$

which implies

$$CV \geq c(S^0, b^0) - c(S^0, b^1) = DS(S^0).$$

This is just a rewritten version of Bartik's result in equation (8.30): CV exceeds the savings in defensive expenditures (holding sick days constant) along the real number line. Again, this means that for increases in the public 'bad', changes in defensive expenditures will overstate losses; for decreases it will understate gains.

Note that there are no cost of illness terms in this bound. Intuitively this results because, using the savings in defensive expenditure approach, we are counting only the costs of moving the individual back to the level of health enjoyed before an increase in pollution. Once having done so there is no change in the cost of illness because there is no change in the level of health (or the number of sick days).[15] CV will differ from $DS(S^0)$ because incurring the costs necessary to return to the same level of health may not be the least cost way of achieving the original level of utility. The expenditure minimizing individual will take into account the changed relative prices of the goods entering utility and will choose to reallocate expenditures and change the composition of consumption. The only case when this will not occur is when changes in b do not change the marginal cost of health, but instead represent a change in real income. This is the linear, separable production case discussed earlier. The results from that section do not change with the addition of cost of illness as long as the per unit cost of sick days does not change with the level of b.

8.5.3 Mitigation and Costs of Illness

Several papers suggest that the costs of illness could serve as a welfare measure in its own right, while others suggest that costs of illness need to be added to defensive expenditures to obtain welfare measures (often supported by the misleading result in equation (8.43)). The intuition given is that defensive expenditures are costs of increases in pollution, as are the lost wages and increased medical expenses. Therefore, they should all be counted. The key difference, of course, is that costs of illness are by definition exogenous costs imposed on the individual while defensive action is a choice variable optimally selected.

In attempting to sort this out, we need first to define the cost of illness measure of relevance. As always we are interested in the effects of a change–this time, a change in the public 'bad', b. So, unless either b^0 or b^1 is zero, the relevant measure is really a *change* in the cost of illness brought about by a change in b. Also we need to sign the term correctly. To be consistent with our definition of CV in equation (8.48) and with our interpretation of DS and ADS, the cost of illness measure is most usefully signed according to the direction of the welfare change. Thus,

$$COI = [wS(b^0) + gM(S(b^0))] - [wS(b^1) + gM(S(b^1))].$$

Now let's consider two clear cases. First, let us ask: are there any conditions under which COI would be equal to the compensating variation of a change in b? Start by assuming as before that increases in b cause exogenous increases in medical costs and lost wages, but now assume that they do not cause any disutility, directly or indirectly through illness, nor is there any way for the

[15]Strictly speaking, this suggests that the DS measure is only relevant for preventive rather than curative defensive expenditures. If medical treatment is the defensive behavior, then it is difficult to argue that one can instantaneously return the individual to the initial health status. Some ill-effects of the disease must certainly occur before the individual is cured.

individual to mitigate the effects of changes in b. In other words eliminate the household production for health status from the problem. If this is an adequate description of the situation, then the increase in medical costs and the lost wages will be the individual's CV measure for the change in b. The reason is simple: H (health status) and therefore S (sick days) are not choice variables in this case, and a change in b translates directly into a change in exogenous income with no behavior involved.

When health changes are felt only through the changes in sick days and medical expenses, the expenditure function is simply

$$\begin{aligned} m(b,w,g,T,u) &= \min\, z_2 + wL + wS(b) + gM(S(b)) - wT \\ &\quad +\mu(u - u(z_2, L)) \qquad (8.49)\\ &= wS(b) + gM(S(b)) + \hat{m}(w,T,u), \qquad (8.50) \end{aligned}$$

where $\hat{m}(w,T,u)$ is the expenditure function ignoring the exogenous costs of illness. The marginal utilities of z_2 and L are not affected by b, by definition. The only thing that b affects is S (sick days), but S does not affect utility nor is it a choice variable, because there is no way for the individual to alter S. The compensating variation of a change in b will equal (minus) the amount of money that is needed to achieve the same utility level after b changes. That is simply the loss from increased medical expenses and lost wages:

$$CV = wS(b^0) - wS(b^1) + gM(S(b^0)) - gM(S(b^1)) = COI. \qquad (8.51)$$

More generally, if S enters utility (naturally reducing it) but there is still no way to mitigate the effects of b, then the COI measure will fall short of the losses due to increases in b. Now, in order to keep utility at its previous level, more z_2 and/or L will need to be consumed, since an increase in b will cause an unmitigatable increase in the 'bad' S. In this case, the COI measure will be a lower bound on the losses.

The second question is: should COI be added to defensive expenditures to get a closer bound for CV? If, by 'defensive expenditures', we mean Bartik's DS measure that holds output constant, then the answer is definitively 'no'. As several researchers have recognized, there can be no increase in costs of illness because, by definition, S has been returned to its initial level (i.e. the individual is returned to his original health status). This point is not inconsistent with Bartik's bounds. In fact, we know we should not be adding anything to DS when there is a degradation, because in this case the absolute value of DS is already an overestimate of the CV measure of the losses.

Having addressed the two clear cases, we are left with two contentions often found in the literature: a) that COI is a legitimate bound on CV, and b) that $COI + ADS$ forms a better bound on CV than does ADS alone. We will consider these contentions in the context of increases in pollution, as this is the literature in which the cost of illness approach generally appears.

In the absence of costs of illness, equation (8.36) helped us determine the ordering of CV and ADS in the face of an increase in b. We can use the same reasoning, but need to add costs of illness and deal with the translation between sick days (S), which enters utility negatively, and z_1. To make the mathematical arguments and notation clearer, we make two simplifying assumptions. First, we subsume leisure into the composite commodity and, second, we define medical costs as being linear in sick days. Define compensating variation for this model as:

$$\begin{aligned} CV &= m(g, w, b^0, u^0) - m(g, w, b^1, u^0) \\ &= [z_2^0 - z_2^{h1}] + [c(S^0, b^0) - c(S^{h1}, b^1)] + [(g+w)(S^0 - S^{h1})]. \end{aligned} \tag{8.52}$$

This expression must necessarily be negative when the environment is degraded ($b^1 > b^0$).[16] It is easy to show that ADS will be a lower bound on the losses from the degradation. The ADS measure is given by:

$$ADS = c(S^0, b^0) - c(S^{m1}, b^1),$$

so the difference between CV and ADS is now:

$$CV - ADS = [z_2^0 - z_2^{h1}] + [(g+w)(S^0 - S^{h1})] + [c(S^{m1}, b^1) - c(S^{h1}, b^1)] < 0. \tag{8.53}$$

Unless the linear, separable production function assumptions hold, the expenditure minimizing individual will choose more z_2 and more S after the degradation in the environment. So either the first two bracketed terms are zero or they are negative. For the same reasons given with respect to (8.36), the last bracketed term will be negative. The negative Marshallian response to the degradation will be bigger than the Hicksian response, resulting in lower levels of H and higher levels of S in the Marshallian regime than in the Hicksian and therefore lower costs in the former than the latter.[17] Because both CV

[16] We should mention again the ambiguity that can arise between defensive expenditures and some elements of the typical cost of illness formulation. For example, medical costs have been viewed by some as mitigating behavior and by others as exogenous costs or outcomes of being ill. If incurring medical expenses leads to a reduction in sick days then this item more appropriately belongs in the defensive expenditure function. We define our notation so that $c(H, b)$ includes efforts that affect H, and $(g+w)$ includes consequences only. The problem can become more complicated if some actions (such as seeking medical help) do not alter sick days (and therefore have no effect on loss of income) but do reduce the disutility from illness.

[17] The Hicksian and Marshallian slopes are related in the following way: $\partial S^m/\partial b = \partial S^h/\partial b - m_b \partial S^m/\partial y$, but since S is a 'bad' $\partial S^h/\partial b > 0$, $m_b > 0$, and $\partial S^m/\partial y < 0$, implying that $S^{m1} \geq S^{h1}$. Therefore the costs of achieving S^{m1} are lower than the costs of achieving S^{h1} at any given level of b.

and ADS are negative, the actual defensive expenditures will underestimate the true losses from the degradation. In other words the ordering will still be as depicted in Figure 8.6.

The cost of illness results are more ambiguous. We restate the COI measure for the simplified model, taking care to denote the fact that S^1 is usually measured as the observable or Marshallian level of S:

$$COI = (g + w)(S^0 - S^{m1}).$$

The difference between the CV (equation (8.52) and COI measures is

$$CV - COI = [z_2^0 - z_2^{h1}] + [(g+w)(S^{m1} - S^{h1})] + [c(S^0, b^0) - c(S^{h1}, b^1)]. \quad (8.54)$$

The first and third bracketed terms are negative as in (8.53), but since S^{m1} is greater than S^{h1}, the second term in (8.54) is positive. For many commodities it might be safe to assume that differences between Marshallian and Hicksian responses will be small, but income effects could potentially be large when dealing with health effects and serious illness. If so, the last term may overwhelm the first two and COI could potentially be an overestimate of losses. This can only occur if defensive behavior is possible; otherwise there is no behavior that can alter the ultimate health state and therefore no meaningful distinction between Marshallian and Hicksian chosen levels of health.

If we could estimate the *Hicksian* costs of illness, then this COI^H measure would be a lower bound on CV, just as ADS is a lower bound. What is more, adding the Hicksian cost of illness to ADS would provide a better approximation to CV. Defining the Hicksian cost of illness as:

$$COI^H = (g + w)(S^0 - S^{h1}),$$

we see that

$$CV - [ADS + COI^H] = [(z_2^0 - z_2^{h1})] + [c(S^{m1}, b^1) - c(S^{h1}, b^1)] < 0.$$

$ADS + COI^H$ still underestimates true losses but forms a closer approximation to CV.

Unfortunately, Hicksian measures are rarely observed, and the applications that argue for using COI invariably use observable Marshallian measures. Most papers suggest using Marshallian costs of illness, but since $S^{m1} > S^{h1}$ adding a Marshallian COI measure to the ADS measure confounds the ordering as we would expect. The difference is now

$$CV - [ADS + COI^M] = \quad (8.55)$$
$$[z_2^0 - z_2^{h1}] + [(g + w)(S^{m1} - S^{h1})] + [c(S^{m1}, b^1) - c(S^{h1}, b^1)],$$

which could easily be either positive or negative. So, for the same reasons that the Marshallian cost of illness measure may not be a lower bound on losses, adding the Marshallian costs of illness measure to ADS moves us in the direction of CV but may or may not overshoot, preventing the use of $ADS + COI$ as any sort of bound on damages.

8.6 Conclusions

In this chapter we treat welfare measurement when the household can defend against or attenuate the pollution that threatens household well-being. Efforts to value changes in public inputs in these circumstances are of two sorts. The first produces neat theoretical solutions that depend on special properties of the household production technology–such as constant marginal costs or the existence of essential inputs. While conceptually appealing these solutions have not often been exploited and are quite often not applicable. The second includes an array of simple uses of expenditure data that produce bounds on welfare estimates.

Health professionals have long attempted to estimate the value of public health programs as the reduction in medical costs and lost work days. This approach, called the 'cost of illness' approach, has intuitive appeal to non-economists, but it can be a good approximation only under some very specific and unlikely conditions. When no defensive actions can be undertaken by the individual and when the only impact of changes in pollution is changes in monetary costs, then the savings in the cost of illness is a valid welfare measure. This however is an extreme and unlikely case because it ignores mitigating behavior and omits the disamenities of being ill or having impaired health.

More useful approximations can be found. Bounding results exist for defensive expenditures holding output constant (DS) and actual changes in defensive expenditures by utility maximizing individuals (ADS). Although in principle, DS might appear to require less information to calculate as it depends solely on household technology, ADS has tended to have greater appeal to economists. This is at least in part due to the fact that for cases of environmental degradation, ADS supplies a lower bound on losses which is far more useful from a damage assessment perspective than the upper bound on losses produced by the DS measure. The empirical examples in the literature have tended to use ADS to bound losses on acute pollution events, such as drinking water contamination. Defensive behavior in response to environmental degradation is easier to detect when the degradation is acute, sudden, and well-publicized. If it is too short-lived there may be little hope of accumulating data on systematic changes in behavior. Nonetheless, these results are less well-known than others in the non-market valuation literature and there may exist considerable potential for using household production models of this sort in regulatory analysis or damage assessment.

8.7 Appendices

8.7.1 Appendix A: Comparative Statics for Cost Minimization

We begin by determining how the Hicksian level of the household produced commodity, z_1, responds to changes in the public bad, b. The household chooses commodities to minimize the costs of attaining a given utility level:

$$m(b, u) = \min_{z_1, z_2} c(z_1, b) + z_2 + \mu(u - (z_1, z_2))$$

which yields first order conditions

$$\begin{aligned} c_{z_1} - \mu u_{z_1} &= 0 \\ 1 - \mu u_{z_2} &= 0 \\ u - u(z_1, z_2) &= 0. \end{aligned}$$

Construct the bordered Hessian and solve the following system:

$$\begin{bmatrix} c_{z_1 z_1} - \mu u_{z_1 z_1} & -\mu u_{z_1 z_2} & -u_{z_1} \\ -\mu u_{z_1 z_2} & -\mu u_{z_2 z_2} & -u_{z_2} \\ -u_{z_1} & -u_{z_2} & 0 \end{bmatrix} \begin{bmatrix} \partial z_1^h / \partial b \\ \partial z_2^h / \partial b \\ \partial \mu / \partial b \end{bmatrix} = \begin{bmatrix} -c_{z_1 b} \\ 0 \\ 0 \end{bmatrix}.$$

Solving this system, we find that

$$\partial z_1^h / \partial b = u_{z_2}^2 \cdot c_{z_1 b} / \Delta \tag{8.56}$$

where Δ, the determinant of the bordered Hessian, is negative for a minimum, implying that $\partial z_1^h / \partial b$ and $c_{z_1 b}$ will be of opposite sign.

Intuitively we might expect the sign of $c_{z_1 b}$ to be positive (or at least non-negative)–the marginal cost of producing z_1 would usually be expected to rise as pollution increases. For example, as disease carrying insects increase, the cost of achieving any marginal increase in health protection using purchased inputs such as pesticides and repellents will increase. As Bartik points out, this is likely but not necessarily the case. To determine the sign of $c_{z_1 b}$ look more carefully at the derivation

$$c(z_1, b) = \min_x x \cdot r + \gamma(z_1 - f(x, b)),$$

where γ is a Lagrangian multiplier. The envelope result gives $c_{z_1} = \gamma$ and it follows that $c_{z_1 b} = \partial\gamma/\partial b$. To interpret this result, start with the first order conditions:

$$\begin{aligned} r - \gamma f_x &= 0 \\ z_1 - f(x, b) &= 0. \end{aligned}$$

Differentiating this set of equations and solving for $\partial\gamma/\partial b$ we find

$$c_{z_1 b} = \frac{\partial \gamma}{\partial b} = \frac{\gamma f_{xx} f_b}{(f_x)^2} - \frac{\gamma f_{xb}}{f_x}. \tag{8.57}$$

The first term on the right hand side is clearly positive, since $f_{xx} \leq 0, f_b < 0, \gamma \geq 0$, and $f_x^2 > 0$. If f_{xb} is negative or if it is small relative to $f_{xx} f_b / f_x$ then the entire expression is positive. But if f_{xb} is positive and sufficiently large, then it is possible for $c_{z_1 b}$ to be negative. This could be true, for example, if one unit of x was capable of reversing any amount of the public bad. Suppose a filter system completely mitigates any amount of contamination in the public water system. Then the marginal product of moving from no filter system to one unit of the filter system will increase dramatically as a function of the level of contamination. For the most part, however, we would probably expect $c_{z_1 b} > 0$, as Bartik argues in his paper.

8.7.2 Appendix B: Alternative Motivation for Bartik's Bounds

In the main part of this chapter, Bartik's results concerning the relationship between savings in defensive expenditures and compensating variation are developed. Here we provide an alternative proof–one that offers some additional insight. Turn back to the first result, that at the optimum $-m_b(b, u) = -c_b(z_1^0, b)$. This is an envelope theorem result and will be true at the optimum level of the choice variables. The expenditure function, $m(b, u)$, and the defensive expenditure function, $c(z_1^0, b)$, will have the same slopes at the optimum, although they will not be tangent as they do not share any points in common as long as some nonzero level of the composite commodity is consumed.

Moving away from the optimum, these two functions have different curvature. The second derivative of the defensive expenditure function (holding output constant) is c_{bb} while the second derivative of the expenditure function is

$$m_{bb} = c_{bb} + c_{bz_1} \frac{\partial z_1^h}{\partial b}. \tag{8.58}$$

Whether the expenditure function or the defensive expenditure function is more concave depends on the sign of $c_{bz_1} \partial z_1^h / \partial b$. From (8.56) we know that this term

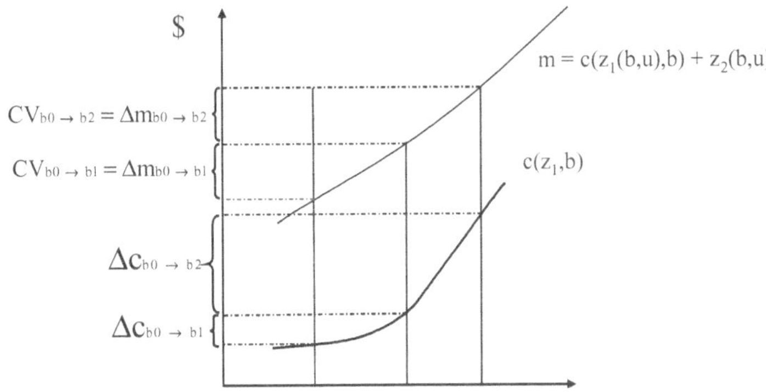

FIGURE 8.8. Bartik's Bounds

must be negative because c_{bz_1} and $\partial z_1^h/\partial b$ have opposite signs. Thus, $m_{bb} < c_{bb}$; the defensive expenditure function will be more concave or less convex from above than the regular expenditure function.

A possible configuration is represented in Figure 8.8. From this picture, one can see that an improvement (that is, a decline in b from b^0 to b^1) will generate savings in defensive expenditures that underestimate the true CV. A degradation (that is, an increase in b from b^0 to b^2) will generate negative savings in defensive expenditures (increases in expenditures) and this negative number will be larger in absolute value than the negative CV associated with the change. This is Bartik's result, shown in a different way. Only when c_{bz_1} equals zero will the savings in defensive expenditures equal compensating variation. This is the linearly separable case discussed in the first part of this chapter.

Earlier in this chapter we saw that the absolute value of the savings in actual defensive expenditures will always be smaller than the analogous measure using savings in defensive expenditures holding output constant. The former will measure smaller gains and smaller losses than the latter. This same result can be illustrated by considering the relative slopes of the two defensive expenditure functions. At the optimum, the two will coincide. That is $c(z_1^m(b^0, y), b^0) = c(z_1^0, b^0)$. As b changes, the two functions diverge as indicated by their different slopes. The slope of the actual defensive expenditure function is given by

$$\frac{dc(z_1(b,y),b)}{db} = c_b + c_{z_1}\frac{\partial z_1^m}{\partial b},$$

where the first term, c_b, is of course the slope of the defensive expenditure function with respect to b holding output constant, and the second term is negative. The graph of the functions in Figure 8.9 shows the defensive expenditure function holding output constant as steeper than the actual defensive expenditure

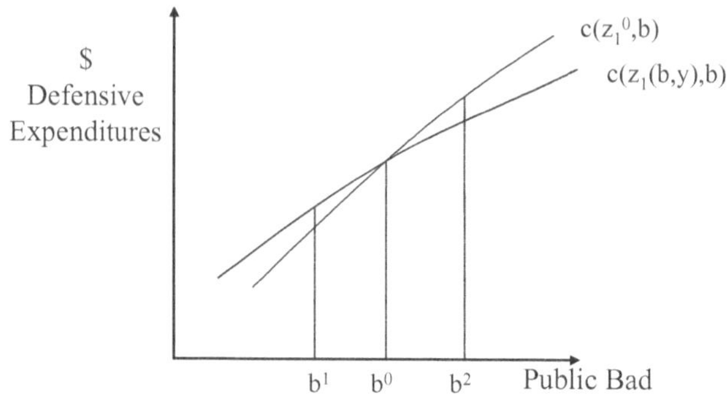

FIGURE 8.9. Output Constant and Output Varying Cost Functions

function. The intuition is straightforward. For environmental degradation, the *DS* measure requires that expenditures be made to return to the original level of z_1 before the increase in pollution. The *ADS* measure reflects a countervailing substitution effect. With a higher implicit price for the commodity, the individual will choose less z_1 and will choose less defensive expenditures then he would need to return to z_1^0. An environmental improvement will typically make z_1 cheaper causing the individual to substitute towards z_1, choosing a new level of z_1 higher than before. As a result, the savings in defensive expenditures made possible by the environmental improvement will not be as great as would have occurred had the individual returned to z_1^0.

Chapter 9

The Environment as an Input into Firms' Production

9.1 Introduction

In Chapters 3 through 8 we explored the problem of measuring the economic gains and losses experienced by households affected by environmental improvements and degradation. These chapters addressed the ways in which households respond when such things as their health or recreational opportunities are affected by the environment. In this chapter we investigate environmental effects on production opportunities for enterprises that sell some part of their output.

The environment can influence production in a number of ways–by changing the productivity of inputs, by altering the quality of the output that is produced, by reducing the effective supply of inputs, etc. These effects can be modeled, conceptually, by treating the environment as an uncontrolled input in the production function. As examples, dissolved oxygen and other environmental inputs, combined with fishing effort, produce fish harvests. Localized reductions in dissolved oxygen lead to fish mortality and effectively shrink the area of habitat suitable for fish. Acid deposition slows tree growth; increasing ozone levels reduce agricultural yields. Our examples highlight primary industries for a reason. Environmental quality is most likely to affect firm and household production possibilities when natural resources such as soil, fish stocks, or forests are directly exploited. In addition to the usual pathways through which the environment affects these resources, the potential for adverse consequences due to climate change has also been considered.

Previous chapters have focused on the effect of the environment on the value of consumption services. The examples are largely from the industrialized countries. In this chapter examples will be drawn from both the developed and developing worlds, but the issues addressed seem more compelling to the latter. In poor countries, households with low incomes and limited time have meagre demand for environmental services as amenities. Production opportunities in poor countries are principally in the primary sectors of agriculture, forestry, fisheries and other activities directly dependent on the vitality and abundance of natural and environmental resources. Further, a greater proportion of households is affected through this route because a larger share of the population continues to live in rural areas in the developing world. The conceptual basis for valuing the environment as an input does not differ whether one investigates low-level ozone damage to crops in the U.S. or losses of mangrove swamps that serve as fish habitat in Thailand. But the scarcity of data and more tenuous market forces make the practice in developing countries different and more difficult. Empirical applications are limited by these constraints, and approximations abound.

9.2 Welfare Measures for Firm Owners

Our interest lies with the welfare effects of environmental and natural resource changes as they emerge through the pathway of production. Naturally this starts with welfare measurement for the firm. But it is important to remember that people, not firms, are the final claimants for all economic gains and losses. As we have emphasized all along, economic welfare is only defined in terms of the individual and, therefore, all costs and benefits must ultimately accrue to individuals. In models of consumer choice this is obvious. In considering firms, we must bear in mind that individuals own the firms and the capital used in production and hence are the claimants on returns to the firm.[1]

We first address the situation in which production decisions are made independent of the consumption decisions of the owners of firms and capital. This is typical of many firms in developed countries, especially when owners have no operational influence on the firm but are merely stockholders. But it is more broadly applicable as we will see in a later section of this chapter. When changes in exogenous circumstances–such as output or input prices–alter net returns to the firm but do not change the firm owners' decisions on consumption

[1] In socialist systems, the state may own the firms and capital. This does not, in principle, preclude the application of welfare economics but it does make it necessary to determine how the returns from production are redistributed in the population given the particular regime.

of goods or supply of factors, then these net returns translate into changes in exogenous income to firm owners. Consumption and labor supply decisions of these individuals have no bearing on what they earn from their partial ownership of the firm, and ownership influences consumption decisions only through the budget constraints of share owners. In these cases, welfare measures in the firm context turn out to be exogenous income changes to individuals, eliminating many of the complications encountered in previous chapters. This is true when consumption and production are separable and net returns are not random. In this section we define the welfare concept of interest in the context of the firm in this usual way. In a subsequent section we investigate under what conditions the same results hold when the distinction between households and producers is blurred.

9.2.1 Defining Welfare for the Firm

Firms are viewed as choosing variable inputs to maximize profits, subject to the available technology and fixity of other inputs. Let the technology for producing a vector of outputs, $\mathbf{z}$, be denoted by the implicit function $T(\mathbf{z}, \mathbf{x}; \mathbf{k})$ where $\mathbf{x}$ is the vector of variable input use available to the firm at fixed prices $\mathbf{w}$, and $\mathbf{k}$ is a vector of fixed inputs including the firm's capital stock. Fixed costs, c^0, are the rental payments for the levels of fixed inputs, $\mathbf{k}$. These may take the form of opportunity costs if the firm owns the fixed factors, but may be actual rental payments for inputs such as land owned by others.

We assume that firms attempt to maximize profits. This objective is reasonable for small, competitive firms such as those found in agriculture and fisheries in most countries and even in industries such as forestry in many.[2] Given the profit maximization objective, the firm faces the following problem:

$$\begin{aligned} \max_{\mathbf{x}} \pi &= \mathbf{p} \cdot \mathbf{z} - \mathbf{w} \cdot \mathbf{x} - c^0(\mathbf{k}) \textit{ subject to } T(\mathbf{z}, \mathbf{x}; \mathbf{k}) = 0 \qquad (9.1) \\ &= \max_{\mathbf{z}} \mathbf{p} \cdot \mathbf{z} - c(\mathbf{z}, \mathbf{w}; \mathbf{k}) - c^0(\mathbf{k}) \\ &= \pi(\mathbf{p}, \mathbf{w}; \mathbf{k}) \end{aligned}$$

where $\mathbf{p}$ is a vector of output prices, $\mathbf{w}$ is a vector of input prices, $c(\mathbf{z}, \mathbf{w}; \mathbf{k})$ is the firm's minimum variable cost function conditioned on $\mathbf{k}$, and $\pi(\mathbf{p}, \mathbf{w}; \mathbf{k})$ is the profit function. The length of the planning horizon determines the mix of variable and fixed inputs. In general, the longer the temporal unit of observation, the larger the share of inputs that are variable.

[2] Welfare measurement in the context of imperfectly competitive, or game theoretic, behavior is addressed in Just, Hueth, and Schmitz (2004, Chapter 10) but will not be dealt with in this book.

Just, Hueth and Schmitz (2004) define an additional welfare-related term, *quasi-rent,* as including both profits and returns to fixed inputs. This concept encompasses the returns to both the owners of the firm and the owners of the fixed factors. From equation (9.1), quasi-rent equals

$$\widehat{\pi} = \pi + c^0 = \mathbf{p} \cdot \mathbf{z} - c(\mathbf{z}, \mathbf{w}; \mathbf{k}). \tag{9.2}$$

Although quasi-rents accrue to owners of firms and fixed factors, changes in quasi-rent accrue to firm owners by definition. This is because, during the length of time in which certain factors are defined as fixed, no change can take place–by definition–in the returns paid to their owners. To the extent that policies and other exogenous events fail to alter fixed costs, there will be no distinction between *changes* in profits and *changes* in quasi-rents, even when the firm is forced to shut down, as long as the firm owner remains responsible for paying fixed costs. With profits defined as in (9.1), changes in π will equal changes in $\widehat{\pi}$.[3] These statements might suggest that the concept of quasi-rent offers little added value, but we will appreciate its usefulness when we turn to the concept of essentiality.

Given our assumption that production and consumption decisions are separable, we can safely treat any changes in quasi-rents as a change in exogenous income to firm owners. The indirect utility function of such an individual would depend on $\widehat{\pi}$, as well as income from other sources, y, and consumer prices, p. The compensating variation (CV) for a change in quasi-rents from $\widehat{\pi}^0$ to $\widehat{\pi}^1$ satisfies the condition

$$v(y + \widehat{\pi}^1 - CV, p) = v(y + \widehat{\pi}^0, p), \tag{9.3}$$

which implies that[4]

$$CV = \widehat{\pi}^1 - \widehat{\pi}^0. \tag{9.4}$$

So the change in quasi-rents equals compensating variation to owners of firms. It is also equal to equivalent variation, since monetary valuations of direct exogenous income changes are independent of utility level. The definitional task is simple as long as production decisions are separable from household decisions, as we assume above.

---

[3]We differ from Just, Hueth and Schmitz in this regard although our difference is purely semantic. As long as profit is defined as equalling $-c^0(\mathbf{k})$ when the firm shuts down, the equivalence between $\Delta\widehat{\pi}$ and $\Delta\pi$ holds. Just, Hueth and Schmitz define profits as zero in the shut down state.

[4]Note that because $\hat{\pi} = \pi + c^0$, when fixed costs are not altered by exogenous changes CV equals the difference in profits as well as the difference in quasi-rents.

9.2.2 Exact Welfare Measures for Price Changes

Our ultimate task is to define empirical welfare measures for changes in non-priced environmental inputs, but we will first need welfare results for price changes. In this section we develop well-known results that link empirically obtainable measures to the desired CV measure, change in quasi-rents. Even in its empirical representation we will continue to use the term 'change in quasi-rents' because it is a generally recognized concept. The popular term 'producer surplus' is less useful because, strictly speaking, it applies only to areas behind supply curves and not areas behind input demand functions. On the other hand, the use of the term 'consumer surplus' for areas behind input demands is troublesome because it can be confused with household's consumer surplus' which has usefulness only as an approximation of CV. To the extent that the areas behind input demand functions measure welfare effects for firm owners, they are exact measures and not approximations. Given the potential for confusion, we will discuss welfare measures solely in terms of changes in quasi-rents and avoid the use of the term 'producer surplus'.

Empirical Measurement of Changes in Quasi-rents

Defining welfare measures for firm owners from price changes is straightforward, but the problem of empirical measurement remains. If data on revenues and costs are observable before and after an exogenous change, welfare calculation is purely an accounting exercise. Such data are rarely available, and never when evaluating potential policy alternatives, *ex ante*. Researchers sometimes estimate a profit function (with or without accompanying output supply and input demand functions). But this is no small undertaking, and so welfare economics of the firm often involves a search for alternative methods that require less information and fewer challenges in estimation. Although still data intensive and problematic, estimation of only an output supply or an input demand function is sometimes all that is required.

It is well known that the change in quasi-rents associated with an output price change can be measured as the change in the area behind the firm's supply curve for that output. The result holds as long as the firm is a price taker–as long as the supply curve of the firm is its marginal cost function. To demonstrate, consider a price change for a single output. Beginning with the expression for quasi-rents in (9.2), the change in quasi-rents associated with a price change will be

$$\Delta\widehat{\pi}_{\Delta p} = \int_{p^0}^{p^1} \frac{\partial\widehat{\pi}(p, \mathbf{w}; \mathbf{k})}{\partial p} dp$$

where

$$\widehat{\pi}(p, \mathbf{w}; \mathbf{k}) = p\, z(p, \mathbf{w}; \mathbf{k}) - c(z(p, \mathbf{w}; \mathbf{k}), \mathbf{w}; \mathbf{k}). \tag{9.5}$$

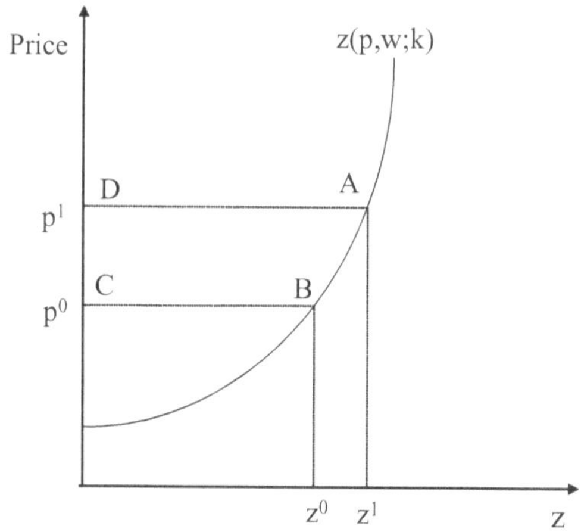

FIGURE 9.1. Welfare Measure of Output Price Change

From the envelope theorem,

$$\frac{\partial \widehat{\pi}(p, \mathbf{w}; \mathbf{k})}{\partial p} = z(p, \mathbf{w}; \mathbf{k}),$$

where $z(p, \mathbf{w}; \mathbf{k})$ is the firm's supply function. The welfare measure will be

$$\Delta \widehat{\pi}_{\Delta p} = \int_{p^0}^{p^1} z(p, \mathbf{w}; \mathbf{k}) dp. \tag{9.6}$$

The change in quasi-rents, described in (9.6) as the integral of the supply function over the change in price, is illustrated in the usual way by area $ABCD$ in Figure 9.1.

Again assuming the firm is a profit maximizer and a price taker in the input market, a similar result is available for a change in an input price. Rewriting (9.5) as $\widehat{\pi} = pf(\mathbf{x}, \mathbf{k}) - \mathbf{w} \cdot \mathbf{x}$ implies that

$$\frac{\partial \widehat{\pi}(p, w_j, \mathbf{w}_{-j}; \mathbf{k})}{\partial w_j} = -x_j(p, \mathbf{w}; \mathbf{k}),$$

so that the change in quasi-rents of a change in the price of the jth input is

$$\Delta \widehat{\pi}_{\Delta w_j} = -\int_{w_j^0}^{w_j^1} x_j(p, w_j, \mathbf{w}_{-j}; \mathbf{k}) dw_j. \tag{9.7}$$

This is the change in the area behind that firm's input demand function as depicted by area $EFGH$ in Figure 9.2.

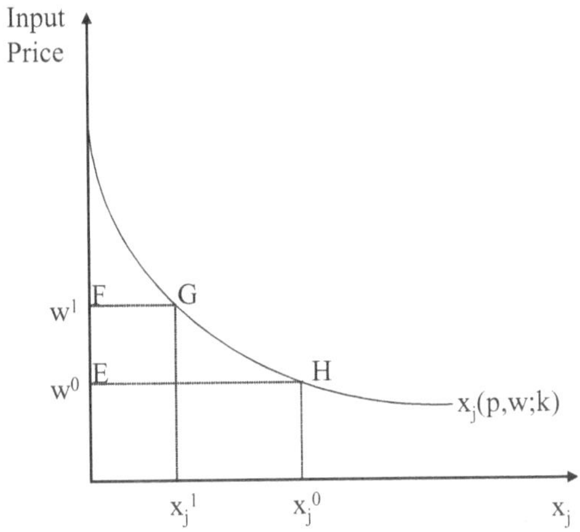

FIGURE 9.2. Welfare Measure of Input Price Change

The change in quasi-rents associated with a multiple price change, for example (p^0, w_j^0) to (p^1, w_j^1), is easy to establish. This can be evaluated as a line integral of the form:

$$\Delta\widehat{\pi} = \int_{\mathcal{L}} \frac{\partial\widehat{\pi}(p, \mathbf{w}; \mathbf{k})}{\partial p} dp + \frac{\partial\widehat{\pi}(p, w_j, \mathbf{w}_{-j}; \mathbf{k})}{\partial w_j} dw_j \tag{9.8}$$

where $\mathcal{L}$ is some path between (p^0, w_j^0) and (p^1, w_j^1). Because the integrand of (9.8) is an exact differential of the quasi-rent function, the expression in (9.8) will be path independent. This expression can be rewritten as

$$\Delta\widehat{\pi} = \int_{\mathcal{L}} z(p, \mathbf{w}; \mathbf{k})dp - x(p, w_j, \mathbf{w}_{-j}; \mathbf{k})dw_j,$$

so that the line integral is equal to the sum of areas such as those portrayed in Figures 9.1 and 9.2, *except* that the price changes must be sequenced. The measures in these two figures can be added together only if either a) $x_j(p, \mathbf{w}; \mathbf{k})$ is conditioned on p^1 or b) $z(p, \mathbf{w}; \mathbf{k})$ is conditioned on w_j^1.

Using Essentiality to Measure Welfare for Price Changes

Just, Hueth, and Schmitz demonstrate an additional result using these same output supply and input demand functions. The result hinges on the definition of essentiality in production–a concept we address below. The crucial point for our purposes is that once such an output or input is identified, the area behind the essential output supply or essential input demand curve evaluated

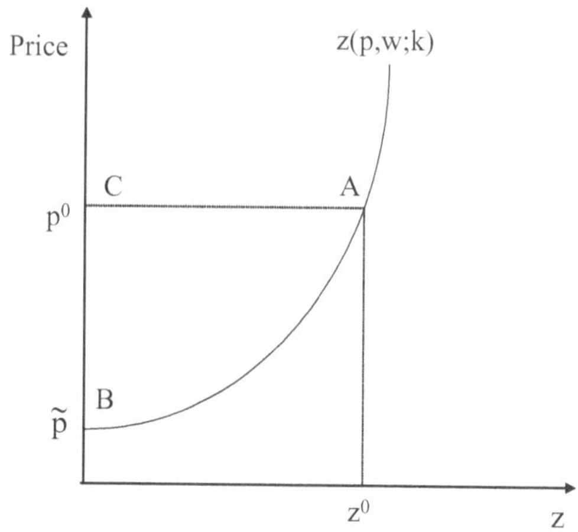

FIGURE 9.3. Quasi-rent as Measured by the Area Behind an Essential Output's Supply Curve

at the existing price measures the *entire* quasi-rent at that price. This result opens the door to measuring the welfare effects of changes of any parameters that influence quasi-rents and is thus a key to valuing environmental changes. Here we outline their basic results.

The first result is that the entirety of quasi-rent can be measured as the area behind a firm's supply curve for a given output and below its price if that output is 'essential' to the firm's operation. This is illustrated in Figure 9.3 as area ABC. An output is essential if at an output price low enough for the firm to cease production of this output (i.e. at any price $\leq \tilde{p}$ in the graph) the firm shuts down. When a firm shuts down, it ceases to hire variable inputs or to produce output so quasi-rents, $pz - wx$, equal zero. This is trivially true for the single output firm, but may also be true for multiple output firms, especially if other outputs are by-products in the production process. Although owners of fixed factors must still be paid, the payment is a transfer from firm owners. The sum of returns to the two groups is therefore zero.

The proof that area ABC equals total quasi-rents is straightforward. Area ABC in Figure 9.3 equals

$$\begin{aligned}\int_{\tilde{p}}^{p^0} z(p, \mathbf{w}; \mathbf{k})dp &= \int_{\tilde{p}}^{p^0} \frac{\partial \widehat{\pi}(p, \mathbf{w}; \mathbf{k})}{\partial p} dp \\ &= \widehat{\pi}(p^0, \mathbf{w}; \mathbf{k}) - \widehat{\pi}(\tilde{p}, \mathbf{w}; \mathbf{k}) \\ &= \widehat{\pi}(p^0, \mathbf{w}; \mathbf{k}) - 0,\end{aligned} \tag{9.9}$$

where $\tilde{p}$ is the price at which the firm ceases to produce z. By definition,

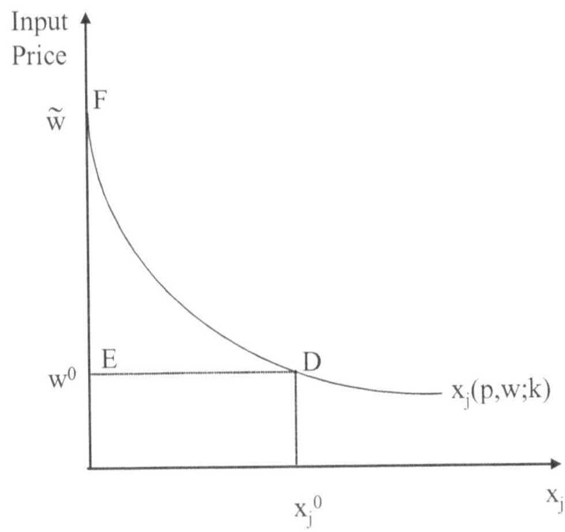

FIGURE 9.4. Quasi-rent as Measured by the Area Behind an Essential Input's Demand Curve

$\widehat{\pi}(\tilde{p}, \mathbf{w}; \mathbf{k}) = 0$, because at output price $\tilde{p}$ for the essential output, the firm shuts down. Thus, the whole area behind the supply function and below price is the measure of quasi-rent at price p^0.[5]

Likewise, an input is essential if, at an input price high enough to cause the firm to cease purchasing the input, the firm shuts down. For example, fuel is an essential input into the commercial harvesting of most fish stocks in developed countries; water will be an essential input into irrigated crops in arid areas. The area above price and to the left of the essential input demand function equals total quasi-rents. This is illustrated in Figure 9.4 as area DEF, where the 'choke price' for the input is labeled $\widetilde{w}$. Similar to (9.9),

$$\begin{aligned} \int_{w_j^0}^{\widetilde{w}_j} x_j(p, \mathbf{w}; \mathbf{k}) dw_j &= -\int_{w_j^0}^{\widetilde{w}_j} \frac{\partial \widehat{\pi}(p, w_j, \mathbf{w}_{-j}; \mathbf{k})}{\partial w_j} dw_j \\ &= \widehat{\pi}(p, w_j^0, \mathbf{w}_{-j}; \mathbf{k}) - \widehat{\pi}(p, \widetilde{w}_j, \mathbf{w}_{-j}; \mathbf{k}) \\ &= \widehat{\pi}(p, w_j^0, \mathbf{w}_{-j}; \mathbf{k}) - 0, \end{aligned} \tag{9.10}$$

because at $\widetilde{w}_j$ the firm shuts down and quasi-rent is zero.

For price changes, equations (9.9) and (9.10) become useful if data or other

[5]Strictly speaking, the correct measure is price times quantity minus the area below the supply (marginal cost) function. This could conceivably differ from the area below the price line and above supply (marginal cost) if marginal cost declines over the initial units of production. In that case the marginal cost at small levels of output could actually be greater than price. Such an example is depicted in Just, Hueth and Schmitz.

problems prevent the direct application of the results in (9.6) or (9.7). For example, if an input price changes but input demand cannot be estimated because input levels are not observed, the change in quasi-rents can be measured as the change in the area behind the shifting supply curve of an essential output (if one exists):

$$\begin{aligned}\Delta\widehat{\pi}_{\Delta w_j} &= \int_{\tilde{p}}^{p^0} z(p, w_j^1, \mathbf{w}_{-j}; \mathbf{k})dp - \int_{\tilde{p}}^{p^0} z(p, w_j^0, \mathbf{w}_{-j}; \mathbf{k})dp \\ &= \widehat{\pi}(p^0, w_j^1, \mathbf{w}_{-j}^0; \mathbf{k}) - \widehat{\pi}(p^0, w_j^0, \mathbf{w}_{-j}^0; \mathbf{k}),\end{aligned}$$

where, if the choke price is a function of w_j, then $\tilde{p}$ is defined as the minimum of the two choke prices, i.e. $\tilde{p} = \min[\tilde{p}(w_j^0, w_j^1)]$. Similar results are possible using essential input demands–that is changes in output price or other input prices can be evaluated as changes in the area behind an essential input demand function.

From an econometric perspective, using changes in the entire area behind behavioral functions is, in general, not as desirable as the more usual measures presented in (9.6) or (9.7), because the former requires accuracy in estimation over a larger range of prices than does the latter. For our problem in which the change is in an exogenous environmental input, we will find that there is no direct approach and this result of Just, Hueth, and Schmitz becomes invaluable.

The welfare measures described to this point are broadly applicable when markets are competitive and complete, but we have dealt only with policies or events that affect *prices*. Our problem is different. No observable function results from the derivative of profits or quasi-rents with respect to the environmental input, q. The same type of problem was encountered in Chapter 3 where no analogy to Shephard's lemma existed for the environmental amenity.

9.2.3 Valuing Changes in an Environmental Input

Now we turn to the real task, the valuation of a change in an exogenous environmental input. To do so, we introduce a generic model in which environmental quality or an ecological service is viewed as an input in the production function. For example, the input may be a natural resource or the quality of that resource, or it may be a flow or stock pollutant in which case its marginal product will be negative. Given the abstract nature of our development, it is worth bearing in mind that the connection between a pollutant or a natural resource on the one hand and the production process of interest on the other may be quite complex. Some pollutants (e.g. acid deposition) have a cumulative effect on the production process, while others (e.g. ozone) are episodic in nature. Some pollutants must exceed a threshold before their effects are felt, and in some cases there is a range beyond which additional levels of the pollutant have no impact. The connection between a pollutant and the production process may involve several steps. As a result, the ultimate effect of the environmental damage may be a function of other ambient circumstances, and a temporal lag may ensue before the effect of the damage is felt in production. As an example, a pollutant might destroy habitat used as a breeding ground for certain fish species. Depending on the life cycle of the species, fishermen may not feel the effects for some years. This suggests that environmental degradation can take different pathways, but it can also produce different types of ultimate effects. A reduction in environmental quality may depress growth rates of fish, timber or crops; it may alter the appearance or nutritional value of the output; or it may increase susceptibility of plants or animals to disease.

Several empirical welfare studies have attempted to model the ecological chain of effects connecting the initial environmental injury to the ultimate effect on production (e.g., Kahn and Kemp, 1985; Swallow, 1994). Others have argued that a reduced form analysis avoids compounding of errors that are the natural outcome of many layers of ecological modeling (e.g. Garcia, Dixon, Mjelde, and Adams, 1986), but these papers run the risk of missing the essential nature of the environment-production interaction. In pervasive and complex cases, econometric evidence of a clear relationship between the environment and an identifiable output may be especially difficult to establish, and careful modeling of ecological connections may be necessary. In this chapter, we focus on the welfare measures appropriate in such cases and not on modeling the environmental linkages. However, no empirical study can proceed without giving this latter aspect careful thought.

In our general model, output depends on purchased inputs and on the environmental quality input, q. For concreteness, let q to be a 'good', representing the level of environmental or ecological services. For example, q could represent depth of top soil on a farm subject to soil erosion. On occasions, however, the most natural concept for discussion is a 'bad'–a measure of pollution or the degradation of the environment. Examples include soil salinization from contaminated irrigation water or loss of wetlands from development. Increased pollution or degradation of resources will be treated as declines in q.

Let the transformation function now be written $T(\mathbf{z}, \mathbf{x}; \mathbf{k}, q) = 0$ where $\mathbf{z}$, $\mathbf{x}$, and $\mathbf{k}$ are as defined before, and q is the environmental factor. The firm's cost function, defined as

$$c(\mathbf{z}, \mathbf{w}; \mathbf{k}, q) = \min\ \mathbf{w} \cdot \mathbf{x} \ \ \textit{subject to}\ T(\mathbf{z}, \mathbf{x}; \mathbf{k}, q) = 0,$$

depends on the exogenous environmental input, and quasi-rent is given by

$$\hat{\pi}(\mathbf{p}, \mathbf{w}; \mathbf{k}, q) = \pi + c^0 = \mathbf{p} \cdot \mathbf{z} - c(\mathbf{z}, \mathbf{w}; \mathbf{k}, q).$$

Now suppose that the environmental input increases from q^0 to q^1. As long as production and consumption are separable, the compensating variation of the change will equal the change in quasi-rents:

$$CV = \hat{\pi}(\mathbf{p}, \mathbf{w}; \mathbf{k}, q^1) - \hat{\pi}(\mathbf{p}, \mathbf{w}; \mathbf{k}, q^0). \tag{9.11}$$

We will suppress **k** as it plays no direct role in the conceptual analysis, although conditioning on **k** in a cross-sectional empirical analysis would be necessary.

Empirically evaluating the welfare effects of a change in q introduces a number of complications. Of course, it is still possible, in concept, to estimate changes in quasi-rents using predicted changes in a profit function, but estimating a profit function poses more difficulties than it does in the conventional setting because now output must be estimated as a function of the environmental input as well as prices. Obtaining observations on profits under circumstances of varying levels of q may often be difficult, and predicting profits when the level of the environmental input is altered through regulation, for example, will typically depend on accurate specification of the relationship between profits and q beyond the range of observed historical data. Nonetheless, a few attempts at estimating profit functions can be found (e.g. Garcia, Dixon, Mjelde and Adams, 1986; Pattanayak and Kramer, 2001; Young and Aidun, 1993).

More commonly, empirical studies attempt to measure welfare effects of changes in environmental inputs as areas between shifting supply functions, where these functions are obtained from econometric estimation or mathematical programming simulations. Researchers regularly assume, without discussion, that the area between the two supply curves conditioned on initial and subsequent levels of q is the appropriate welfare measure. A sufficient condition for establishing this as the correct measure is that the output for which the supply function is drawn is an essential output to the firm. To show this, first re-write the quasi-rent function as a function of q as well as prices and fixed inputs, $\widehat{\pi}(p_1, \mathbf{w}; \mathbf{k}, q)$, and define $\tilde{p}_1$ as a price low enough such that at this price z_1 ceases to be produced at both initial and subsequent levels of q. Specifically, $\tilde{p}_1 \leq \min[\tilde{p}_1(q^0), \tilde{p}_1(q^1)]$. If output z_1 is an essential output (which it will be, trivially, for the single output firm), then the change in quasi-rents associated with a change in q can now be measured as the change in the area behind the supply curve, as q shifts that supply curve because

$$\begin{aligned}\Delta\widehat{\pi}_{\Delta q} &= \int_{\tilde{p}_1}^{p_1^0} z(p_1, \mathbf{w}, q^1)dp_1 - \int_{\tilde{p}_1}^{p_1^0} z(p_1, \mathbf{w}, q^0)dp_1 \\ &= [\hat{\pi}(p_1^0, \mathbf{w}, q^1) - \hat{\pi}(\tilde{p}_1, \mathbf{w}, q^1)] - [\hat{\pi}(p_1^0, \mathbf{w}, q^0) - \hat{\pi}(\tilde{p}_1, \mathbf{w}, q^0)] \\ &= [\hat{\pi}(p_1^0, \mathbf{w}, q^1) - \hat{\pi}(p_1^0, \mathbf{w}, q^0)].\end{aligned} \tag{9.12}$$

Given that the entire measure of quasi-rent can be found behind the supply curve for an essential output, the change in that area induced by a change in q

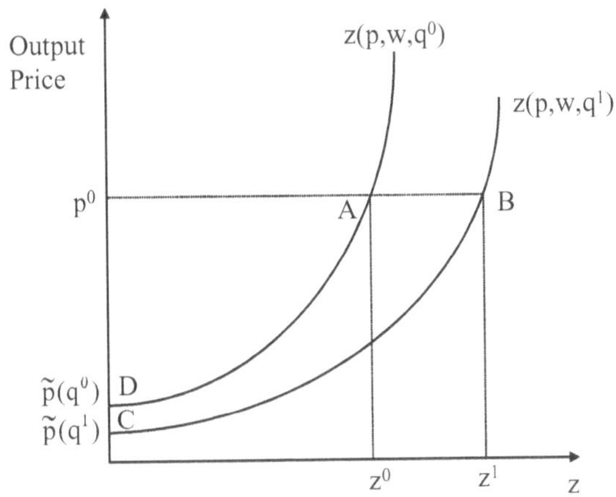

FIGURE 9.5. Welfare Measure Using an Essential Output

measures the change in quasi-rents. If the environmental input has increased, the welfare measure will be positive as depicted by area $ABCD$ in Figure 9.5.

The result in (9.12) holds, trivially, for the single output firm, but it also holds if there are multiple outputs and either a) one of these outputs can be identified as essential or b) only one output uses q.[6] In the more general case, total quasi-rent can be obtained by adding all areas equivalent to (9.12) for each output, but only if the price changes are sequenced correctly. To illustrate, consider joint production of two outputs. Total quasi-rents equal the sum of the areas behind the two supply curves, but the second supply curve must be

[6]Huang and Smith (1998) consider the cases outlined in this section but label these 'weak complementarity in production'. This seems an odd term to apply to a relationship between the output of a production process and an environmental amenity, but an example will prove to explain why they chose to use the term. Suppose the firm produces many outputs, all of which require water as an input and none are essential. However, one output requires high quality water (perhaps because humans ingest it) and we will designate that one z_1. If q is water quality, then the firm does not care about changes in q unless output z_1 is produced. Huang and Smith show that integrating between the two output supply functions for z_1 conditioned on different levels of q will provide the proper welfare measure. The mathematics matches that in (9.12) because $\hat{\pi}(\tilde{p}_1, p_{-1}, \mathbf{w}, q^1) = \hat{\pi}(\tilde{p}_1, p_{-1}, \mathbf{w}, q^0)$. This condition holds when q enters only the production function for z_1. The authors' use of the term 'weak complementarity' has led others to believe that non-essentiality is required (similar to weak complementarity in consumption), but this is not the case. Because of the confusion over essentiality and non-essentiality suggested by this term, we avoid using it in the context of production.

conditioned on the choke price of the first:

$$\int_{\tilde{p}_1(p_2^0)}^{p_1^0} z_1(p_1, p_2^0, \mathbf{w}, q)dp_1 + \int_{\tilde{p}_2(\tilde{p}_1)}^{p_2^0} z_2(\tilde{p}_1, p_2, \mathbf{w}, q)dp_2. \tag{9.13}$$

We can show that this expression equals total quasi-rent. Rewriting in terms of the quasi-rent function gives:

$$\hat{\pi}(p_1^0, p_2^0, \mathbf{w}, q) - \hat{\pi}(\tilde{p}_1, p_2^0, \mathbf{w}, q) + \hat{\pi}(\tilde{p}_1, p_2^0, \mathbf{w}, q) - \hat{\pi}(\tilde{p}_1, \tilde{p}_2, \mathbf{w}, q) = \hat{\pi}(p_1^0, p_2^0, \mathbf{w}, q).$$

The fourth term on the left hand side must equal zero because at prices $(\tilde{p}_1, \tilde{p}_2)$, the firm is producing no output. After canceling the second and third terms which are of equal but of opposite sign, we are left with the first term–the measure of total quasi-rent.

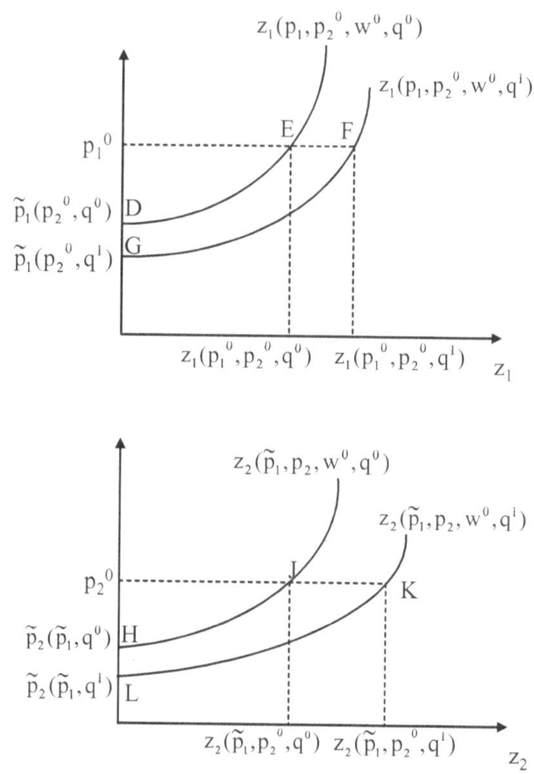

FIGURE 9.6. Changes in Producer Surplus for a Multiproduct Firm

Now consider measuring the change in quasi-rents from a change in q using these two supply functions. The measure equals the change in (9.13) with a change in q:

$$\begin{aligned}\Delta\hat{\pi}_{\Delta q} &= \int_{\tilde{p}_1}^{p_1^0} z_1(p_1, p_2^0, \mathbf{w}, q^1)dp_1 + \int_{\tilde{p}_2}^{p_2^0} z_2(\tilde{p}_1, p_2, \mathbf{w}, q^1)dp_2 \qquad (9.14)\\ &\quad - \int_{\tilde{p}_1}^{p_1^0} z_1(p_1, p_2^0, \mathbf{w}, q^0)dp_1 - \int_{\tilde{p}_2}^{p_2^0} z_2(\tilde{p}_1, p_2, \mathbf{w}, q^0)dp_2\\ &= \hat{\pi}(p_1^0, p_2^0, \mathbf{w}, q^1) - \hat{\pi}(p_1^0, p_2^0, \mathbf{w}, q^0),\end{aligned}$$

where $\tilde{p}_1 \leq \min[\tilde{p}_1(p_2^0, q^0), \tilde{p}(p_2^0, q^1)]$ and $\tilde{p}_2 \leq \min[\tilde{p}_2(\tilde{p}_1, q^0), \tilde{p}_2(\tilde{p}_1, q^1)]$. This measure is depicted in Figure 9.6 as the sum of areas $DEFG$ and $HJKL$. If the supply (marginal cost) curves are independent, then output decisions become independent and the sequencing does not matter. Then, as we argued above, only those outputs that depend on q are relevant.

Sometimes the estimation of supply functions is not feasible. Output may not be well-defined or easily measured. An alternative approach employs input demand functions. Parallel to the results above, if x_j is an essential output then the change in quasi-rents due to a change in q can be measured as the change in the area behind the demand curve for the essential input, as q shifts that demand curve. Again, the mathematical statement is obvious:

$$\begin{aligned}\Delta\hat{\pi}_{\Delta q} &= \int_{w_j^0}^{\tilde{w}_j} [x_j(p, w_j, \mathbf{w}_{-j}, q^1) - x_j(p, w_j, \mathbf{w}_{-j}, q^0)]dw_j \qquad (9.15)\\ &= \hat{\pi}(p, w_j^0, \mathbf{w}_{-j}, q^1) - \hat{\pi}(p, w_j^0, \mathbf{w}_{-j}, q^0),\end{aligned}$$

where $\tilde{w}_j \geq \max[\tilde{w}_j(q^0), \tilde{w}_j(q^1)]$. This area is depicted as area $MNOP$ in Figure 9.7. Pattanayak and Butry (2005) use this result to calculate the welfare effects of watershed protection. Their measure is based on estimated labor demand as a function of soil erosion and hydrological baseflows.

An alternative production function restriction can assure that expression (9.15) is the proper measure, without appealing to the essentiality argument. For the area between the two demand functions to measure the change in quasi-rents, all we really need is that $\hat{\pi}(p, \tilde{w}_j, \mathbf{w}_{-j}, q^0) = \hat{\pi}(p, \tilde{w}_j, \mathbf{w}_{-j}, q^1)$ even if neither term equals zero. Suppose that the environmental input is a quality dimension of x_j, so that when x_j is not employed in production, changes in q have no impact on production. The environmental quality only affects the firm's production when x_j is positive. We can think of the transformation function as $T(z, h(x_j, q), \mathbf{x}_{-j})$ with the property that $h(0, q) = 0$. As an example, suppose that the firm has two sources of irrigation water, so that no single source is essential. Assume that one source of water is polluted with an industrial effluent, where the water quality of that source is denoted by our environmental variable, q. The quasi-rents associated with changes in the

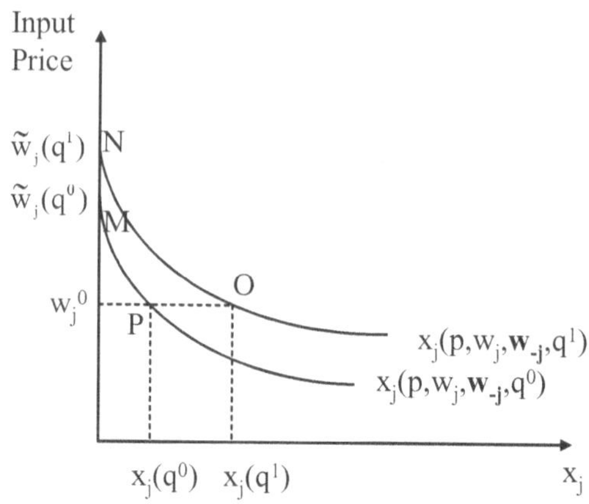

FIGURE 9.7. Welfare Measurement Using An Essential Input

effluent can be estimated for the firm as the change in the area under the derived demand for the effluent-contaminated water. The welfare measure is the same as that depicted in Figure 9.7, but the restriction on technology is different. This case is one of complementarity between the environmental good and a purchased input rather than essentiality of the input. These are simply alternative reasons why $\hat{\pi}(p, \tilde{w}_j, \mathbf{w}_{-j}, q^0)$ might equal $\hat{\pi}(p, \tilde{w}_j, \mathbf{w}_{-j}, q^1)$. Any rationale for this condition that meets the facts of the problem will achieve the same result.

9.2.4 Aggregation and Market Interactions

In some instances the environmental change will be pervasive enough to affect many firms. This causes no added problems in welfare measurement unless the agents' actions are interdependent. In the absence of interdependence of any form (except interdependency through markets), we need only add the welfare measures in (9.11) over affected firms to get aggregate welfare estimates.

Where many firms are affected by a change in q, it is typically easier to analyze industry level output supply and/or input demand functions. Because adding firm level functions horizontally across otherwise independent competitive firms yields industry level functions, the areas behind these industry curves can be shown to be the sum of the areas behind the individual firm supply or input demand curves. Thus, as long as firms' production functions are independent, we have the satisfying result that to obtain aggregate welfare measures we need only substitute industry supply and/or demand functions for firm level functions in all the previous results.

An extensive environmental change that affects multiple firms in an industry may result in price changes, especially where markets are localized. Once prices change, there are additional welfare effects to be taken into account. To calculate the full welfare effects, the additional challenge is to estimate any relevant price changes. Figure 9.8 describes the problem in the output market, where the curves marked $S(p, w, q)$ are industry supply curves conditioned on different levels of q, and $D(p)$ is the industry demand curve. As long as z is an essential output, total quasi-rent to the industry before the change in q is measured by area ABp^0. The full welfare effect for the industry of the environmental change will be $CDp^1 - ABp^0$ but calculating this requires knowledge, not only of the shift in the industry supply curve due to the change in q, but also the location of the industry demand function. With this information, not only can the new price level be predicted and the welfare effects for firms estimated, but the welfare effects for consumers can be calculated as well. In this example, the consumer surplus gain due to the environmental improvement is given by Ap^0p^1D, and the total gain for all agents equals $ABCD$.

Price changes may also be induced in input markets, if the industry in question is of sufficient size relative to any of these markets. Once again, additional information is necessary to calculate the full welfare effects. Now we need industry supply functions of input producing firms to complete the analysis. The environmental change could lead to impacts on primary factor markets as well–that is, labor and land markets. Price changes in factor markets mean changes in quasi-rents for factor owners.

If an exogenous change in an environmental good that is an input in firm production ultimately induces changes in consumer goods' prices or wage rates, then welfare measures developed for price changes for households will become part of the analysis. This is obvious when consumer prices are affected, but less obvious for wages. If the household labor supply function is not perfectly inelastic, wage rate changes will lead to endogenous income effects that involve labor-leisure choice.

Market effects signal pecuniary interdependencies, but interdependencies can also be technological, in which case externalities exist between agents. Where production functions are interdependent, welfare measures cannot simply be added across firms. In order to evaluate welfare effects in the presence of technological interdependence, the nature of the externality must be explicitly accounted for and the behavior of agents modeled appropriately, as we will see in an example in the final section of this chapter. When the interdependency or externality *is* the environmental good of interest, for example if q is a pollutant emitted by one firm or industry and affecting other economic agents, then special care must be taken in evaluating changes. For example, suppose an agricultural sector's farming methods cause downstream sedimentation, thus reducing fish populations and quasi-rents to commercial fishermen. Evaluation of an effluent regulation would involve predicting the new level of q and

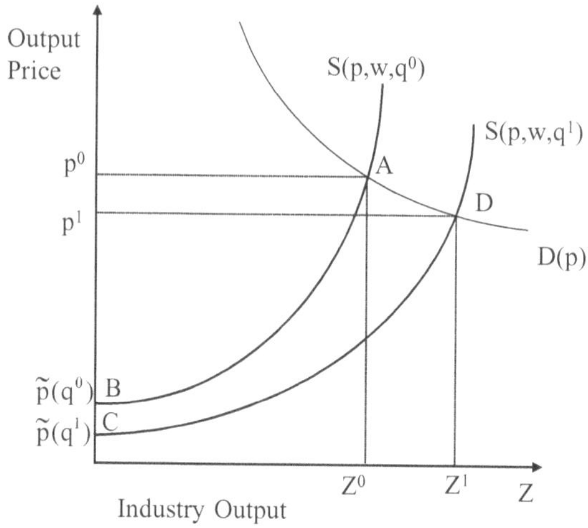

FIGURE 9.8. Welfare Measurement for the Industry

estimating the resulting gains and losses in both sectors.

The message is that changes in the environment or in the status of natural resources are likely to affect more than one producer. The more agents involved, the more likely are complications in welfare analysis. The researcher needs to assess whether interdependencies exist–through the market by affecting prices, through production function interdependencies in the form of technological externalities, or both–and take them into account in setting up the underlying behavioral model that forms the basis of welfare measurement.

9.3 The Household as Producer

Owners of large incorporated firms are usually non-participating stock-holders, separated from the day to day operations of the enterprise. This is less true of firms in primary industries such as fishing and farming, even in developed countries, and in developing countries the household is often the production unit. In such settings, output might be distributed between household consumption and market sales. Household labor might be allocated between production and alternative uses of household time, and a household might use labor on the farm but also sell labor to others. In this section we investigate the conditions that must prevail for production to be treated as separable and the quasi-rents from production viewed as simply exogenous changes in income.

9.3.1 Separability of Production and Consumption

To relate the discussion to producing and consuming households, the household production model is recast in a way consistent with the literature (e.g. Singh, Squire, and Strauss, 1986; deJanvry, Fafchamps, and Sadoulet, 1991). Initially we ignore the environmental input. We replace the more general transformation function used earlier with a production function, and assume that the household produces one output, z_1, using labor, L, land, A, and purchased inputs, $\mathbf{x}$, according to production function $z_1 = F(L, A, \mathbf{x})$. Labor used in production is composed of own household labor, L_h, and hired labor, L_r. Likewise, land used in production includes land owned by the household, A_h, and rented land, A_r. Monetary returns to the household from the production activity depend on how much of the produced good is directly consumed by the household (z_1^C) and how much is sold on the market at price p_1. The monetary returns are

$$p_1[F(L_h + L_r, A_h + A_r, \mathbf{x}) - z_1^C] - wL_r - rA_r - \mathbf{s} \cdot \mathbf{x}, \tag{9.16}$$

where w is the market wage rate, r is the market rental rate for land, and $\mathbf{s}$ is the vector of prices of other inputs.[7] While accurate as far as it goes, the statement of profits in expression (9.16) ignores the opportunity costs of owned land and household labor. Presumably owned land could be rented out (A_s). Household labor could be sold on the labor market (L_s) or used in other household production activities including leisure (T_h). The general case allows for the household to consume or sell output, to buy or sell labor, and to rent or rent out land.

To capture these alternatives, we write the entire utility maximization problem. Utility is expressed as a function of two consumption goods (z_1 and z_2) and time left over for household use (T_h). Good z_2 can only be acquired through purchase on the market at price, p_2. Good z_1 is produced by the household, and some or all of this production can be consumed. The household could also purchase z_1 on the market at price, p_1. The optimization problem is

$$\max\ u(z_1, z_2, T_h),$$

subject to

$$\begin{aligned} p_1[F(L_h + L_r, A_h + A_r, \mathbf{x}) - z_1^C] - wL_r - rA_r - \mathbf{s} \cdot \mathbf{x} + wL_s + rA_s & \\ = p_1(z_1 - z_1^C) + p_2 z_2 \ \ (\textit{money constraint}) & \\ \bar{T} - T_h - L_h - L_s = 0 \ \ (\textit{time constraint}) & \\ \bar{A} - A_h - A_s = 0 \ \ (\textit{acreage constraint}), & \end{aligned} \tag{9.17}$$

[7]To be consistent with the usual statement of the problem, we assume here that the 'length of run' of the analysis is long enough so that land is a variable factor and land rents are included as a cost in the calculation of quasi-rents.

where $\bar{T}$ and $\bar{A}$ are the household's time and acreage endowments.[8] In the first constraint, income is earned from three potential sources: the sale of household labor, the rental of household land, and returns from production.

Written in this way, the constraints imply some important embedded assumptions. First, there exist exogenous market prices for goods, labor, and land, and the household can buy and sell at the same price. Also, produced and purchased z_1 are assumed equally desirable in consumption; and hired and own labor are assumed equally productive, as are rented and owned land. This is a picture of complete and smoothly working markets with no significant transactions costs. It is easy to show that if these features characterize the household's problem, then the production decision is separable from the utility maximization decision.

The separability becomes obvious if we rewrite the maximization problem. Substitute $L_s = \bar{T} - T_h - L_h$ and $A_s = \bar{A} - A_h$ into the money constraint and eliminate $p_1 z_1^C$ from both sides of the money constraint, so that the new problem becomes

$$\begin{aligned} &\max u(z_1, z_2, T_h) \\ &\textit{subject to} \\ &p_1 F(L_h + L_r, A_h + A_r, \mathbf{x}) - wL_r - rA_r - \mathbf{s}\cdot\mathbf{x} + \\ &w(\bar{T} - T_h - L_h) + r(\bar{A} - A_h) - p_1 z_1 - p_2 z_2 = 0. \end{aligned} \tag{9.18}$$

Regrouping terms and writing in Lagrangian form yields

$$\begin{aligned} &\max u(z_1, z_2, T_h) \\ &+\lambda\{[w\bar{T} + r\bar{A}] + [p_1 F(L, A, \mathbf{x}) - wL - rA - \mathbf{s}\cdot\mathbf{x}] - p_1 z_1 - p_2 z_2 - wT_h\} \end{aligned} \tag{9.19}$$

where $L = L_h + L_r$ and $A = A_h + A_r$. The first term in square brackets in the constraint represents household endowments and can be treated as exogenous 'full' income in the Becker (1965) sense. The second term in square brackets is quasi-rent from the production activity. While L and A are choice variables, they can be optimally determined independently of the remainder of the problem. The resulting quasi-rent becomes income to the household. Finally the last three terms simply charge market prices for the goods that enter the utility function. Because a market exists for both the produced good and labor (or household time), both z_1 and T_h have per unit opportunity costs equal to market prices.[9]

---

[8]The household's time endowment is determined by the household's size, gender and age composition.

[9]If land not used for production can be used for leisure or other utility generating uses, then $\bar{A} - A_h$ also enters the utility function and has an opportunity cost equal to r. This is less likely to characterize developing than developed country settings.

First order conditions for the problem in (9.19) are simply:

$$\begin{aligned} u_{z_1}(z_1, z_2, T_h) - \lambda p_1 &= 0, & (9.20a)\\ u_{z_2}(z_1, z_2, T_h) - \lambda p_2 &= 0, & (9.20b)\\ u_{T_h}(z_1, z_2, T_h) - \lambda w &= 0, & (9.20c)\\ p_1 F_L(L, A, \mathbf{x}) - w &= 0, & (9.20d)\\ p_1 F_A(L, A, \mathbf{x}) - r &= 0, & (9.20e)\\ p_1 F_x(L, A, \mathbf{x}) - \mathbf{s} &= 0, & (9.20f)\\ w\bar{T} + r\bar{A} + \hat{\pi} - p_1 z_1 - p_2 z_2 - wT_h &= 0. & (9.20g) \end{aligned}$$

Equations (9.20d, e, and f) are the household's conditions for quasi-rent maximization, where quasi-rent equals $\hat{\pi} = p_1 F(L, A, \mathbf{x}) - wL - rA - \mathbf{s} \cdot \mathbf{x}$. The production decision is not dependent on the utility maximization problem and, except for the budget constraint, consumption is independent of production. The household, in effect, chooses how much to produce, by maximizing its quasi-rents, and then makes its consumption decisions on the basis of resulting income. Household labor used in the production process is indistinguishable from hired labor because, by assumption, the two are assumed to have equal productivity and because the option of selling household labor on the market implies that household labor has the same opportunity cost as hired labor. Given the choices of L in production and T_h in consumption, the various amounts of L_r, L_h and L_s can be determined.[10] The same is true for the input land.

Researchers argue that circumstances in developing countries rarely provide complete and smoothly operating markets (deJanvry, Fafchamps, and Sadoulet, 1991). Equation (9.19) may not accurately capture household decisions for other reasons as well. Household labor may be more productive than purchased labor or require less supervision, or households may prefer to supply labor for household production rather than sell it on the labor market (Lopez, 1991). Given the casting of the problem in (9.17), impediments or imperfections in the labor market will cause difficulties. If land markets are missing, households are constrained to use only their own land for production and, as a consequence, may produce less. However, since in our formulation land does not enter the utility function, the consumption and production decisions remain separable. The only consequence is that $\hat{\pi}$ (and therefore exogenous income) can be no

[10] Note that in this model a household would not hire labor and sell labor at the same time. Also, own time is contrained by $\bar{T}$ and own land by $\bar{A}$. Given hese conditions a full solution of the problem is possible if L, A, $\mathbf{x}$, z_1, z_2, and T_h are choice variables. It must be true that if $A \geq \bar{A}$ then $A_h > 0$, $A_s = 0$, and $A_r \geq 0$. Otherwise $A_h > 0$, $A_s \geq 0$, and $A_r = 0$. Likewise, if $L \geq \bar{T} - T_h$, then $L_h > 0$, $L_s = 0$, and $L_r \geq 0$. Otherwise, $L_h > 0$, $L_s \geq 0$, and $L_r = 0$.

larger when land markets are missing than when they exist and work smoothly.

Consider what happens if land markets work, but labor markets are non-existent, so that all labor used in production must come from the household and no market exists to sell labor.[11] The model now becomes

$$\begin{aligned} \max \quad & u(z_1, z_2, T_h) \\ & \textit{subject to} \\ & r\bar{A} + p_1 F(L_h, A, \mathbf{x}) - rA - \mathbf{sx} - p_1 z_1 - p_2 z_2 = 0 \\ & \bar{T} - T_h - L_h = 0. \end{aligned} \tag{9.21}$$

There is now no way to avoid the fact that T_h and L_h are related. To see this, note that the problem can be rewritten by substituting $T_h = \bar{T} - L_h$ into the utility function or $L_h = \bar{T} - T_h$ into the production function. Choosing the former, the problem becomes:

$$\max\ u(z_1, z_2, \bar{T} - L_h) + \lambda(r\bar{A} + p_1 F(L_h, A, \mathbf{x}) - rA - \mathbf{sx} - p_1 z_1 - p_2 z_2).$$

First order conditions reveal the snag:

$$\begin{aligned} & u_{z_1} - \lambda p_1 = 0, \\ & u_{z_2} - \lambda p_2 = 0, \\ & p_1 F_L - u_{T_h}/\lambda = 0, \\ & p_1 F_A - r = 0, \\ & p_1 F_x - \mathbf{s} = 0, \\ & r\bar{A} + \tilde{\tilde{\pi}} - p_1 z_1 - p_2 z_2 = 0, \end{aligned} \tag{9.22}$$

where $\tilde{\tilde{\pi}} = p_1 F(L_h, A, \mathbf{x}) - rA - \mathbf{sx}$. The third of these first order conditions links production and consumption inseparably: labor is used in production up to the point where the marginal value of the output of labor equals the value of labor in other household uses.

Several authors have shown that even if labor markets do exist, the absence of credit markets can prevent poorer farmers from making use of them. Cash flow problems may make it impossible to hire labor because returns from production are earned after labor must be paid. The absence of insurance markets also causes problems if returns are uncertain. When firm owners face substantial undiversifiable risk and cannot insure against it, we should account for the owner's preferences over multiple moments of the probability distribution, making the nature of the preference function relevant in describing production decisions. This is most germane for the small firm with no internal means of diversifying its risk portfolio.

---

[11] For a general model of market imperfections, see Thornton and Eakin (1992).

Pitt and Rosenzweig(1986) and Benjamin (1992) have suggested an empirical test for production and consumption separability. If markets are complete then household composition should not affect production decisions (and quasi-rents). In the environment as an input literature, Pattanayak and Kramer (2001) and Pattanayak and Butry (2005) use this test and reject the notion of incomplete markets in their study of the effects of drought mitigation (through watershed protection) on agricultural production in Indonesia. Nevertheless non-separability is likely to characterize many developing country problems.

9.3.2 Welfare Effects of Price Changes with Separability

It is worthwhile to consider the implications of this separability for welfare measurement. Here we consider the welfare effects of price changes.

The household optimization problem, given in (9.19), implies the expenditure function:

$$\begin{aligned} &m(p_1, p_2, w, r, \mathbf{s}, \bar{T}, \bar{A}, u^0) = \\ &\min\ p_1 z_1 + p_2 z_2 + wT_h - \\ &[p_1 F(L, A, \mathbf{x}) - wL - rA - \mathbf{s} \cdot \mathbf{x}] - w\bar{T} - r\bar{A} + \\ &\mu(u^0 - u(z_1, z_2, T_h)). \end{aligned} \tag{9.23}$$

Compensating variation of a change in p_1, for example, is defined as

$$CV = m(p_1^0, p_2, w, r, \mathbf{s}, \bar{T}, \bar{A}, u^0) - m(p_1^1, p_2, w, r, \mathbf{s}, \bar{T}, \bar{A}, u^0). \tag{9.24}$$

But the applied welfare question remains: how do we reveal the value of this conceptual measure using observable data?

Begin by noting that

$$\frac{\partial m}{\partial p_1} = z_1^h - z_1^P. \tag{9.25}$$

For clarity, we introduce the notation of z_1^h is the Hicksian demand for good z_1 (i.e. the demand for consumption of z_1 obtained from own production and/or market purchases), and $z_1^P = F(L, A, \mathbf{x})$ is the amount the household produces. According to our assumptions the good can be bought and sold on the market at the same price, p_1. It is clear from the first order conditions of the expenditure minimization problem that the Hicksian demand for z_1 will be a function of p_1, p_2, w, and u^0, while the supply function, z_1^P, will be a function of p_1, w, r, and $\mathbf{s}$. If we integrate $\partial m / \partial p_1$ over the change in p_1 we must now take into account both demand and supply. That is,

$$CV = -\int_{p_1^0}^{p_1^1} \frac{\partial m}{\partial p_1} dp_1 = -\int_{p_1^0}^{p_1^1} z_1^h(p_1, p_2, u^0) dp_1 + \int_{p_1^0}^{p_1^1} z_1^P(p_1, w, r, \mathbf{s}) dp_1, \tag{9.26}$$

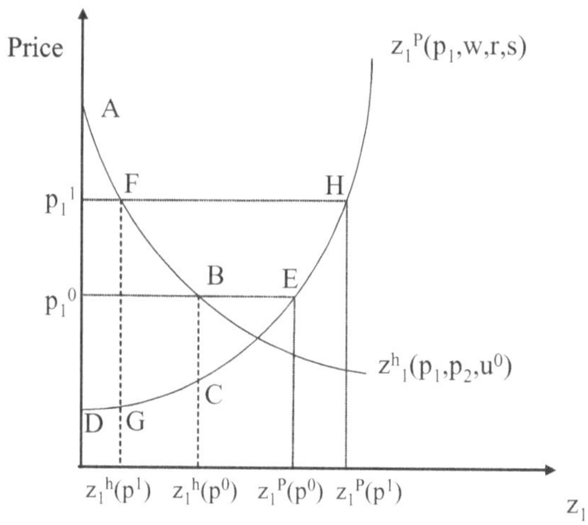

FIGURE 9.9. Welfare Measure of Output Price Change for Consuming and Producing Household

which is the change in the area behind the supply (or marginal cost) curve as price changes minus the change in the area behind the Hicksian demand curve as price changes.

Figure 9.9 illustrates the case in which the household produces more than it consumes, selling the remainder on the market. At p_1^0 the first $z_1^h(p_1^0)$ units of output are dedicated to household consumption. *Area ABCD* measures willingness to pay to consume this amount minus the costs of producing this amount. The remaining output, equal to $z_1^P(p_1^0) - z_1^h(p_1^0)$ on the graph, is sold on the market, generating quasi-rents equal to area BCE. However, we could have gotten the same total area by adding two measures, the sum of which identically equals *area* $ABCD$ + *area* BCE:

a) the compensating variation that any household with this Hicksian demand function would get from having access to good z_1 at price p_1^0 (a measure described by *area* ABp_1^0); and

b) the quasi-rent that any firm with this supply (marginal cost) function would earn given price p_1^0 (a measure described by *area* p_1^0ED).

Now consider the welfare effects of a *change* in p_1 for this household. At a new price of p_1^1, the household consumes only $z_1^h(p_1^1)$ and sells the remainder on the market because p_1^1 is higher than the household's willingness to pay (as demonstrated by its Hicksian demand) for any more than this amount of the good. As a result of the price change, the consuming household loses *area* $p_1^1FBp_1^0$, but as a producer the household gains *area* $p_1^1HEp_1^0$, resulting in a net gain of *area* $HEBF$. Even though a change in output price affects the household in both its consuming and producing roles, the welfare effect can be calculated separately, and in the usual way, for each of these roles.

This story has been told in terms of compensated demand, so that an exact correspondence between CV and areas behind curves can be established. Marshallian rather than Hicksian demands are more likely to be observed, raising the usual questions about approximation. The conditions specified by Willig (1976) are arguably less likely to hold in this setting. Households may be poor, and they may consume few goods making the demand for z_1 relatively inelastic and the ratio of the consumer surplus of a change in the price of z_1 for this good relative to income quite large. In any event, Willig's results provide bounds on CV and EV measures that can be calculated with information derived from an ordinary demand curve.

Evaluating a change in output price for the producing household involves two terms, as in expression (9.26), because output price affects the household in both its consuming and producing roles. Evaluating a change in the price of an element of the input vector, denoted $\mathbf{x}$ in the above model, would not involve consumption, only production, so that the welfare measure would be identical to that stated in expression (9.7). However, the welfare measure of a change in w will be more complex, since time valued at the wage rate enters utility as well as production.

To evaluate the effects of a wage change in the separability case, begin with expression (9.23). The compensating variation measure is

$$CV = m(p_1, p_2, w^0, r, \mathbf{s}, \bar{T}, \bar{A}, u^0) - m(p_1, p_2, w^1, r, \mathbf{s}, \bar{T}, \bar{A}, u^0).$$

Welfare measurement begins with the partial derivative of the expenditure function:

$$\frac{\partial m}{\partial w} = T_h + L - \bar{T} = L - (\bar{T} - T_h). \tag{9.27}$$

The first term on the right hand side of (9.27) is the household's demand for labor in production (irrespective of its source) while the second term is the supply of household labor (irrespective of whether it is used in own production or sold on the labor market). By assumption, labor can be bought and sold at the same wage rate, w. Integrating $\partial m/\partial w$ over the change in w and taking into account both demand and supply, yields

$$CV = -\int_{w^0}^{w^1} \frac{\partial m}{\partial w} dw = \int_{w^0}^{w^1} [\bar{T} - T_h^h(p_1, p_2, w, u^0)]dw - \int_{w^0}^{w^1} L(p_1, w, r, \mathbf{s})dw, \tag{9.28}$$

which is the change in the area behind the household's Hicksian labor supply curve $(\bar{T} - T_h^h)$ as wage changes minus the change in the area behind the household's demand curve for labor to be used in production as wage changes. Household labor supply is $\bar{T} - T_h$, which equals household labor used in production (L_h) plus household labor sold on the market (L_s).

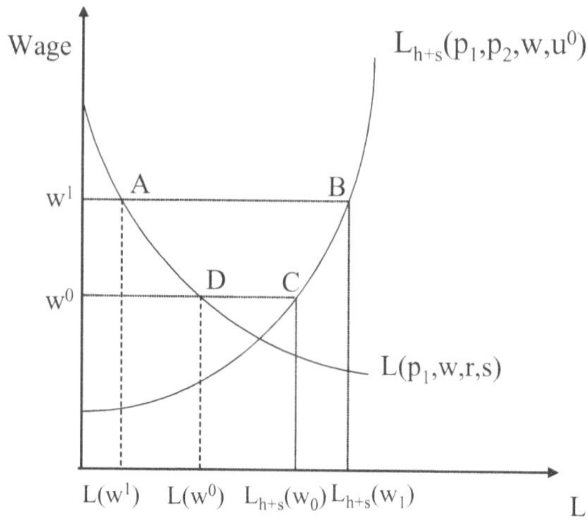

FIGURE 9.10. Welfare Measure of Wage Change for Producing Household that Sells Labor

Equation (9.28) is illustrated in Figure 9.10 where labor supply, $\bar{T} - T_h$, is denoted L_{h+s}. At the initial and subsequent wage rates, wages are sufficiently high so that the household supplies more labor than it uses in its own production. As a result of the wage change from w^0 to w^1, the producing household loses *area* w^0w^1AD, but as a supplier of labor the household gains *area* w^0w^1BC, resulting in a net gain of *area* $ABCD$. Just as with the output price change, the welfare effect can be calculated separately, and in the usual way, for each of these roles. This holds as long as production and consumption are separable.

9.3.3 Welfare Effects for the Environment as an Input

Now we turn to the central task–obtaining measures of the welfare effects of a change in q when the household is a producer. When production and consumption are separable and the environmental input affects households only through the production function, welfare measurement of a change in q should be straightforward, at least conceptually. Because production is separable from consumption, this should reduce to a problem of welfare measurement in the context of the firm, and all of the earlier results in this chapter should apply.

Let us take the case where z_1 is an essential output for the household's production activities. The household's expenditure minimization problem, including the environmental input, q, is given by

$$m(p_1, p_2, w, r, \mathbf{s}, \bar{T}, \bar{A}, q, u^0) = \\ \min\ p_1 z_1 + p_2 z_2 + wT_h - [p_1 F(L, A, \mathbf{x}, q) - wL - rA - \mathbf{sx}] \\ -w\bar{T} - r\bar{A} + \mu(u^0 - u(z_1, z_2, T_h)). \tag{9.29}$$

The compensating variation of a change in q is

$$CV = m(p_1, p_2, w, r, \mathbf{s}, \bar{T}, \bar{A}, q^0, u^0) - m(p_1, p_2, w, r, \mathbf{s}, \bar{T}, \bar{A}, q^1, u^0). \tag{9.30}$$

If production and consumption are indeed separable, then we expect to find that the methods for measuring welfare change for the firm will work here. Specifically, let's see whether expression (9.30) reduces to the area between two supply curves (conditioned on different levels of q) for the produced good.

To prove this we need to establish two results. First, we need to show that the area between $z_1^P(q^1)$ and $z_1^P(q^0)$ equals

$$\int_{\tilde{p}_1}^{p_1^0} \frac{\partial m(p_1, q^0, u^0)}{\partial p_1} dp_1 - \int_{\tilde{p}_1}^{p_1^0} \frac{\partial m(p_1, q^1, u^0)}{\partial p_1} dp_1, \tag{9.31}$$

where we suppress the irrelevant arguments. In (9.31), $\tilde{p}_1 \leq \min[\tilde{p}_1(q^0), \tilde{p}_1(q^1)]$, where $\tilde{p}_1(q^i)$ is the supply choke price when $q = q^i$. As in the last section, $\partial m/\partial p_1$ is equal to the Hicksian demand curve (z_1^h) minus the supply curve (z_1^P):

$$\frac{\partial m}{\partial p_1} = z_1^h(p_1, u^0) - z_1^P(p_1, w, r, \mathbf{s}, q). \tag{9.32}$$

Integrating this expression over a price change yields two terms reflecting the effect of the price change on the household in its role as consumer and producer. However, when production and consumption decisions are separable, the change in the integral with a change in q will reduce to the expression we are looking for. Expanding (9.31):

$$\int_{\tilde{p}_1}^{p_1^0} \frac{\partial m(p_1, q^0, u^0)}{\partial p_1} dp_1 - \int_{\tilde{p}_1}^{p_1^0} \frac{\partial m(p_1, q^1, u^0)}{\partial p_1} dp_1 \tag{9.33}$$

$$= [\int_{\tilde{p}_1}^{p_1^0} z_1^h(p_1, u^0) dp_1 - \int_{\tilde{p}_1}^{p_1^0} z_1^P(p_1, q^0) dp_1] \\ -[\int_{\tilde{p}_1}^{p_1^0} z_1^h(p_1, u^0) dp_1 - \int_{\tilde{p}_1}^{p_1^0} z_1^P(p_1, q^1) dp_1], \tag{9.34}$$

but this equals

$$= \int_{\tilde{p}_1}^{p_1^0} z_1^P(p_1, w, r, \mathbf{s}, q^1) dp_1 - \int_{\tilde{p}_1}^{p_1^0} z_1^P(p_1, w, r, \mathbf{s}, q^0) dp_1,$$

because Hicksian demand is not a function of q and the first and third terms in (9.34) cancel.

The second result we need requires that

$$\int_{\tilde{p}_1}^{p_1^0} \frac{\partial m(p_1, q^0, u^0)}{\partial p_1} dp_1 - \int_{\tilde{p}_1}^{p_1^0} \frac{\partial m(p_1, q^1, u^0)}{\partial p_1} dp_1 = m(p_1^0, q^0, u^0) - m(p_1^0, q^1, u^0).$$

For this to be true, $m(\tilde{p}_1, q^1, u^0) - m(\tilde{p}_1, q^0, u^0)$ must equal 0. Evaluating these terms:

$$\begin{aligned} & m(\tilde{p}_1, q^1, u^0) - m(\tilde{p}_1, q^0, u^0) && (9.35) \\ = \; & [\tilde{p}_1 z_1 + p_2 z_2 + wT_h - \widehat{\pi}(\tilde{p}_1, q^1)] && (9.36) \\ & -[\tilde{p}_1 z_1 + p_2 z_2 + wT_h - \widehat{\pi}(\tilde{p}_1, q^0)], \end{aligned}$$

where z_1, z_2, and T_h are at their optimal levels, given $\tilde{p}_1$ and the initial levels of p_2 and w. Because none of these choice variables is a function of q, the expression in (9.36) equals

$$\widehat{\pi}(\tilde{p}_1, q^0) - \widehat{\pi}(\tilde{p}_1, q^1) = 0. \qquad (9.37)$$

This difference in quasi-rents equals zero because, by definition, quasi-rent from the production of z_1 is zero at prices less than or equal to the shut down price. Thus, the requirement that $m(\tilde{p}_1, q^1, u^0) - m(\tilde{p}_1, q^0, u^0) = 0$ holds.

It is not surprising that the area between two supply curves, conditioned on different levels of q, is the welfare measure of a change in q. It simply says that when the household is a producer but consumption and production are separable, measuring welfare effects of a change in an environmental input involves only the production decision and in the usual way. This is true, even though we are using the mechanism of integrating over p_1 which enters the consumption decision. It's not that changes in p_1 have no effect on consumption but that, evaluated at any given level of p_1, *changes in* q have no effect on consumption.

The above derivation requires that the Hicksian consumption decision is not a function of exogenous variables affecting production and vice versa. With incomplete markets, this condition is violated and the results unravel. Suppose the circumstances described in model (9.21) were relevant. Inserting q into the production function and setting the problem up as an expenditure minimization would give:

$$\begin{aligned} & m(\mathbf{p}, \mathbf{s}, r, \bar{A}, \bar{T}, q, u^0) && (9.38) \\ = \; & \min \; \mathbf{p} \cdot \mathbf{z} - [p_1 F(L_h, A, \mathbf{x}, q) - rA - \mathbf{s} \cdot \mathbf{x}] - r\bar{A} \\ & \textit{subject to } u^0 - u(z_1, z_2, \bar{T} - L_h) \end{aligned}$$

where we have included all the arguments for future reference. First order conditions for this problem would look much like those in (9.22) except that the budget constraint would be replaced by the utility constraint. A condition exactly equivalent to the third first order condition in (9.22) would persist, leading to the complicating connection between production and consumption decisions. As a result, demands for commodities, demands for inputs, and supply would all be functions of all exogenous variables, including q. In attempting to execute the procedure described in (9.33) to (9.37), the Hicksian demands would be found to be functions of q and therefore would not cancel out in the expression in (9.33).

In principle, recovering welfare measures in the absence of complete markets is still possible. For example, we could pursue the strategy that begins with equations (9.32) and (9.33). The expression in (9.26) is still relevant, but both the demand and supply functions must be specified as functions of all exogenous variables associated with the household's roles as consumer and producer. Now demand will shift with a change in q, so that the correct welfare answer will equal the change in the area between the supply and demand functions as q changes. Admitting practical difficulties of implementation, the result is still theoretically correct because at $p_1 = \tilde{p}_1$, production of z_1 ceases and the household no longer cares about changes in q.

The welfare result could alternatively be obtained by measuring the change in the area behind an essential input, x_j. Using input demand has the advantage of requiring the estimation of only one behavioral function, although it too must be specified as a function of all the exogenous variables in the problem. Denoting the price of x_j as s_j and $\tilde{s}_j$ as its choke price, the change in the area to the left of the demand for x_j can be written as

$$\begin{aligned} & \int_{s_j^0}^{\tilde{s}_j} x_j(\mathbf{p}, s_j, \mathbf{s}_{-j}, r, \bar{A}, \bar{T}, q^1, u^0) ds_j \qquad (9.39) \\ & - \int_{s_j^0}^{\tilde{s}_j} x_j(\mathbf{p}, s_j, \mathbf{s}_{-j}, r, \bar{A}, \bar{T}, q^0, u^0) ds_j \\ = & \int_{s_j^0}^{\tilde{s}_j} \frac{\partial m(\mathbf{p}, s_j, \mathbf{s}_{-j}, r, \bar{A}, \bar{T}, q^1, u^0)}{\partial s_j} ds_j \\ & - \int_{s_j^0}^{\tilde{s}_j} \frac{\partial m(\mathbf{p}, s_j, \mathbf{s}_{-j}, r, \bar{A}, \bar{T}, q^1, u^0)}{\partial s_j} ds_j \\ = & \; m(\mathbf{p}, \tilde{s}_j, \mathbf{s}_{-j}, r, \bar{A}, \bar{T}, q^1, u^0) - m(\mathbf{p}, s_j^0, \mathbf{s}_{-j}, r, \bar{A}, \bar{T}, q^1, u^0) \\ & - m(\mathbf{p}, \tilde{s}_j, \mathbf{s}_{-j}, r, \bar{A}, \bar{T}, q^0, u^0) + m(\mathbf{p}, s_j^0, \mathbf{s}_{-j}, r, \bar{A}, \bar{T}, q^0, u^0). \end{aligned}$$

We carry the cumbersome set of arguments throughout as a reminder that the estimation of the demand function will not be easy under the absence of separability, and that input demand depends on technology and tastes. Expression (9.39) equals CV if we can be sure that $m(\mathbf{p}, \tilde{s}_j, \mathbf{s}_{-j}, r, \bar{A}, \bar{T}, q^1, u^0) = m(\mathbf{p}, \tilde{s}_j, \mathbf{s}_{-j}, r, \bar{A}, \bar{T}, q^0, u^0)$. Essentiality of x_j assures this, because at $\tilde{s}_j$ the household ceases to produce z_1 and changes in q no longer matter. An alternative means of ensuring this condition requires that q be viewed as a quality characteristic of input, x_j. If so, then at $s_j = \tilde{s}_j$ changes in q will not matter because x_j is not purchased. In agriculture, an example might be one in which the quality of irrigation water changes, where irrigation water, x_j, is purchased at a price, s_j. Even if this particular source of irrigation water is not essential to production, the welfare effects of its contamination can be estimated using shifts in the demand for the input.

All of these alternative means of obtaining a welfare measure exist in principle, but each requires estimating a complex function with accuracy up to the relevant choke price. Even under separability, strategies require measuring the change in entire areas behind behavioral functions. This is troublesome, as estimation accuracy will be poorer over ranges of prices where data are scarce. But there is a more fundamental problem. Where there are few alternative sources of income and only one crop, household-producers are unlikely to be observed shutting down production operations. And shut-down is required in every one of our strategies except when q is a quality characteristic of an input. In that case alone, we need not rely on the mechanism that q no longer matters when the household ceases production. Whether even this strategy is practicable may be questioned. The difficulties inherent in trying to estimate the input demand function in (9.39) are likely to be formidable, especially in a developing country setting.

9.4 Welfare Bounds and Approximations

Using the results developed so far in the chapter requires estimating a supply or demand function as a function of prices and the environmental resource. There are many examples where these techniques have been used for analyzing price changes, especially in agricultural policy analysis. But obtaining data to estimate supply or demand functions, especially under varying levels of q, is difficult. In developing countries, where environmental degradation is a common threat to production, research efforts to value losses have frequently depended on much cruder approximations of the theory.

9.4.1 Approximations by Pricing Output Changes

The guiding spirit of revealed preference models entails the use of *behavioral* models to measure welfare. Yet much of the literature that attempts to capture the welfare effects of environmental changes on production begins with either field experiments or production function estimation intended to capture the *physical* relationship between q and yields. Approximations using the production function provide useful information as long as the researcher remains aware of the likely behavioral adjustments that are absent in experimental data.

Dose Response or Damage Functions

Collecting data on outputs and inputs and estimating the production function, $z = F(\mathbf{x}; \mathbf{k}, q)$, directly would *appear* to be advantageous. For example, Acharya (1998) estimates a relationship between crop output in the Hadejia-Nguru Floodplain in Nigeria and the inputs: land, labor, fertilizer and irrigation water. Irrigation water is extracted from aquifers, and pumping costs are affected by groundwater levels that are in turn affected by wetlands–the environmental resource of interest. In another example, Lynne, Conroy and Prochaska (1981) estimate a fisheries production function where harvest of blue crabs is a function of fishing effort and acreage of marshlands, which serves as a natural spawning ground for the crab. Narain and Fisher (1994) provide a third example, one in which the relationship between agricultural output and the population of the Anolis lizard (a natural pest predator) is estimated.

Direct estimation of production functions is not, in general, a reliable way to estimate welfare measures. In contrast to the estimation of behavioral functions (i.e. supply or demand) that include only exogenous factors such as prices, production function estimation is plagued by endogeneity problems. In the typical production problem, input levels will be endogenously determined and estimated coefficients will be biased. This occurs because unobserved heterogeneity–omitted from the researcher's analysis but visible to the decision maker–affects the endogenous choice of input levels. As a result input levels are correlated with omitted variables, causing bias in parameter estimates.

Consider an agricultural example. The researcher estimates a relationship between yield and inputs, including pesticide use. If observations include enterprises with varying pressure from agricultural pests, one could easily find more pesticide use associated with lower yields leading to the estimation of a production function in which increased pesticide use is a deterrent to production. This is because the level of pesticide use is determined by the farmer and is related to the magnitude of the pest threat. Higher pest threats lead to lower yields, all else equal, and lower yields may be only partially mitigated by pesticide use. Correlation between unobserved pest threats and endogenously chosen pesticide use will lead to a downward bias in the estimated effect of this input. Results such as this can occur with fertilizers where more is applied to soils that are naturally less fertile, for water use where more irrigation is used in more drought-prone areas, etc.

Now consider a production function in which q is included, one that will be used for a dose-response type analysis. The consequences of attempting to estimate the production function directly depend on the relationship between q and the other inputs in the model. If q is uncorrelated with the chosen inputs, then the estimate of the effect of q on yields may be safely considered unbiased. However, if q is related to an input, for example if q is pest level and one of the inputs is pesticide use, then correlation between q and the input use will make it difficult to estimate either coefficient with accuracy and will further point up the error of including the input use as an explanatory variable in estimation, as the input is both endogenous and central to the problem. Where an input is used directly to mitigate declines in q, a damage function approach that holds inputs constant is particularly inappropriate.

Occasionally, dose-response or damage functions are available from theoretical or experimental science, precluding the necessity of estimating the production (response) function. This at least avoids the endogeneity problem in estimating production functions, if not the problems in determining the optimal level of inputs when q changes. These dose-response functions may come from agricultural experiments in which researchers subject different plots of land to different levels of environmental degradation as in the case of the National Crop Loss Assessment Network (NCLAN) or from engineers who have developed such constructs as the Universal Soil Loss Equation (USLE) for predicting soil erosion. Brekke, Iversen, and Aune (1999) adopt the USLE model in a complicated, dynamic framework to model soil degradation in Tanzania. As we will see in a later section, several authors have used NCLAN to assess the effects on US crops of changes in ambient ozone levels.

The damage or dose-response functions commonly available from natural scientists usually assess the change in output with a change in the environmental input, holding other inputs constant. Agronomic field tests are classic examples. In these tests, the level of the environmental input is randomly assigned across experimental plots or chambers, while other inputs are held constant. But, in practice, if producers have any discretion over input levels, they may be able to change their use of inputs thus reducing experimentally predicted losses or increasing gains. If it were possible to recover the entire production surface, any number of functions useful for welfare evaluation could be derived. But typically experimental dose-response functions hold inputs constant and will be too restrictive to yield exact welfare measures.

Welfare Approximation Using Predicted Output Changes

Even when dose-response function parameters are estimated consistently, use of these functions to obtain welfare results has a known potential bias. The use of dose-response functions is akin to having knowledge of a production relationship but assuming that the input vector is fixed. Hence one finds researchers approximating damages from a degradation in the environmental input as

$$app_1\ \Delta\hat{\pi} = pF(\bar{\mathbf{x}}, q^1) - pF(\bar{\mathbf{x}}, q^0) \tag{9.40}$$

where the input vector is assumed fixed, often at the level of $\mathbf{x}$ that was optimal when $q = q^0$, but not necessarily optimal when $q = q^1$. This is to be compared with true changes in quasi-rents assuming no induced changes in either input or output prices, given by:

$$\Delta\hat{\pi}_{\Delta q} = [pF(\mathbf{x}(q^1), q^1) - pF(\mathbf{x}(q^0), q^0)] - [\mathbf{w} \cdot \mathbf{x}(q^1) - \mathbf{w} \cdot \mathbf{x}(q^0)].$$

Expression (9.40) will be accurate if changes in the production process brought about by changes in q do not elicit changes in input use. Lack of an input response may occur if changes in q are not easily perceived by producers, such as in production processes subject to many uncontrollable environmental and climatic factors. It will also occur if all inputs are indeed fixed inputs over the time span of the problem. But the longer this time span, the less likely the assumption of fixed inputs holds.

A situation that would serve this approximation well would be the release of a toxic chemical that destroys part of a farmer's crop late in the growing season. Because the inputs have, for the most part, already been committed to the production process, the assumption of fixed inputs may be approximately correct. Further, the costs of these inputs (with the exception of harvesting costs) will have already been incurred. The value of the change in quasi-rents would approximately equal the change in revenues because the input costs would not be changed appreciably. Acharya uses this argument in her evaluation of the damages from reduced groundwater in Nigeria, because groundwater levels do not become known until after farmers have already committed other inputs. As a result, no immediate (within season) adjustment to groundwater change is possible.

Some studies of the returns to proposed investment projects have also used the approach of valuing predicted changes in output, often based on restrictive assumptions about how land will be used. Consider the analysis of a watershed management project by Fleming (1983). This study fixes by assumption the amount of land that would be used in agriculture, grazing, pasture, scrubland, forest and plantations with and without the management project, with further assumptions made about how productivity of the land in these different uses will change with the project. Valuation of the grazing land component, for example, involves calculating predicted changes in the yield of milk per hectare of all land assumed to be used for grazing, and valuing this at milk prices. This approach will provide a reasonable approximation if assumptions about land use and productivity are reliable, and if technology is so restrictive that there is little further discretion left to land users. This study is superior to many in that it does consider the change in expenditures on one variable input–livestock feed. However most studies of this sort are forced to ignore the costs of labor, as measurement of these is extremely difficult, especially where formal labor markets are absent. In an evaluation of a soil management project in Lesotho, Bojø (1990) values the project as the crop price times the change in yield minus changes in variables costs, and describes the difficulty of pricing labor. In evaluating the benefits of afforestation in Nigeria, Anderson (1987) provides a sensitivity analysis based on different means of calculating the opportunity cost of labor.

In general, firms will be able to respond to the change in the environmental input, either to take advantage of the change if it is a productive one or to take defensive actions if the change is deleterious. This will cause both revenues and costs to change, but if the firm chooses to take the action then the change in revenues minus costs must be positive. The principle of Le Châtelier, which implies that optimizing agents cannot be worse off when constraints are relaxed, helps us assess the direction of bias in the welfare measure when changes in inputs are ignored. If adjustments are possible, the damage function approach will overestimate the welfare losses from an undesirable change in the environment and underestimate the gains from an improvement in the environment. This will be true as long as individuals are price takers and aggregate behavior does not lead to induced changes in input or output prices. If sufficient numbers of producers are affected, then induced price changes may occur and this will tend to dampen any welfare effects on producers. It will also introduce other agents, such as those who purchase output or sell inputs to this industry, whose welfare effects must now be measured.

Welfare Approximations Using Observed Changes in Output

Approximations of welfare measures can be derived when researchers observe, *ex post*, the *actual* change in output due to the environmental change. A typical approximation might be

$$app_2\ \Delta\hat{\pi} = p\Delta z = p[z(p, \mathbf{w}, q^1) - z(p, \mathbf{w}, q^0)] \tag{9.41}$$

where Δz is the actual, observed change in output due to the change in the environmental input. Expression (9.41) differs from $app_1\ \Delta\hat{\pi}$ in (9.40) which depends on an *approximation* of the change in z equal to $F(\bar{\mathbf{x}}, q^1) - F(\bar{\mathbf{x}}, q^0)$. Thus, (9.41) includes the effects of any optimal changes in inputs that the firm is able to make in response to the change in q.

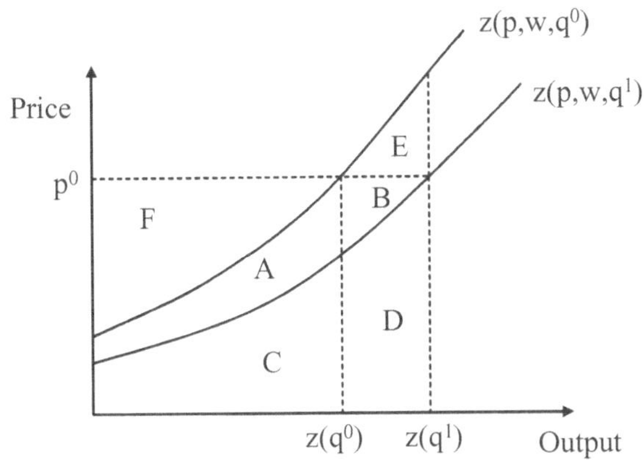

FIGURE 9.11. Commonly Used Approximations

Expression (9.41) is still an approximation because it ignores the change in costs unless input levels, and thus costs of production, do not change in response to the change in q. To investigate the accuracy of the approximation in (9.41), consider the formula for the welfare effect associated with a change in q, given that quasi-rent equals $\widehat{\pi} = pz - c(z, \mathbf{w}, q)$. We can rewrite the change in quasi-rents with a change in q as the sum of three effects: the change in revenues due to the change in output, the change in costs of producing the original level of output after q changes, and the change in costs due to the expansion of output once q has changed. These can be written as

$$\begin{aligned} \Delta\widehat{\pi} &= p[z(p, \mathbf{w}, q^1) - z(p, \mathbf{w}, q^0)] \\ &\quad -[c(z(q^0), q^1) - c(z(q^0), q^0)] \\ &\quad -[c(z(q^1), q^1) - c(z(q^0), q^1)]. \end{aligned} \tag{9.42}$$

Assuming the firm to be a price taker, the supply curve conditioned on q^0 represents marginal costs, so that the area below this curve equals total variable costs. To illustrate the approximation, some assumptions about the nature of the cost function are necessary. Specifically, we assume what we believe to be the typical case–that marginal costs decline with increases in q (i.e. $c_{zq} < 0$).[12] Now consider such an increase in q from q^0 to q^1. The exact value of this environmental improvement is area $A + B$ in Figure 9.11. The nature of the approximation in (9.41) can be appreciated by relating the figure to equation

[12] In Chapter 8 we explored properties of the household's defensive expenditure function, but here we simply make the most common assumption. Note that the contention that c_{zq} is likely to be negative is the same as the assumption made in much of Chapter 8 that c_{zb} was positive, since b was defined as an environmental 'bad'.

(9.42). The first line in (9.42) will be positive and will equal area $B+D$ in Figure 9.11, the second line will be positive as well and will equal area A, and the third line will be negative, equalling area $-D$, so that the total effect is area $A+B$. The approximation in (9.41) equals $B+D$ and may be an overestimate or an underestimate of the true change in quasi-rents, depending on the relative sizes of areas A and D. If the increase in costs from the increase in output is greater than the reduction in costs from the increase in the environmental variable, then the approximation overestimates the welfare effect from the environmental improvement.

Just as with the damage function approach, the approximation here will be more accurate the less flexibility the firm has to adapt, although unlike the damage function approach, *ex post* output is observed and need not be estimated. If no inputs can be adjusted, there are no changes in costs and the change in quasi-rents is simply $p(\Delta z/\Delta q)$, holding all other inputs constant. As an example, consider the case in which the only variable input in production is labor, and labor has few alternatives so that it is not responsive to changes in productivity. Hanayama and Sano (1986) make this assumption in valuing a loss in fish habitat. They characterize fishermen as working the same hours irrespective of catch rates, so that changes in variable costs are not relevant and the change in revenues equals the change in quasi-rents due to the change in fish stocks. Welfare effects are calculated, *ex post*, by multiplying the actual decrease in fish catch by fish price and ignoring any changes in inputs.

9.4.2 Approximating Welfare Changes Using Cost Data

Estimating welfare gains or losses by calculating the change in the cost of producing the same output levels in the presence of the changed environment is particularly popular among engineers, but examples can also be found in the economics literature. Here we make an important distinction between two quite different approaches. One relates to *firm* production decisions, calculating cost changes in the context of the firm's production technology. The other, commonly referred to as the 'replacement cost' approach to valuation, refers to public projects that could reverse environmental damage.

Changes in Defensive Expenditures in the Context of the Firm

In Chapter 8 we explored two bounds (developed by Bartik) on the welfare effects of a change in an environmental input for a household: the savings in defensive expenditures holding output constant (DS) and the savings in actual defensive expenditures (ADS). In the context of the firm we find similar results.

The firm's cost savings with a change in q, holding z constant (typically at the level that was optimal before the change in q) is

$$app_3\ \Delta\hat{\pi} = c(\bar{z}, \mathbf{w}, q^0) - c(\bar{z}, \mathbf{w}, q^1) = DS.$$

These cost savings will be a lower bound on the welfare measure along the real number line. Figure 9.11 illustrates this. In the figure, the reduction in costs, maintaining output at its original level, with an environmental improvement from q^0 to q^1 is given by area A. This is an underestimate of the exact gains from environmental enhancement, which equals $A + B$. With declines rather than increases in q, $app_3\ \Delta\hat{\pi}$ would overestimate the losses from environmental deterioration. For example, were q to fall from q^1 to q^0, the true welfare loss would equal $-[A + B]$ but the change in cost keeping output at its original level (z^1 in this case) would be $-[A + B + E]$. Regulated public utilities that are required to provide a given level of service offers one example in which institutions force firms to remain at a fixed level of output. Other examples include quotas and production limits, as long as the institutional restrictions are binding both before and after the change in q.

Calculating the *actual* change in costs as

$$app_4\ \Delta\hat{\pi} = c(z(p, \mathbf{w}, q^0), \mathbf{w}, q^0) - c(z(p, \mathbf{w}, q^1), \mathbf{w}, q^1) = ADS$$

ignores the fact that revenues change as well and will be accurate only if output returns to its original level. Consider Figure 9.11 once again. In the case where q increases, initial costs are $A + C$ and final costs are $C + D$. The decrease in actual costs equals $A - D$, an underestimate of the true gains $A + B$. When environmental quality declines, the actual increase in costs underestimates the true welfare losses. The latter is, of course, $-[A + B]$, and the former (the actual change in costs) is given by $-A + D$. These results assume no induced price changes, but if a sufficient number of firms is affected additional analysis will be required.

The Concept of Replacement Costs

Some of the more serious mistakes in environmental valuation are made by attempting to measure the social cost of environmental degradation as the engineering cost of replacing the damaged resource. In discussing replacement costs, it is important to make the distinction between this concept and the change in a firm's costs necessary to produce the same level of output after a change in q (as we discussed in the last section) by employing more of other inputs. The concept of replacement costs relates to costs incurred by the public sector to reproduce a lost environmental resource or return a damaged resource to its previous quality level.

To crystallize the argument, consider a simple and highly stylized example. Suppose there is an oil spill on a remote island off the California coast. All parties agree that the injury from the spill is limited to oil on the beach, with no effects, ecological or otherwise, lasting longer than a year. The economic loss is the foregone consumer surplus (CS) from eliminating boat access for one year, and very few boats use the island. Restoration would entail a highly sophisticated engineering operation involving considerable manpower and specialized equipment to remove the oil. The replacement costs would be quite high, but the damages averted by complete restoration almost negligible. One would be forced to conclude that replacement costs would be a poor measure of the damage from the oil spill. What is obvious from this example also holds for the general case. Replacement or remediation costs are related in no particular way to the lost value from pollution.

There is one sense in which the cost of the remediation effort has a useful interpretation. Were the perpetrator of the oil spill held liable for the damage, it would make sense to limit that liability to the cost of remediation because *by our assumptions* the entire effect of the spill could be reversed through this engineering solution. But this is only true if complete mitigation is possible and if it would actually be undertaken by the public sector.

For this reason remediation costs are often referred to as upper bounds on losses due to damage events. But this interpretation is misleading. Actually, complete remediation implies the *exchange* of the losses due to the oil spill for the social costs of remediation. Remediation would eliminate the direct losses that would arise from the damage incident and replace them by the costs of the remediation action. If complete remediation does take place, then direct losses are not experienced and remediation costs exactly equal the losses to society due to the damage event. If they do not take place, they have no bearing on the social losses at all.[13]

If we were to observe a profit maximizing producer freely undertaking a mitigation project, then we would know that the present value of loss *to this producer* would be at least as great as the cost of the remediation program. If it were not, he would not undertake the project. This information is revealed by the firm's optimization behavior, taking into account his losses from the environmental degradation and his gains from remediation. Likewise, if a public agency responsible for a resource knows the true losses of benefits, as well as the costs of alternatives, and chooses remediation, then the researcher might (reasonably) assume that remediation represents the least cost solution. But engineering costs cannot *substitute* for knowledge of the losses from environmental degradation when the public sector has no information about the magnitude of these losses. It is this latter circumstance in which we often find

[13]In reality remediation generally takes time and therefore can not entirely prevent losses. If so, the cost of remediation, if it takes place, will underestimate ultimate losses because the direct, but transitory, losses must be added in as well.

ourselves and 'replacement' or remediation costs can do little to help.

In a study of the value of a program to reduce soil erosion in Korea, Kim and Dixon (1986) use what they refer to as the 'replacement-cost approach'. They calculate the cost of the manpower that would be needed to dig up the lost sediment from its deposits downstream, the truck rental costs of hauling the sediment back upstream, and the costs of spreading the sediment in its original location. They take this total cost to be the 'minimum estimate of the value' of a project that will prevent the damage. There is no correspondence between the cost of this remediation project and either the social cost of soil erosion or the benefit of a program that prevents it. Even if this specific remediation project were *known* to be the least cost method of completely mitigating the soil loss process, the cost calculation would still do nothing to inform our estimate of the *value* of avoiding the soil loss.

9.5 Examples of the Environment as an Input

In this section we explore two literatures in which welfare measurement of the effects of a change in an environmental input is central. These examples are chosen to highlight the complications that can arise in the presence of externalities or policy-induced market distortions. Welfare measurement for firms and households is simplest when agents act independently of each other and when the sole role of government is enforcement of contracts. In natural resources, these conditions are often violated either because of market failures, government intervention, or both. Government intervention in agricultural commodity markets is pervasive, leading to complications in welfare measurement. In fisheries, extraction externalities in the face of open access also complicate welfare measurement, as firms are no longer independent agents.

9.5.1 The Welfare Effects of Changes in Ozone Levels

In the face of potential changes in air quality standards for low level ozone (O_3), several studies in the 1980's and early 1990's attempted to measure the welfare effects in the agricultural sector of changes in ambient ozone levels. Two different approaches to capturing the effect of ozone can be found in this literature.

By far the most common approach and one used in Adams and McCarl (1985), Adams, Hamilton and McCarl (1986), and Kopp, Vaughan, Hazilla and Carson (1985) is to embed the results from dose-response functions, such as EPA's National Crop Loss Assessment Network (NCLAN), in a broader economic model of welfare assessment. NCLAN provides a link between ozone levels and yields for various crops in different regions based on experiments that attempt to simulate commercial field conditions. Using experimental data rather than estimated relationships has the advantage of helping us evaluate the effects of ozone levels that are higher or lower than those currently or historically observable. In addition, it is sometimes possible to test whether the levels of other environmental factors (such as rainfall or temperature) affect the relationship between ozone levels and yields. The chief disadvantage, as we have argued earlier, is that optimal adjustment in inputs is not generally taken into account.[14] Inputs are either held constant in experiments or adjusted in fixed proportions. The latter is relevant to the ozone case only if a change in ozone levels is equivalent to a neutral technological change.

Only a few papers have taken a dual approach and attempted to directly estimate profit functions. As an example, Garcia, Dixon, Mjelde and Adams (1986) estimate a farm level profit function for Illinois corn and soybean producers using cross-section/time-series data over the period 1979 through 1981. The profit function includes prices of variable inputs (normalized on output price) and levels of fixed inputs including ozone levels which vary over space and over time. For this approach to be viable, data must be available over circumstances in which prices, fixed inputs and ambient ozone levels vary. Including ambient ozone levels directly in the profit function, thus avoiding costly experiments and compounding of modeling errors, is seen as an advantage by the authors. The principal strength of the approach, however, is that production decisions consistent with profit maximizing behavior are incorporated into the estimation, thus accounting for optimal input adjustment. In general, an approach that ignores adaptations available to the decision maker will over-estimate losses and under-estimate gains from changes in environmental circumstances. Although the Garcia *et al.* paper implicitly accounts for input adjustments, like most other analyses it ignores the possibility of mitigating the effects of high ozone levels by switching to other cultivars or other crops that are less sensitive to this pollutant.

Any advantage of the 'dual' approach due to accounting for input adjust-

[14]The damage function approach is also frequently adopted in assessing the damages to agriculture from global warming. In a study of the costs of global warming to agriculture in the US, Mendelsohn, Nordhaus and Shaw (1994) argue that the damage function approach is equivalent to the 'dumb farmer', because it assumes that farmers do not know enough to adjust their inputs. These authors assume that farmers can adjust fully and instantaneously to changes in climate, leading to estimates of global warming damages that are lower than the damage function approach produces. Quiggin and Horowitz (1999), in commenting on the damage estimates for global warming, note that the damage function approach and the Ricardian or instantaneous adjustment approach are on opposite extremes in terms of adjustment costs. The damage function assumes that adjustment costs are infinite, but instantaneous adjustment assumes that adjustment costs are zero. Estimates using these different assumptions may bound true welfare measures.

ments can easily be lost if the effects of changes in the environmental input are not accurately perceived by farmers and taken into account in their production decisions. Some authors have argued that while environmental inputs may affect average yields, individual farmers may be incapable of detecting systematic differences due to the environmental variable of interest when viewed against a backdrop of other causes of uncontrolled variation in yields. A comparison of the damage function and profit function approaches clearly turns on the likelihood and importance of input adjustments and the complexity of the relationship between the environmental variable and yields.

There is another shortcoming in analyses such as Garcia *et al.*'s. Given that changes in ozone standards would likely change ambient ozone levels nationally, their implicit assumption of constant prices is probably unwarranted. Adams *et al.* (1985, 1986) and Kopp *et al.* (1985) incorporate the yield response to changing ozone levels into a market framework in which predicted changes in output are reflected in shifting supply curves, inducing changes in output price. These authors calculate both the consumer surplus and changes in quasi-rents associated with changing ozone levels.

This method of calculating welfare effects is appropriate in a perfectly competitive market, but McGartland (1987) points out that the US grain market is far from that. Agricultural price support and deficiency payment policies existed during the period of these analyses, introducing distortions into the market. That is, for some crops the government paid the difference between a target price and market price to farmers, while for others a price floor was set and the government bought up sufficient stocks so that market price did not drop below that floor. With these sorts of interventions, price-quantity pairs along the aggregate supply curve differ from price-quantity pairs consistent with aggregate demand and any attempt to define demand and supply curves in the usual way will produce incorrect estimates of these curves. Welfare measurement requires the separate recovery of the market demand and supply functions, as well as careful treatment of government subsidy payments. Presumably any increase in the latter represents a loss to taxpayers.[15] McGartland shows that ignoring these distortions vastly overestimates the gains from ozone reduction. A subsequent paper by Kopp and Krupnick (1987) further argues that the benefits from environmental regulation of ozone will depend on the changing nature of agricultural policy.

---

[15] If the subsidies take the form of income supports this will be more straightforward than if the policy involves price supports. The latter generates increases in stocks of surplus agricultural products that cannot be easily valued. For a complete treatment of the welfare effects of agricultural policy instruments see Just, Hueth and Schmitz (2004), chapter 8.

9.5.2 Welfare Effects in Fisheries

An obvious pathway through which changes in environmental quality or ecosystem health affects humans is through fishery stocks. Declines in water quality can reduce survival rates of fish at various life-cycle stages and can change the mix of species in an area. Degradation of habitat, such as coral reefs, mangrove swamps, or wetlands, can lead to similar effects. Evidence that nitrogen deposition is a major contributor to nitrogen loadings in the Chesapeake Bay and acid rain a major factor in making aquatic life unsupportable in Adirondack lakes suggests that even air quality has a significant effect on fish. Given that environmental quality will be a factor in fish production, how does one best measure the welfare effects of changes in quality that arise through this pathway? In this section the problems that arise in measuring welfare effects in fishery stocks are discussed. We illustrate using a few papers that have attempted to measure the welfare effect of changes in the amount of wetlands and other types of fish spawning and nursery grounds.

In one of the earliest attempts to measure the welfare effects of changes in wetlands, Ellis and Fisher (1987) drew on work by Lynne, Conroy, and Prochaska (1981) that investigated the effect of wetlands on the blue crab fishery off the Florida Gulf Coast. Borrowing parameter estimates from the latter, Ellis and Fisher derived an industry supply function from an assumed Cobb-Douglas production function where harvest is a function of fishing effort and wetlands acreage. Assuming a constant elasticity market demand function, equilibrium is specified as the price at which industry marginal cost equals demand. With these two functions in hand, we measure the welfare effect of a change in wetlands acreage as an area such as $ABCD$ in Figure 9.8.

Freeman (1991) argues that the model set up by Ellis and Fisher was essentially incorrect for an open access fishery. Given the usual result that entry occurs in open access fisheries until all resource rent is dissipated, moving from one steady state equilibrium to another should generate no change in returns to fishermen. The only welfare consequence from a change in wetlands, Freeman argues, will be the change in consumer surplus, where this is calculated as the change in the area behind the demand curve and above price. This argument together with the characterization of equilibrium as the point at which price equals industry average, not marginal, cost led Freeman to challenge the Ellis and Fisher results and recalculate welfare effects.[16]

The above problem arises because interdependencies link fishermen's produc-

[16] A similar argument is presented by McConnell and Strand (1989), who note that pollution control aimed at increasing the productivity of commercial fisheries will have zero value in the long run with open access because of the zero rent condition. When there are price effects from the increased productivity, there will be gains in consumer surplus.

tion functions. Fishermen harvest the same stock and each fisherman's harvest lowers current stock for everyone. It also affects future stocks, if 'recruitment' is stock dependent (i.e. if spawning and survival rates are a function of stock size.) If the amount of effort needed to harvest a given amount of fish is a function of the size of the stock, which it will be for any stock subject to a sufficient amount of fishing effort, then each fisherman's harvest has an impact on the production function (and therefore the cost function) of others–both today and in the future.

In the standard model of the fishery, we assume that fishermen are profit maximizers just like any other type of firm. Typically fishing effort is viewed as the only variable input and can be purchased at a fixed price per unit. The resource stock is a fixed input, but as it is not owned in an open access fishery, no rent is paid for its use. Starting from an equilibrium in which there are zero excess profits to the competitive industry, either increases in price (due to shifts in demand) or decreases in costs (due to exogenous improvements in fish stocks) will induce the entry of new fishing effort. An increase in aggregate effort in the fishery will shift upward each fisherman's cost curve and will continue until any potential gains in quasi-rents from the change are eliminated.

Because of the importance and nature of intertemporal stock dependence in the fishery story, most models cast the problem in terms of long run steady state equilibrium. Freeman is implicitly assuming this when he states that price will equal industry average cost rather than marginal cost. Moving between two steady state equilibria means moving between solutions in which firm interdependencies are accounted for. The resulting industry level cost functions will reflect the long run path along which the industry can expand. This path is described by price equaling average cost (given that all firms are identical) because it assumes that entry will shift individual firm's cost functions and will continue until all excess profits and resource rents are driven to zero.[17] Freeman's measure of the welfare effect of a degradation in the environmental input is depicted in Figure 9.12 where $D(p)$ is the industry demand curve and $AC(q^i)$ is the industry average cost function evaluated at environmental quality equal to q^i. The welfare effect of a change in q from q^0 to q^1 is denoted by *area* p^0ABp^1. This measure includes only the change in consumer surplus because in the steady state the open access fishery generates no excess profits or rents. When the demand for fish from this fishery is perfectly elastic, this result implies that the social value of the fish stock is zero. This means that environmental changes that damage the fish stock as well as those that improve

[17] A similar situation would occur, although for a different reason, if we were to evaluate the effects of some exogenous change on an industry that was characterized by constant returns in the long run. Quasi-rents would not exist because of free entry and the absence of any fixed factors.

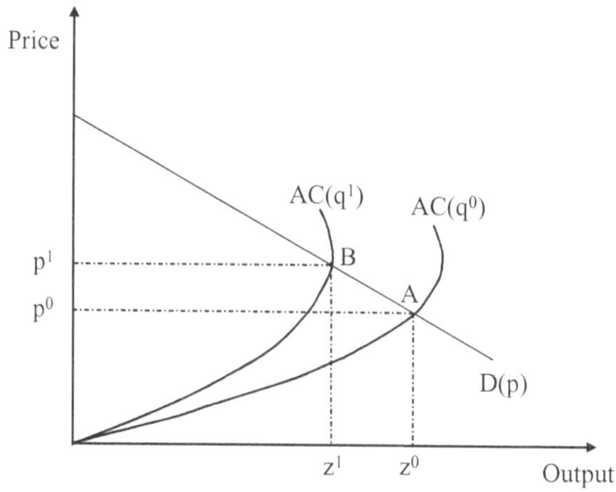

FIGURE 9.12. Welfare Measurement for Steady State Equilibrium in an Open Access Fishery

the fish stock have no social cost or value through the fishery. Naturally if there are other uses of the fish stock, including its support of ecological functions, this result does not hold.

If we are to analyze the fishery problem in terms of steady state equilibria, then the static Cobb-Douglas production function used by Ellis and Fisher is inappropriate. Stock dependent recruitment implies that future catch rates will be a function of current stock levels and current catch rates. Empirical specifications consistent with dynamic models that describe the intertemporal connections between stocks and harvests would seem far more desirable. Most of these view changes in stocks as having the following general form:

$$X_{t+1} - X_t = F(X_t) - h(X_t, E_t)$$

where X_t is the biomass of the fish stock in period t, $F()$ describes how stocks change in the absence of harvesting, and $h()$ expresses harvest as a function of stocks and fishing effort, E_t, in period t.

Some researchers have given more structure to the problem. For example Barbier, Strand, and Sathirathai (2002) specify a Schaefer type model in which $F(X)$ is a logistic growth function of the form: $F(X) = rXk - rX^2$, and harvests are given by $h = aXE$. In these models r is the intrinsic growth rate and k the environmental carrying capacity, while a denotes a catchability coefficient. An environmental factor can be introduced into this model as affecting either the intrinsic growth rate or the carrying capacity or both. Kahn (1987) has suggested that pollutants might more likely affect the former, while changes in habitat might affect the latter.

Barbier *et al.* investigate the welfare effects associated with the loss of mangrove swamps that serve as nursery and spawning grounds for coastal fisheries in the Gulf of Thailand and the Andaman Sea. They introduce mangrove swamp area into the model through the carrying capacity parameter, k, and ultimately estimate a derived steady state model of harvests as a function of mangrove area and fishing effort. Given that this fishery is open access, welfare effects of changes in mangrove area are calculated as changes in consumer surplus only, where a constant elasticity demand function is assumed and welfare estimates are calculated for different elasticities. They show that losses due to mangrove destruction are greater the less elastic is demand.

Ignoring quasi-rents in the calculation of welfare effects depends on the assumption that the fishery is in long run equilibrium and that all fishermen are alike. In the short run, before entry/exit adjusts to changing conditions, there would likely be transitory gains or losses to producers.[18] One might think of models such as Ellis and Fisher's as being short run models, but the problem with that interpretation is that the effects of changes in the environmental input (particularly when it is spawning and nursery habitat) are unlikely to be immediate, so that the short run has little meaning. To our knowledge, however, no welfare analyses have attempted to capture the adjustment process either in terms of firm entry/exit or stock adjustment to changing ecological conditions.

9.6 Conclusions

Valuing changes in environmental services that influence enterprises producing goods and services, whether households or firms, is an underdeveloped area in environmental economics. This is principally because the institutional structure for valuation has arisen in western Europe and North America where valuation is used for policy, rule making and litigation. In these regions, environmental changes affect households in activities that are not chiefly oriented towards employment and production. In the developing world, where reliance on environmental and natural resources is greater, improvements in valuation methods that measure the cost of pollution to firms and production-oriented households would be particularly rewarding.

Valuation of environmental changes that influence production is especially sensitive to the institutional structure. How a resource is managed or how an

[18]There could be quasi-rents in an open access fishery, even in the long run, if fishermen differ in skills. These intra-marginal rents would accrue to earlier entrants into a fishery, perhaps because of more experience. There is no reason to expect that a change in environmental circumstances would necessarily *change* those rents to experience, however.

industry is regulated will affect welfare measurement procedures and impact in a systematic way the size of welfare effects. This will be true in all types of economies. Yet the greatest need for welfare analysis may arise in developing country settings where households play the role of both consumers and producers and operate in institutional settings where labor markets and/or land markets are imperfect. These circumstances tend to involve households that are poorer and more vulnerable to environmental changes, but potentially able to be aided by policies that recognize the nature of their losses.

Chapter 10

Some Broader Considerations

10.1 The Territory Covered

A book on conceptual issues in revealed preference approaches to valuation does not lend itself to linear reading from beginning to end. Nevertheless, we conclude the book with some ideas about more general issues surrounding valuation, and some thoughts about challenging and potentially fruitful research directions.

Our approach in each chapter has been to connect the welfare measure that researchers typically want with observed behavior. By proceeding in this manner, we have covered the most prominent revealed preference approaches to environmental valuation. Two points emerged from this discourse that we would like to reiterate. The first concerns conventional practice in welfare economics. Economists are likely to think of welfare measures as derived from the standard prices-as-parameters model. This practice, which connects utility functions, expenditure functions, and indirect utility functions so neatly with demand functions, is not often a good guide for devising welfare measures when the change is some dimension of the environment or some type of public good. Rather, when changes in circumstances involve non-price influences, one must search for plausible restrictions on preferences that will allow behavior to reveal or approximate the conceptually correct welfare measure.

The second theme concerns marginal values. For almost all valuation problems, it is easy to derive the marginal value of a public good, and in many such cases, it may even be possible to connect the marginal value directly with the slope of an estimated behavioral function. This has led some researchers to compute marginal values in benefit cost exercises, especially those depending on hedonic techniques or averting behavior models. In the case of hedonic wage models, the marginal value of risk would seem the exact measure needed. But many policy actions or exogenous events induce non-marginal changes in outcomes. In cases where marginal values are not constant, which is the situation in most economic problems, multiplying a marginal value by a discrete underlying change will typically produce conceptually misleading and inaccurate welfare assessments.

10.2 When *Not* To Do Valuation

Throughout the book we have addressed valuation as if the purpose were clear. Valuation has a clear role in correcting market failures. When there is no market failure, it is not in society's interest to devote resources to valuation.[1] This is obvious and noncontroversial. When the rationale for valuation does exist, the undertaking must also be responsive to a well-formed question. This almost always relates to *changes* in policies or to exogenous events that change the amount and/or quality of environmental goods and services. The most obvious uses of environmental valuation include the evaluation of regulatory policy and damage assessment. Estimation of individual and aggregate values is required to evaluate rule making in the design phase or in assessing the effectiveness of past policy. Environmental valuation is essential for the estimation of compensation for natural resource damage cases.

Employing valuation where it is not answering a well-formed question will not in general lead to useful results. A good case of the misuse of valuation occurs in the several attempts made by non-economists to value the services of the world's ecosystems. Notable among these is the frequently sited paper by Costanza, d'Arge, de Groot, Farber, Grasso, Hannon, Limburg, Naeem, O'Neill, Paruelo, Raskin, Sutton and van den Belt (1997).[2] The aim of this paper is to answer an ill-formed question: what is the value of the world's ecosystems? To answer such a question we would need to compare the current state of the world with a well-defined description of what the world would look like in the absence of these ecosystems. It takes little imagination to see the impossibility

[1]However, there is substantial, and no doubt more profitable, opportunity for private firms to employ some valuation tools in market research. See for example Louviere, Hensher and Swait (2000) who have used conjoint analysis for a variety of market-related tasks.

[2]Among the numerous critiques of this paper are Bockstael, Freeman, Kopp, Portney, and Smith (2000), Freeman (2002), Pearce (1998), Smith (1997b), and Toman (1998). Other ecologists' attempts at valuation include Ehrlich and Ehrlich (1996), and Pimentel, Wilson, McCullum, Huang, Dewn, Flack, Tran, Salmon, and Cliff (1997).

of this task and a great deal to imagine an economic world without ecosystems. If nothing replaces current ecosystems (whatever that notion of 'nothingness' might imply), then it is fairly safe to say that individuals would be willing to pay everything they had to avoid it and could not be compensated by any finite amount to be willing to accept this change. These are trivial answers to a meaningless question.

The authors are able to come up with empirical 'measures' in apparent answer to this question by taking from the existing literature value estimates of small localized changes in given ecosystem services and multiplying these implied losses per area by the world's total area of that type of ecosystem. For example, a previous study estimated the loss of several hectares of a local wetland at \$15,000/hectare. These results were then used by Costanza *et al.* to value the world's 330 million hectares of wetlands at about \$5 trillion ((\$15,000×330 million hectares). Ultimately aggregated values of many different types of ecosystems, each extrapolated in this way, are added together to get the authors' final figure of \$33 trillion/year for the world's ecosystems and natural assets. This is so obviously a misuse of these valuation estimates as to warrant little comment. As we discussed in the context of marginal valuation, the welfare effects of small changes cannot be simply scaled up to measure large changes and values for the loss of one localized ecosystem could never be simply added to that of another, because the initial estimates are conditioned on the existence of these substitutes. From an ecological perspective this procedure is equally senseless, because concepts of substitution and interdependence are as relevant in biological as economic regimes.

Even if these technical problems were satisfied, efforts to value the services of the world's ecosystems answer no plausible policy or compensation question. One would find virtual unanimity that if the world's ecosystems were to be destroyed, the amount of compensation that would be required for the world's population would be infinite. The world's economies would reasonably be expected to collapse, making measures in any monetary system irrelevant. Further, no policy change would ever be considered that had this expected outcome and few events can be imagined, short of a nuclear holocaust, that would wipe out all the world's ecosystems.

There are other environmental valuation questions which may be well-formed welfare questions *in concept* but offer little hope of being empirically measurable. When the environmental change is sufficiently small, we can have little confidence in the precision of valuation analysis. In such cases, trust in empirical results will be misplaced. The notion of 'too small' an effect may be relative and depend on the valuation approach used. For example, a change in recreational fishing catch rates may induce fishermen to alter their recreational trip behavior and thus be valued in a recreational demand model, but variation in catch rates are unlikely to be detectable through analysis of housing markets, even though some avid fishermen may choose their housing location on this basis. These sorts of mistakes are especially dangerous in hedonic type models where omitted variable bias is always a threat. For many reasons, regional fish catch patterns may be sufficiently correlated with things that do affect buyers to produce *statistically* significant results even though they have no reasonable behavioral basis. Statistical significance must be combined with good economic sense to provide convincing evidence about the value of attributes.

10.3 Stated or Revealed Preference?

In many situations both stated and revealed preference approaches can be fruitfully used. These are typically circumstances in which individuals actually use the resources or respond to the environmental changes in question, and a means of recovering observations on this use or response exists. Carson, Flores, Martin and Wright (1994), in a comprehensive effort to compare estimates of values for changes in public goods, find 83 studies that estimate values from both revealed and stated preference approaches. The Carson *et al.* survey provides a good illustration of the possibility of using both methods for a given valuation question.

There are clearly cases where only stated preference approaches can be used, however. The most obvious are those that involve non-use (or 'passive' use) value–that is, an individual's compensating variation for a change in the status of a resource that is enjoyed and appreciated without *in situ* use.[3] The Exxon Valdez oil spill in Prince William Sound provides an illustration. Although individuals used Prince William Sound for kayaking, fishing, and wildlife watching, the numbers of such recreators were tiny compared with the numbers of individuals who cared about the health of the untouched natural environment. The lion's share of damage estimates was derived from households in the lower 48 states, almost all of whom had never, nor would ever, visit Alaska, much less the more inaccessible Prince William Sound. The ultimate case rested on values obtained from stated, not revealed, preference studies (see Carson, Mitchell, Hanemann, Kopp, Presser, and Ruud, 1992).

In occasional circumstances, it may be impossible to deduce use value from behavior if the resource has been sufficiently degraded as to discourage use. This is a particular drawback when attempting to value the benefits of clean-up when no recent history of use is available. A good example of this arose in

---

[3]It may be that the demand for certain existence-type goods can be partially revealed through market transactions when consumers buy higher-priced goods with ecolabels. An example is dolphin-safe tuna. The difficulty with this approach is that its incentive properties are not clear, resembling most closely the voluntary contribution mechanism for public goods.

the PCB contamination of New Bedford harbor. This would have been an ideal opportunity for applying a discrete choice model of beach choice, had it not been for the historic pollution of the beaches by waste water from sewage treatment plants. In the absence of recent use, researchers were forced to adopt a form of stated preference that elicited changes in hypothetical behavior (McConnell, 1986).

This discussion should not leave the impression that stated preference techniques can be used in any situation. It is true that stated preference approaches can be employed to estimate both use and non-use values, while there is no role for revealed preference approaches in measuring the latter. Yet caution is warranted. Stated preference methods may reveal how much people value a well-defined, initial exogenous change, but that is not always the complete answer to the policy or damage assessment question. A change that induces behavioral adjustment may have more complicated outcomes. The behavioral changes themselves might induce market effects or might themselves change the quality of the resource. In such cases, stated preference approaches cannot be expected to yield correct valuation results. Revealed preference approaches, based on models that explain changes in behavior, can be adapted to produce the necessary information for valuation.

The most obvious example comes from the housing hedonic discussion in Chapter 6. The response to a stated preference question would give an answer equivalent to what we have called 'the pure willingness to pay' effect. However, this is not always the correct answer to the valuation problem because markets adjust (and prices change) as individuals attempt to alter their behavior in response to the exogenous stimulus. Equally obvious are examples such as that provided by Foster and Just (1989). The contamination of the Hawaiian milk supply led to behavioral changes and subsequent induced price changes that must be taken into account in assessing the true welfare effects of the contamination incident.

Behavioral changes can induce further effects even in the absence of markets if there are interdependencies among individuals. For example, an improvement in the water quality at a beach will cause more people to use it and may ultimately lead to a less desirable experience than expected because of increased congestion. Without knowing how demand responds to water quality, the levels of congestion cannot be predicted. Similarly, an increase in water quality might increase fish stocks, but subsequent increases in recreational fishing effort may deplete stocks, reducing catch rates, and limiting the ultimate gains to fishermen. Stated preference methods that consider individual responses to the water quality question will overestimate the value of these improvements.

There is at least one more setting in which stated preference techniques can encounter difficulties that are overcome (at least partially) in revealed preference analysis: when the environmental good or service in question does not enter preference functions directly. Chapters 8 and 9 covered cases in which the environmental good served as an input into production. Households that use the input in household production activities may be able to take into account the household production process and the availability of substitutes in answering direct valuation questions, but households are unlikely to be able to give meaningful answers to stated preference questions relating to firm production processes. Asking the average individual how much he would value a reduction in soil erosion in Kansas would, of course, be nonsensical, since the primary reason this is likely to matter to him is if reductions in soil erosion lead to changes in food prices. But the economist working with an agronomist is far better suited to answer this question, by modeling the positive economics of the problem and then evaluating market welfare effects. This complaint carries over to problems in which ecosystem services are threatened. Asking an individual his value for an increase in wetlands habitat, for example, is likely to produce a confused response, since the individual is unlikely to understand the complex biology and economics that connects wetland habitat to fish reproduction to fish harvests to market prices and will surely be ignorant of the role of a given wetland in providing less obvious but potentially more important ecosystem services. It is not unreasonable to assume that individuals can value some of the services of wetlands, but there is no basis for believing that individuals know how ecological processes produce these services.

The economics literature contains 'ill-advised' examples of both stated and revealed preference analysis. The former tend to be problems in which the representative respondent can not be expected to have the proper knowledge about the resource in question and whose feelings about it are irrelevant. The latter arise when researchers rely on spurious statistical significance despite the lack of a convincing story about how people know about the environmental effect and why they change their behavior in response.

10.4 Some Concluding Thoughts

The principal goal of the book has been to elucidate models that connect behavior with welfare measures. We have not dwelt on some of the types of applications that have posed challenges to existing revealed (and often stated) preference approaches. To conclude the book, we mention two of these valuation problems that will require advances in the methods.

10.4.1 Behavioral Economics

Over the past several decades, researchers have amassed a significant body of evidence that challenges the standard neoclassical model of consumer and producer behavior. This research has formed a new field, behavioral economics.[4] The initial investigations in this area were experimental but recently field evidence has provided further support for the idea that economic agents are not always rational optimizers. The field has developed into a significant research area in two ways. First, there is recognition that there are several broad areas of economic behavior and phenomena that contradict the neoclassical model. These areas include time discounting, decisions under uncertainty, reference dependence and labor market outcomes.[5] Second economists have constructed more general theories that explain non-neoclassical outcomes but accommodate neoclassical preferences.

The systematic development of behavioral economics seems likely to provide a strong challenge to valuation. Researchers using revealed preferences for measuring the value of environmental amenities have taken full advantage of the assumption that households are cost-minimizing, utility-maximizing agents who plan well and react rationally. This assumption will require strengthening to withstand non-optimizing evidence of the sort found in behavioral economics. Perhaps more important, the use of benefit cost analysis is predicated on the idea of rational choice. Behavioral economic models may challenge this means of resource allocation as well as models of private choice. We consider a few examples of non-neoclassical behavior that could play a role in valuation studies.

A discrete choice model estimated under the assumption that the only systematic influence on choice is the set of amenities for each alternative uses this assumption to assess the welfare effects of changes in amenities. However, if the choice problem is frequently repeated and an 'anchoring' effect exists, welfare measurement that implicitly assumes complete reoptimization on every choice occasion may produce biased estimates of the gains or losses from changes in amenities. While it may be difficult statistically to discriminate between anchoring and habit formation, both concepts pose a challenge to the simplest model of choice among alternatives.[6] In making discrete choices, households change behavior less frequently than models predict.

The distinction between subjective and objective measures of amenities may be further distorted by the tendency to inflate small probabilities and compress

[4] These developments are surveyed in the essay by Camerer, Lowenstein and Rabin (2004).

[5] Within environmental economics, researchers are likely to be familiar with the idea of reference dependence and the closely related idea of status quo bias, which leads to the observed discrepancy in willingness to pay and willingness to accept in experimental and stated preference settings.

[6] Adamowicz (1994) developed a model of habit formation for discrete choice settings. Because current choices influence future preferences when habit formation prevails, this phenomenon poses significant challenges to welfare analysis.

large ones. When the amenity is a risk of exposure to a hazardous substance, for example, a household might overreact to a risk with a small reported probability of a health effect and under-react to one with a large probability. When welfare evaluation entails finite changes in risk, the difference between an expected utility model that is linear in probabilities and an alternative that employs a probability weighting function can be substantial. The impact of changes in risk will depend strongly on the baseline risk.

The phenomenon of 'choice bracketing' refers to the breadth of alternatives considered for a given choice. Results from behavioral economics suggest that individuals may keep separate mental accounts of spending, for example. Hence money used for clothing might not be considered fungible with money used for recreational activities. In contrast, standard welfare economics assumes that there is no constraint on income allocation and no phenomenon such as choice bracketing. In an attempt to take account of this empirically-observed phenomenon, Baerenklau and Provencher (2005) argue for a 'mental accounts' approach to the seasonal choice of recreational trips by assuming that the time spent on trips is fixed, and that individuals may substitute to change the distribution of those trips over time but not substitute the recreational activity for some other type of activity. Changes in underlying assumptions of this sort should have implications for welfare calculations.

The prevalence of these and other behavioral anomalies is well known. Yet researchers have not investigated their implications for revealed preference models nearly as much as for stated preference approaches. Accommodating behavior that appears to contradict the axioms of rational choice is likely to be an important future task for revealed preference research.

10.4.2 Valuing Ecosystem Services

The criticisms by environmental economists of the Costanza *et al.* effort, such as those outlined in an earlier section, should not be interpreted as a lack of appreciation for the importance of ecosystems. Increasingly economists and ecologists are working together to improve our understanding of the interdependencies between ecological and economic systems–each of which is a complex and adaptive system driven by interacting agents adjusting their 'behavior' at multiple spatial and temporal scales. In such a world, the pathways through which changes in ecosystems alter human well-being are difficult to define.

Finding meaningful ways to value changes in ecosystem services and functions has been a high priority at the U.S. EPA for some years. Yet environmental valuation has not advanced much beyond the default practice of valuing ecosystem services piece-meal, focusing on only those outcomes that we already have conventional techniques to address (such as fishery productivity or storm water protection). The limited success in 'ecosystem valuation' has two basic causes.

First, measuring the welfare effects of an insult to an ecosystem requires understanding a good deal more about ecosystem functioning than we do. It also requires understanding how human behavior responds to ecosystem changes and how economic and ecological systems interact. These are difficult questions but not initially welfare questions. They are really problems that require good ecological science and 'positive' economics. Indeed if the problem is to value conventional outcomes such as fish, timber, water quality, etc., that arise from complex ecosystem interactions, then the real challenge in ecosystem valuation is understanding the joint system dynamics, not in rethinking methods for estimating welfare effects.

There is a second dimension to the valuation challenge. Recent work by ecologists and economists (e.g. Levin, 1998; Carpenter, Brock and Hanson, 1999; Ludwig, Walker and Holling, 1997; Dasgupta, Levin, and Lubchenco, 2000) suggest that the most important aspects of an anthropogenically induced disturbance to an ecological system may be its effect on important properties of the system. Resilience is one such property, describing a system's ability to absorb disturbance without fundamental change. Systems tend to be more resilient the more heterogeneous and redundant they are, where heterogeneity refers to diversity of species and redundancy is related to multiple species fulfilling similar functions. These features increase the ability of the system to 'substitute' away from compromised individuals and therefore sustain their current functions. The most important effects of policies or exogenous events on ecosystems may therefore be a change in system properties such as resilience. Persistent pollution, even at low levels, may change system properties in ways that have no immediate outcomes for humans, but change the probabilities of future catastrophic consequences by making the system less resistant to collapse in the face of even minor perturbations. The combination of high levels of uncertainty, path dependence, and the possibility of small changes causing catastrophic events, makes this type of problem difficult to conceptualize in the current welfare effects framework.

References

Abdalla, C.W., B.A. Roach and D.J. Epp. 1992. 'Valuing environmental quality changes using averting expenditures: an application to groundwater contamination', *Land Economics*, **68**: 163-69.

Acharya, G. 1998. 'Valuing the environment as an input: the production function approach', in M. Acutt and P. Mason, eds., *Environmental Valuation, Economic Policy and Sustainability*, Cheltenham, UK: Edward Elgar.

Adamowicz, W.L. 1994. 'Habit formation and variety seeking in a discrete choice model of recreation demand', *J. of Agricultural and Resource Economics*, **19**: 19-31.

Adams, R., S. Hamilton and B. McCarl. 1986. 'The benefits of pollution control: the case of ozone and U.S. agriculture', *American J. of Agricultural Economics*, **67**: 886-893.

Adams, R. and B. McCarl. 1985. 'Assessing the benefits of alternative ozone standards on agriculture: the role of response information', *J. of Environmental Economics and Management*, **12**: 264-276.

Agee, M.D. and T.D. Crocker. 1996. 'Parental altruism and child lead exposure', *J. of Human Resources*, **31:** 677-91.

Akerman, J., F.R. Johnson and L. Bergman. 1991. 'Paying for safety: voluntary reduction of residential radon risks', *Land Economics*, **67:** 435-46.

Alberini, A. and A. Krupnick. 2000. 'Cost of illness and WTP estimates of the benefits of improved air quality in Taiwan', *Land Economics*, **76**: 37-53

2004. 'Valuing the health effects of pollution' in T. Tietenberg and H. Folmer, eds., *The International Yearbook of Environmental and Resource Economics* 2002/2003, Cheltenham, UK: Edward Elgar.

Allen, S. 1995. 'Updated notes on the interindustry wage structure, 1890-1990', *Industrial and Labor Relations Review*, **48:** 305-321.

Anderson, D. 1987. *The Economics of Afforestation,* Baltimore, MD: Johns Hopkins Press.

Arrow, K. 1951. *Social Choice and Individual Values,* New Haven: Yale University Press for Cowles Foundation.

Arrow, K., R. Solow, P. Portney, E. Leamer, R. Radner and H. Schuman. 1993. 'Report of the NOAA panel on contingent valuation', *Federal Register*, **58**: 4601-4614.

Baerenklau, K.A and B. Provencher. 2005. 'Static modeling of dynamic recreation behavior: implications for prediction and welfare estimation', *J. of Environmental Economics and Management*, **50**: 617-636.

Bailer, A.J., L.T. Stayner, N.A. Stout, L.D. Reed and S.J. Gilbert. 1998. *Occupational and Environmental Medicine*, 'Trends in rates of occupational fatal injuries in the United States', **55**: 485-489.

Banzhaf, S. 2003. 'Hedonic pricing in realistic urban structures, or what if Tiebout called and nobody sorted?', working paper, Resources for the Future.

Barbier, E., I. Strand and S. Sathirathai. 2002. 'Do open access conditions affect the valuation of an externality? estimating the welfare effects of mangrove-fishery linkages in Thailand', *Environmental and Resource Economics*, **21**: 343-367.

Bartik, T.J. 1988a. 'Evaluating the benefits of non-marginal reductions in pollution using information on defensive expenditures', *J. of Environmental Economics and Management*, **15**: 111-127.

1988b. 'Measuring the benefits of amenity improvements in hedonic price models', *Land Economics*, **88:** 172-183.. Mills ,ed., *Handbook on Urban Economics,* Amsterdam: North Holland Publishers.

Bateman, I., R. Carson, B. Day, W.M. Hanemann, N. Hanley, T. Hett, M. Jones-Lee, G. Loomes, S. Mourato, E. Özdemiroglu, D. Pearce, R. Sugden and J. Swanson. 2002. *Economic Valuation with Stated Preference Techniques,* Northampton, MA: Edward Elgar.

Bayer, P., N. Keohane, N. and C. Timmins. 2005. 'Migration and hedonic valuation: the case of air quality', Working paper, Duke University, (http://www.econ.duke.edu/~timmins/cv.html).

Bayer, P. and C. Timmons. Forthcoming. 'Estimating equilibrium models of sorting across location', *Economic J.*

Becker, G.S. 1965. 'A theory of the allocation of time', *Economic J.*, **75:** 493-517.

1981. *A Treatise on the Family.* Harvard University Press.

Ben-Akiva, M. and S.R. Lerman. 1985. *Discrete Choice Analysis: Theory and Application to Travel Demand*, Boston, MA: MIT Press.

Benjamin, D. 1992. 'Household composition, labor markets and labor demand: testing for separation in agricultural household models', *Econometrica*, **60**: 287-322.

Biddle, J.E. and G.A. Zarkin. 1988. 'Worker preference and market compensation for job risk', *Review of Economics and Statistics*, **70:** 660-667.

Black, D.A., J. Galdo and L. Liu. 2003. 'How Robust are Hedonic Wage Estimates of the Price of Risk?' Center for Policy Research, Syracuse University, Syracuse, NY. Report presented to National Center for Environmental Economics, Office of Policy Economics and Innovation, USEPA.

Black, D. and T. Kniesner. 2003. 'On the measurement of job risk in hedonic wage models', *J. of Risk and Uncertainty*, **27**: 205-220.

Blackburn, M. 1995. 'Decomposing wage variation: a comment on individual heterogeneity and interindustry wage differentials', *J. of Human Resources*, **30**: 853-860.

Blackburn, M. and D. Neumark. 1992. 'Unobserved ability, efficiency wages and interindustry wage differentials', *Quarterly J. of Economics*, **107**: 1421-1436.

Blomquist, G., M. Berger and J. Hoehn. 1988. 'New estimates of quality of life in urban areas', *American Economic Review*, **78**: 89-107.

Bockstael, N.E., A. Freeman, R. Kopp, P. Portney and V.K. Smith. 2000. 'On valuing nature', *J. of Environmental Science and Technology*, **34**: 1384-1389.

Bockstael, N.E., W.M. Hanemann and C. Kling. 1987. 'Modeling recreational demand in a multiple site framework', *Water Resources Research*, **23**: 951-60.

Bockstael, N.E., M. Hanemann and I.E. Strand. 1986. 'Measuring the Benefits of Water Quality Improvement Using Recreation Demand Models', Environmental Protection Agency, Cooperative Agreement CR-81143-01-1.

Bockstael, N.E. and C.L. Kling. 1988. 'Valuing environmental quality changes when quality is a weak complement to a set of goods', *American J. of Agricultural Economics,* **70**: 654-662.

Bockstael, N.E. and K.E. McConnell. 1983. 'Welfare measurement in the household production framework', *American Economic Review,* **83**: 806-814.

1993. 'Public goods as characteristics of non-market commodities', *Economic J.,* **103**: 1244-1257.

Bockstael, N.E. and I.E. Strand. 1985. 'Distributional issues and nonmarket benefit measurement', *Western J. of Agricultural Economics,* **10**: 162-169.

Bockstael, N.E., I.E. Strand and W.M. Hanemann. 1987. 'Time and the recreation demand model', *American J. of Agricultural Economics,* **69**: 293-302.

Bojø, J. 1990. 'Benefit-cost analysis of the farm improvement with soil conservation project in Mahputseng, Mohale's Hoek District, Lesotho' in J. Dixon, D. James and P. Sherman, eds., *Dryland Management: Economic Case Studies*, London: Earthscan Publications, Limited.

Bourguignon, F. and P.A. Chiappori. 1992. 'Collective models of household behavior: an introduction', *European Economic Review*, **36**: 355-364.

Boyle, K.J., P.J. Poor and L.O. Taylor. 1999. 'Estimating the demand for protecting freshwater lakes from eutrophication', *American J. of Agricultural Economics*, **81**: 1118-22.

Breffle, W.S. and E.R. Morey. 2000. 'Investigating preference heterogeneity in a repeated discrete-choice recreation demand model of Atlantic salmon fishing', *Marine Resource Economics,* **15**: 1-20.

Brekke, K.A. 1997. 'The numeraire matters in cost benefit analysis', *J. of Public Economics*, **64**: 117-23.

Brekke, K.A., V. Iversen and J.B. Aune. 1999. 'Tanzania's soil wealth', *Environment and Development Economics,* **4**: 333-356.

Bresnahan, B. and M. Dickie. 1995. 'Averting behavior and policy evaluation', *J. of Environmental Economics and Management*, **29**: 378-392.

Brookshire, D.S., M.A. Thayer, W.D. Schulze and R.C. d'Arge. 1982. 'Valuing public goods: a comparison of survey and hedonic approaches', *American Economic Review*, **72**: 165-77.

Brookshire, D.S., M.A. Thayer, J. Tschirhart and W.D. Schulze. 1985. 'A test of the expected utility model: evidence from earthquake risks', *J. Political Economy*, **93**: 369-89.

Brown, G. and R. Mendelsohn. 1984. 'The hedonic travel cost method', *Review of Economics and Statistics*, **66**: 427-433.

Brown, J.M. and H. Rosen. 1982. 'On the estimation of structural hedonic price models', *Econometrica,* **50**: 765-768.

Browning, M. and P.A. Chiappori. 1998. 'Efficient intra-household allocation: a general characterization and empirical tests', *Econometrica,* **66**: 1241-1278.

Browning, M., P.A. Chiappori and A. Lewbel. 2004. 'Estimating consumption economies of scale, adult equivalence scale and bargaining power' Boston College Economics Working Paper 88.

Camerer, C. and G. Lowenstein. 2004. 'Behavioral economics, past, present, future' in Camerer, C., G. Lowenstein and M. Rabin, eds., *Advances in Behavioral Economics,* Princeton, NJ: Princeton University Press.

Carpenter, S., W. Brock, and P. Hanson. 1999. 'Ecological and social dynamics in simple models of ecosystem management', *Ecology and Society*, 3: online at: http://www.consecol.org/vol3/iss2/art4/.

Carson, R. forthcoming. *Contingent Valuation: A Comprehensive Bibliography and History*. Northampton, MA: Edward Elgar.

Carson, R., N. Flores, K. Martin and J. Wright. 1996. 'Contingent valuation and revealed preference methodologies: comparing the estimates for quasi-public goods', *Land Economics,* **72**: 113-128.

Carson, R., R. Mitchell, W. M. Hanemann, R. Kopp, S. Presser and P. Ruud. 1992. *A Contingent Valuation Study of Lost Passive Use Values Resulting From the Exxon Valdez Oil Spill.* Report to the Attorney General of the State of Alaska.

Cesario, F. 1976. 'Value of time and recreation benefit studies', *Land Economics,* **52**: 32-41.

Cesario, F. and J. Knetsch. 1970. 'Time bias in recreation benefits estimates', *Water Resources Research,* **6**: 700-704.

Chattopadhyay, S. 1998. 'An empirical investigation into the performance of Ellickson's random bidding model, with an application to air quality valuation', *J. of Urban Economics,* **43**: 292-314.

1999. 'Estimating the demand for air quality: new evidence based on the Chicago housing market', *Land Economics,* **75**: 22-38.

2000. 'The effectiveness of McFadden's nested logit model in valuing amenity improvement', *Regional Science and Urban Economics,* **30**: 23-43.

2002. 'Divergence in alternative Hicksian welfare measures: the case of revealed preference for public amenities', *J. of Applied Econometrics,* **17**: 641-666.

Chen, H. and S. Cosslett. 1998. 'Environmental quality preference and benefit estimation in multinomial probit models: a simulation approach', *American J. of Agricultural Economics,* **78**: 512-520.

Chen, H., F. Lupi and J. Hoehn. 1999. 'An empirical assessment of multinomial probit and logit models for recreation demand' in J. Herriges and C. Kling, eds., *Valuing Recreation and the Environment,* 141-161. Northampton, MA: Edward Elgar.

Clark, D. E., W. Herrin, T. Knapp and N. White. 2003. 'Migration and implicit amenity markets: does incomplete compensation matter?', *J. of Economic Geography,* **3:** 289-307.

Clark, D. E. and J. R. Kahn. 1989. 'The two-stage hedonic wage approach: a methodology for the valuation of environmental amenities', *J. of Environmental Economics and Management,* **16**: 106-120.

Clawson, M. 1959. 'Methods of Measuring the Demand for and the Value of Outdoor Recreation', Reprint No. 10, Resources for the Future.

Cooper, J. and J. Loomis. 1993. 'Testing whether waterfowl hunting benefits increase with greater water deliveries to wetlands', *Environmental and Resource Economics,* **3**: 545-61.

Cornes, R. 1992. *Duality and Modern Economics,* Cambridge: Cambridge University Press.

Costa, D.L. and M.E. Kahn. 2004. 'Changes in the value of life, 1940-1980. *J. of Risk and Uncertainty,* **29**: 159-180.

Costanza, R., R. d'Arge, R. de Groot, S. Farber, M. Grasso, B. Hannon, K. Limburg, S. Naeem, R. O'Neill, J. Paruelo, R. Raskin, P. Sutton and

M. van den Belt. 1997. 'The value of the world's ecosystem services and natural capital', *Nature,* **387**: 253-260.

Courant, P.N. and R.C. Porter. 1981. 'Averting expenditure and the cost of pollution', *J. of Environmental Economics and Management,* **8**: 321-329.

Cragg, M. and M. Kahn. 1997. 'New estimates of climate demand: evidence from locational choice', *J. Urban Economics,* **42**: 261-284.

Cropper, M. and A. Arriaga-Salinas. 1980. 'Inter-city wage differentials and the value of air quality', *J. Urban Economics,* **8**: 236-254.

Cropper, M.L., L.B. Deck and K.E. McConnell. 1988. 'On the choice of functional form for hedonic price equations', *Review of Economics and Statistics,* **70**: 668-75.

Cropper, M., L. Deck, N. Kishor and K.E. McConnell. 1993. 'Valuing product attributes using single market data: a comparison of hedonic and discrete choice approaches', *Review of Economics and Statistics,* **75**: 225-232.

Cummings, R. and G. Harrison. 1995. 'The measurement and decomposition of nonuse values: a critical review', *Environmental and Resource Economics,* **5**: 225-47.

Curtis, R. and R. Hicks. 2000. 'The cost of sea turtle preservation: the case of Hawaii's pelagic longliners', *American J. of Agricultural Economics,* **82**: 1191-1197.

Dasgupta, P. 2004. 'Valuing health damages from water pollution in urban Delhi, India: a health production function approach', *Environment and Development Economics,* **9**: 83-106.

Dasgupta, P., S. Levin and J. Lubchenco. 2000. 'Economic pathways to ecological sustainability', *Bioscience,* **50**: 339-45.

Deaton, A. and J. Muellbauer. 1980. *Economics and Consumer Behavior,* Cambridge: Cambridge University Press.

deJanvry, A., M. Fafchamps and E. Sadoulet. 1991. 'Peasant household behavior with missing markets: some paradoxes explained', *Economic J.,* **101**: 1400-1417.

Dickens, W.T. and L.F. Katz. 1987. 'Inter-industry wage differences and industry characteristics' in K. Lang and J. Leonard, eds., *Unemployment and the Structure of Labor Markets,* Oxford: Basil Blackwell.

Dickie, M. and S. Gerking. 1987. 'Interregional wage differentials: an equilibrium perspective', *J. of Regional Science*, **27**: 571–585.

1991, 'Willingness to pay for ozone control: inferences from the demand for medical care', *J. of Environmental Economics and Management*, **21**: 1-16.

1996. 'Formation of risk beliefs, joint production and willingness to pay to avoid skin cancer', *Review of Economics and Statistics*, **78**: 451-63.

Dillingham, A. and R.S. Smith. 1983. 'Union effects on the valuation of fatal risk' in B. Dennis, ed., *Proceedings of the Industrial Relations Research Association 36th Annual Meeting.*

Dixon, J.A., L.F. Scura, R.A. Carpenter and P. B. Sherman. 1995. *Economic Analysis of Environmental Impacts*, London, UK: Earthscan Publications.

Domencich, T. and D. McFadden. 1975. *Urban Travel Demand-A Behavioral Analysis*, Amsterdam: North Holland Publishers.

Dorman, P. and P. Hagstrom. 1998. 'Wage compensation for dangerous work revisited', *Industrial and Labor Relations Review*, **52**: 116-135.

Dorsey, S. and N. Walzer. 1983. 'Worker's compensation, job hazards and wages', *Industrial and Labor Relations Review*, **36**: 642-654.

Ebert, U. 1998. 'Evaluation of nonmarket goods: recovering unconditional preferences', *American J. of Agricultural Economics*, **80**: 241-254.

Eeckhoudt, L.R. and J.K. Hammitt. 2001. 'Background risks and the value of a statistical life', *J. of Risk and Uncertainty*, **23**: 261-279.

Ehrlich, P. and A. Ehrlich. 1996. *Betrayal of Science and Reason*, Washington, D.C.: Island Press.

Ekeland, I., J.J. Heckman and L. Nesheim. 2004. 'Identification and estimation of hedonic models', *J. of Political Economy*, Part 2 Supplement, **112**: S60-109.

Ellickson, B. 1981. 'An alternative test of the hedonic theory of housing markets', *J. of Urban Economics*, **9**: 56-79.

Ellis, G.M. and A.C. Fisher. 1987. 'Valuing the environment as input', *J. of Environmental Management*, **25**: 149-156.

Englin, J. and R. Mendelsohn. 1991. 'A hedonic travel cost analysis for valuation of multiple components of site quality: the recreational value of forest management', *J. of Environmental Economics and Management,* **21**: 275-290.

Eom, Y-S and D.M. Larson. forthcoming. 'Improving environmental valuation estimates through consistent use of multiple information', *J. of Environmental Economics and Mangement.*

Epple, D. 1987. 'Hedonic prices and implicit markets: estimating demand and supply functions for differentiated markets', *J. of Political Economy,* **95**: 59-80.

Epstein, L.G. 1981. 'Generalized duality and integrability', *Econometrica,* **49**: 655-678.

Evans, A.W. 1990. 'The assumption of equilibrium in the analysis of migration and interregional differences: a review of some recent research', *J. of Regional Science*, **30**: 515-531.

Feather, P. and W.D. Shaw. 1999. 'Estimating the cost of leisure time for recreation demand models', *J. of Environmental Economics and Management,* **38**: 49-65.

2000. 'The demand for leisure time in the presence of constrained work hours', *Economic Inquiry,* **38**: 651-61.

Feenberg, D. and E. A. Mills. 1980. *Measuring the Benefits of Water Pollution Abatement.* New York: Academic Press.

Fisher, F. and K. Shell. 1971. 'Taste and quality change in the pure theory of the true cost-of-living index' in Griliches, Z., ed., *Price Indexes and Quality Change*, Cambridge, MA: Harvard University Press.

Fleming, W.M. 1983. 'Phewa Tal catchment mangement program: benefits and costs of forestry and soil conservation in Nepal' in L.S. Hamilton, ed., *Forest and Watershed Development and Conservation in Asian and the Pacific,* Boulder, Co.: Westview Press.

Foster, W. and R.E. Just. 1989. 'Measuring welfare effects of product contamination with consumer uncertainty', *J. of Environmental Economics and Management,* **17**: 266-283.

Freeman, A.M. 1979. *The Benefits of Environmental Improvement*, Baltimore: Johns Hopkins Press.

1982. *Air and Water Pollution Control: A Benefit-Cost Assessment*, New York: John Wiley.

1991. 'Valuing environmental resources under alternative management regimes', *Ecological Economics,* **3**: 247-256.

1993a. 'Nonuse values in natural resource damage assessment' in R.J. Kopp and V.K. Smith, eds., *Valuing Natural Assets: The Economics of Natural Resource Damage Assessment*, Washington, D.C.: Resources for the Future.

1993b. *The Measurement of Environmental and Resource Values: Theory and Methods*, Washington, DC: Resources for the Future.

2002. 'How much is nature really worth? an economic perspective.' in *Valuing Nature, Papers Resulting from the Shipman Symposium*, Bowdoin College, 19-32. Online at: http://academic.bowdoin.edu/ environmental_studies/dissemination/valnat.pdf.

2003. *The Measurement of Environmental and Resource Values: Theory and Methods*, 2nd editon, Washington, DC: Resources for the Future.

Friedman, M. 1953. 'The methodology of positive economics' in *Essays in Positive Economics,* Chicago: University of Chicago Press.

Garcia, P., B.L. Dixon, J. W. Mjelde and R.M. Adams. 1986. 'Measuring the benefits of environmental change using a duality approach: the case of ozone and Illinois cash grain farms', *J. of Environmental Economics and Management,* **13**: 69-80.

Garen, J. 1988. 'Compensating wage differentials and the endogeneity of job riskiness', *Review of Economics and Statistics*, **70**: 9-16.

Gegax, D., S. Gerking and W. Schulze 1988. 'Perceived risk and the marginal value of safety', *Review of Economics and Statistics*, **70**: 589-596.

Gerking, S. and L.R. Stanley. 1986. 'An economic analysis of air pollution and health: the case of St. Louis', *Review of Economics and Statistics,* **68**: 115-121.

Goldfarb, R. S. and A. Yezer. 1987. 'Interregional wage differential dynamics', *Papers, Regional Science Association,* **62**: 45-56.

Gorman, W. M. 1953. 'Community preference fields', *Econometrica,* **21**: 63-80.

Graves, P.E. 1983. 'Migration with a composite amenity: the role of rents', *J. of Regional Science*, **23**: 541-546.

Graves, P.E. and T. A. Knapp. 1985. 'Hedonic analysis in a spatial context: theoretical problems in valuing location-specific amenities', *The Economic Record,* **16**: 737-743.

Graves, P. E. and P. R. Mueser. 1993. 'The role of equilibrium and disequilibrium in modeling regional growth and decline: a critical reassessment', *J. of Regional Science,* **33**: 69-84.

Greenwood, M., G. L. Hunt, D. S. Rickman and G. I. Treyz. 1991. 'Migration, regional equilibrium and the estimation of compensating differentials', *American Economic Review,* **81**: 1382-1390.

Haab, T.C. and R. Hicks. 1997. 'Accounting for choice set endogeneity in random utility models for recreational demand', *J. of Environmental Economics and Management,* **34**: 127-47.

Haab, T.C. and K.E. McConnell. 2002. *Valuing Environmental and Natural Resources.* Northampton, MA: Edward Elgar.

Hammack, J. and G. M. Brown. 1974. *Waterfowl and Wetlands: Towards Bioeconomic Analysis,* Baltimore, Md.: Johns Hopkins Press.

Hammitt, J.K. and J.D. Graham. 1999. 'Willingness to pay for health protection: inadequate sensitivity to probability?', *J. of Risk and Uncertainty,* **8**: 33-62.

Hanayama,Y, and I. Sano. 1986. (adapted by Maynard M. Hufschmidt and John A. Dixon) 'Valuation of losses in marine product resources caused by coastal development of Tokyo Bay' in J.A. Dixon, M. M. Hufschmidt, eds., *Economic Valuation Techniques for the Environment,* Baltimore, Maryland: Johns Hopkins University Press.

Hanemann, W.M. 1978. A Methodological and Empirical Study of the Recreation Benefits from Water Quality Improvement. Ph. D. dissertation, Department of Economics, Harvard University.

1980. 'Measuring the worth of natural resource facilities: comment', *Land Economics,* **56**: 482-90.

1982. 'Applied welfare analysis with qualitative response models', University of California Berkeley, Agricultural and Resource Economics, Giannini Foundation Working Paper #241, extended and reprinted as Hanemann 1999b).

1984. 'Discrete/Continuous Models of Consumer Demand', University of California Berkeley, Agricultural and Resource Economics, Giannini Foundation Working Paper.

1991. 'Willingness to pay and willingness to accept: how much can they differ?', *American Economic Review*, **81**: 635-47.

1999a. 'The economic theory of WTP and WTA' in I. J. Bateman and K.C. Willis, eds., *Valuing Environmental Preferences: Theory and Practice of the Contingent Valuation Method in the US, EC and Developing Countries*, Oxford: Oxford University Press.

1999b. 'Welfare analysis with discrete choice models' in J. Herriges and C. Kling, eds., *Valuing Recreation and the Environment*, Northampton, MA: Edward Elgar.

Hanemann, W.M. and E. Morey. 1992. 'Separability, partial demand systems and consumer's surplus measures', *J. of Environmental Economics and Management,* **22:** 241-258.

Harrington, W., A. Krupnick and W. Spofford. 1989. 'The economic losses of a waterborne disease outbreak', *J. of Urban Economics*, **25**: 116-137.

Harrington, W. and P.R. Portney. 1987. 'Valuing the benefits of health and safety regulation', *J. of Urban Economics,* **22**: 101-112.

Harrison, D. and D.L. Rubinfeld. 1978. 'Hedonic housing prices and the demand for clean air', *J. of Environmental Economics and Management,* **5**: 81-102.

Hause, J.C. 1975. 'The theory of welfare cost measurement', *J. of Political Economy,* **83**: 1154-78.

Hausman, J.A. 1981. 'Exact consumer's surplus and deadweight loss', *American Economic Review,* **71**: 662-676.

Hausman, J.A., G.K. Leonard and D. McFadden. 1995. 'A utility-consistent, combined discrete choice and count data model: assessing recreational use losses due to natural resource damage', *J. of Public Economics*, **56**: 1-30.

Hegan, R., G. Hauer and M. Luckert. 2003. 'Is the tragedy of the commons likely? factors preventing the dissipation of fuelwood rents', *Land Economics*, **79**: 181-197.

Henderson, J. V. 1982. 'Evaluating consumer amenities and interregional welfare differences', *J. of Urban Economics*, **11**: 32-59.

Herriges, J. and C. Kling. 1999. 'Nonlinear income effects in random utility models', *Review of Economics and Statistics*, **81**: 62-73.

Herriges, J., C. Kling and D.J. Phaneuf. 1999. 'Corner solution models of recreation demand: a comparison of competing frameworks' in J.A. Herriges and C. Kling ,eds., *Valuing Recreation and the Environment*, Northampton, MA: Edward Elgar.

2004. 'What's the use? welfare estimates from revealed preference models when weak complementarity does not hold', *J. of Environmental Economics and Management,* **47**: 55-70.

Herriges, J. and D. Phaneuf. 2002. 'Inducing patterns of correlation and substitution in repeated logit models of recreational demand', *American J. of Agricultural Economics*, **84**: 1076-1090.

Herzog, H.W. and A. M. Schlottmann. 1990. 'Valuing risk in the workplace: market price, willingness to pay and the optimal provision of safety', *Review of Economics and Statistics*, **72**: 463-470.

Hicks, J.R. 1939. 'The foundations of welfare economics', *Economic J.*, **49**: 696-712.

Hicks, R. and I.E. Strand. 2000. 'The extent of information: its relevance for random utility models', *Land Economics*, **76**: 374-85.

Hoch, I and J. Drake. 1974. 'Wages, climate and the quality of life', *J. of Environmental Economics and Management*, **1**: 268-275.

Hoehn, J., M. Berger and G. Blomquist. 1988. 'A hedonic model of wages, rents and amenity values', *J. of Regional Science*, **27**: 605-620.

Hotelling, H. 1947. 'Letter to the National Park Service', reprinted in *An Economic Study of the Monetary Evaluation of Recreation in National Parks*, Washington, DC: US Department of the Interior.

Huang, J-C. and V.K. Smith. 1998. 'Weak complementarity and production', *Economics Letters*, **60**: 329-333.

Hwang, H-S, W. R. Reed and C. Hubbard. 1992. 'Compensating wage differentials and unobserved productivity', *J. of Political Economy*, **100**: 835-858.

Jones-Lee, M.W. 1976. *The Value of Life: An Economic Analysis*, London: Martin Robertson.

Just, R.E., D.L. Hueth and A. Schmitz. 1982. *Applied Welfare Economics and Public Policy*, Englewood Cliffs, N.J.: Prentice-Hall.

2004. *The Welfare Economics of Public Policy: A Practical Approach to Project and Policy Evaluation*, Northampton, MA: Edward Elgar.

Kahn, J.R. 1987. 'Measuring the economic damages associated with terrestrial pollution of marine ecosystems', *Marine Resource Economics,* **4**: 193-209.

Kahn, J.R. and W.M. Kemp. 1985. 'Economic losses associated with the degradation of an ecosystem: the case of submerged aquatic vegetation in Chesapeake Bay', *J. of Environmental Economics and Management,* **12**: 246-263.

Kahn, S. 1987. 'Occupational safety and worker preferences: is there a marginal worker?', *Review of Economics and Statistics,* **69**: 262-268.

Kahn, S. and K. Lang. 1988. 'Efficient estimation in structural hedonic systems', *International Economic Review,* **29**: 157-166.

Kaldor, N. 1939. 'Welfare propositions of economics and interpersonal comparisons of utility', *Economic J.,* **49**: 549-552.

Killingsworth, M. 1983. *Labor Supply.* Cambridge, UK: Cambridge University Press.

Kim, S-H. and J.A. Dixon. 1986. 'Economic valuation of environmental quality aspects of upland agricultural projects in Korea' in J.A. Dixon and M.M. Hufschmidt, eds., *Economic Valuation Techniques for the Environment: A Case Study Workbook,* Baltimore, MD: Johns Hopkins Press.

Kling, C.L. 1992. 'Some results on the variance of welfare estimates from recreation demand models', *Land Economics,* **68**: 318-328.

Kling, C. and J. Herriges. 1995. 'An empirical investigation of the consistency of the nested logit models with utility maximization', *American J. of Agricultural Economics,* **77**: 875-884.

Kling, C.L. and C.J. Thomson. 1996. 'The implications of model specification for welfare estimation in nested logit models', *American J. of Agricultural Economics,* **78**: 103-114.

Knapp, K. and P. Graves. 1989. 'On the role of amenities in models of migration and regional development', *J. Regional Science,* **29**: 71-87.

Kniesner, T.J. and J.D. Leeth. 1988. 'Simulating hedonic labor market models: computational issues and policy applications', *International Economic Review,* **29**: 755-786.

1991. 'Compensating wage differentials for fatal injury risk in Australia, Japan and the United States', *J. of Risk and Uncertainty,* **4**: 75-80.

Kopp, R.J. and A.J. Krupnick. 1987. 'Agricultural policy and the benefits of ozone control', *American J. of Agricultural Economics,* **69**: 956-962.

Kopp, R.J., W.J. Vaughan, M. Hazilla and R. Carson. 1985. 'Implications of environmental policy for U.S. agriculture: the case of ambient ozone standards', *J. of Environmental Management,* **20**: 321-331.

Krueger, A. and L. Summers. 1987. 'Reflections on the inter-industry wage structure' in K. Land and J. Leonard, eds., *Unemployment and the Structure of Labor Markets,* New York: Basil Blackwell.

1988. 'Efficiency wages and the inter-industry wage structure', *Econometrica,* **56**: 259-293.

Krutilla, J. 1967. 'Conservation reconsidered', *American Economic Review,* **57**: 777-786.

LaFrance, J. 1990. 'Incomplete demand systems and semilogarithmic demand models', *Australian J. of Agricultural Economics,* **34**: 118-31.

LaFrance, J.T. and W.M. Hanemann. 1989. 'The dual structure of incomplete demand systems', *American J. of Agricultural Economics*, **71**: 262-274.

Lancaster, K. J. 1966. 'A new approach to consumer theory', *J. of Political Economy,* **74**: 132-57.

Larson, D. 1991. 'Recovering weakly complementary preferences', *J. of Environmental Economics and Management,* **21**: 97-109.

Larson, D.M. and S.L. Shaikh. 2001. 'Empirical specification requirements of two-constraint models of recreation demand', *American J. of Agricultural Economics,* **83**: 428-440.

2004. 'Recreation demand choices and revealed values of leisure time', *Economic Inquiry,* **42**: 264-278.

Laughland, A.S., W.N. Musser, J.S. Shortle and L.M. Musser. 1996. 'Construct validity of averting cost measures of environmental benefits', *Land Economics,* **72**: 100-12.

Leeth, J. and J. Ruser. 2003. 'Compensating wage differentials for fatal and nonfatal injury risk by gender and race', *J. of Risk and Uncertainty,* **27**: 257-277.

Leggett, C. 2002. 'Environmental valuation with imperfect information', *Environmental and Resource Economics,* **23**: 343-355.

Leggett, C. and N.E. Bockstael. 2000. 'Evidence of the effects of water quality on residential land prices', *J. of Environmental Economics and Management,* **39**: 131-144.

Leigh, J. 1995. 'Compensating wages, value of a statistical life and inter-industry differentials', *J. of Environmental Economics and Management,* **28**: 83-97.

Leigh, J. and F.R. Folsom. 1984. 'Estimates of the value of accident avoidance at the job depend on the concavity of the equalizing difference curve', *Quarterly Review of Economics and Business*, **24**: 56-66.

Lerman, S.R. and C.R. Kern. 1983. 'Hedonic theory, bid rents and willingness-to-pay: some extensions of Ellickson's results', *J. of Urban Economics*, **13**: 358-63.

Levin, S. 1998. 'Ecosystems and the biosphere as complex adaptive systems', *Ecosystems* **1**: 431-436.

Lind, R.C. 1973. 'Spatial equilibrium, the theory of rents and benefits from public programs', *Quarterly J. of Economics,* **87**: 188-207.

Liu, J.-T. and J.K. Hammitt. 1999. 'Perceived risk and value of workplace safety in a developing country', *J.of Risk Research*, **2**: 263-275.

Lopez, R. 1991. 'Structural models of the farm household that allow for interdependent utility and profit-maximization decisions' in I. Singh, L. Squire and J. Strauss, eds., *Agricultural Household Models,* Baltimore, MD: Johns Hopkins Press.

Louviere, J., D. Hensher and J. Swait. 2000. *Stated Choice Methods: Analysis and Application*, New York: Cambridge University Press.

Ludwig, D., B.H. Walker and C. S. Holling. 1997. 'Sustainability, stability and resilience', *Conservation Ecology,* **1**: 7.

Lynne, G.D., P. Conroy and F.J. Prochaska. 1981. 'Economic valuation of marsh areas for marine production processes', *J. of Environmental Economics and Management,* **8**: 175-186.

Maddison, D. and A. Bigano. 2003. 'The amenity value of the Italian climate', *J. Environmental Economics and Management*, **45**: 319-332.

Mäler, K.-G. 1974. *Environmental Economics: A Theoretical Inquiry,* Baltimore, MD: Johns Hopkins University Press for Resources for the Future.

1992. 'Production function approach in developing countries', in: J.R. Vincent, E. W. Crawford, J. P. Hoehn, eds., *Valuing Environmental Benefits in Developing Countries,* (Special report 29, Michigan State University).

Manser, M. and M. Brown. 1980. 'Marriage and household decision-making: a bargaining approach', *International Economic Review*, **21**: 31-44.

Manski, C. and D. McFadden. 1981. 'Alternative estimators and sample designs for discrete choice estimators' in C. Manski and D. McFadden, eds., *Structural Analysis of Discrete Data with Econometric Applications,* Cambridge, MA: MIT Press.

Marshall, A. 1930. *Principles of Economics.* London: MacMillan.

Mason, C. and J.M. Quigley. 1997. 'Comparing the performance of discrete choice and hedonic models' in *The Economics of Housing*, J.M. Quigley, ed., Northampton, MA.: Edward Elgar.

Mathur, V.K. and S.H. Stein. 1991 'A dynamic interregional theory of migration and population growth', *Land Economics*, **67**: 292-298.

1993. 'The role of amenities in a general equilibrium model of regional migration and growth', *Southern Economic J.*, **59**: 394-409.

McConnell, K.E. 1983. 'Existence and bequest value' in R. Rowe and L. Chestnut, eds., *Managing Air Quality and Scenic Resources at National Parks and Wilderness Areas*, Boulder, CO: Westview Press.

1986. 'The Damages to Recreation Activities from PCB's in New Bedford Harbor', Report for Ocean Assessment Division, Rockville, MD: National Oceanic and Atmospheric Administration.

1992. 'On-site time in the demand for recreation', *American J. of Agricultural Economics,* **74**: 918-25.

McConnell, K.E. and I.E. Strand. 1981. 'Measuring the cost of time in the demand for recreation', *American J. of Agricultural Economics*, **63**: 153-56.

1989. 'Benefits from commercial fisheries when demand and supply depend on water quality', *J. of Environmental Economics and Management,* **17**: 284-292.

McConnell, K.E., I.E. Strand and L. Blake-Hedges. 1995. 'Random utility models of recreational fishing: catching fish with a Poisson process', *Marine Resource Economics*, **10:** 247-262.

McElroy, M. and M.J. Horney. 1981. 'Nash-bargained household decisions: toward a generalization of the theory of demand', *International Economic Review*, **22**: 333-49.

McFadden, D. 1974. 'Conditional logit analysis of qualitative choice behavior' in P. Zarembka, ed., *Frontiers in Econometrics*, New York: Academic Press, 105-142.

1999. 'Computing willingness-to-pay in random utility models' in Moore, J., R. Riesman and J. Melvin, eds., *Trade, Theory and Econometrics: Essays in Honour of John S. Chipman*, London: Routledge.

1978. 'Modeling the choice of residential location' in A. Karlqvist, L. Lundqvist, F. Snickars and J.W. Weibull, eds., *Spatial Interaction Theory and Planning Models*, 75-96.

McFadden, D. and K.E. Train. 2000. 'Mixed MNL models for discrete response', *J. of Applied Econometrics*, **15:** 447-70.

McGartland, A.M. 1987. 'The implications of ambient ozone standards for U.S. agriculture: a comment and some further evidence', *J. of Environmental Management*, **24**: 139-146.

Mendelsohn R., W. Nordhaus and D. Shaw. 1994. 'The impact of global warming on agriculture: a Ricardian analysis', *American Economic Review*, **84**: 753-771.

Moore, M. and W.K. Viscusi. 1988. 'Quantity-adjusted value of life', *Economic Inquiry*, **26**: 369-388.

Morey E. 1999. 'Two RUM's uncloaked: nested logit models of site choice and nested logit models of participation and site choice', in J. Herriges and C. Kling, eds., *Valuing Recreation and the Environment*, Northhampton, MA: Edward Elgar, 65-116.

Morey, E.R., R. Rowe and M. Watson. 1993. 'A repeated nested-logit model of Atlantic salmon fishing', *American J. of Agricultural Economics*, **75**: 578-92.

Morey, E.R., V.R. Sharma and A. Karlstrom. 2003. 'Incorporating income effects into logit models', *American J. of Agricultural Economics*, **85**: 248-253.

Morrall, J. 1986. 'A review of the record', *Regulation*, **10**: 25-34.

2003. 'Saving lives: a review of the record', *J. of Risk and Uncertainty*, **27**: 221-237.

Mrozek, J. and L. Taylor. 2002. 'What determines the value of life? a meta-analysis', *J. of Policy Analysis and Management*, **21**: 253-270.

Mueser, P. and P. Graves. 1995. 'Examining the role of economic opportunities and amenities in explaining population redistribution', *J. Urban Economics*, **37**: 176-200.

Munro, A. 2005. 'Household willingness to pay equals individual willingness to pay if and only if the household income pools', *Economics Letters*, **88**: 227-230.

Murdoch J.C. and M.A. Thayer. 1990. 'The benefits of reducing the incidence of nonmelanoma skin cancers: a defensive expenditures approach', *J. of Environmental Economics and Management*, **18**: 107-119.

Narain, U. and A. Fisher. 1994. 'Modeling the value of biodiversity using a production function approach', in C. Perrings, K.-G. Mäler, C. Folke, C. Holling and B. Jansson, eds., *Biodiversity Conservation: Problems and Policies*, London: Kluwer.

Neary, J. and K. Roberts. 1980. 'The theory of household behaviour under rationing', *European Economic Review*, **13**: 25-42.

Olson, C. 1981. 'An analysis of wage differentials received by workers on dangerous jobs', *J. of Human Resources*, **16**: 167–185.

Palmquist, R.B. 1984. 'Estimating the demand for the characteristics of housing', *Review of Economics and Statistics*, **66**: 394-404.

1988. 'Welfare measurement for environmental improvements using the hedonic model: the case of nonparametric marginal prices', *J. Environmental Economics and Management*, **14**: 297-312.

1991. 'Hedonic methods', in J.B. Braden and C.D. Kolstad, eds., *Measuring the Demand for Environmental Quality*, Amsterdam: North Holland Publishers, 77-120.

1992. 'Valuing localized externalities', *J. of Urban Economics*, **31**: 59-68.

2005a. 'Weak complementarity, path independence and the intuition of the Willig condition', *J. of Environmental Economics and Management*, **49**: 103-15.

2005b. 'Property value models' in K-G. Mäler and J. Vincent, eds., *Handbook of Environmental Economics*, V. 2, *Valuing Environmental Changes*, Amsterdam: North Holland Publishers. 763-819.

Palmquist, R.B. and A. Israngkura. 1999. 'Valuing air quality with hedonic and discrete choice models', *American J. of Agricultural and Resource Economics*, **81**: 128-133.

Parsons, G. 1986. 'An almost ideal demand system for housing attributes', *Southern Economic J.*, **53:** 347-363.

2001. 'A bibliography of revealed preference random utility models in recreation demand', University of Delaware Department of Economics, Working Paper.

Parsons, G., P. Jakus and T. Tomasi. 1999. 'A comparison of welfare estimates from four models for linking seasonal recreational trips to multinomial logit models of site choice', *J. of Environmental Economics and Management*, **38**: 143-157.

Pattanayak, S. and D. Butry. 2005. 'Complementarity of forests and farms: a spatial econometric approach to ecosystem valuation in Indonesia', manuscript, Research Triangle Institute.

Pattanayak, S. and R. Kramer. 2001. 'Worth of watersheds: a producer surplus approach for valuing drought control in Eastern Indonesia', *Environment and Development Economics,* **6**: 123-45.

Pearce, D. 1998. 'Auditing the earth', *Environment,* **40**: 23-28

Pendleton, L.H. 1999. 'Reconsidering the hedonic vs. RUM debate in the valuation of recreational environmental amenities', *Resource and Energy Economics*, **21**: 167-189.

Pendleton, L.H. and R. Mendelsohn. 1998. 'Estimating the economic impact of climate change on the freshwater sportsfisheries of the northeastern U.S.', *Land Economics*, **74**: 483-96.

2000. 'Estimating recreation preferences using hedonic travel cost and random utility models', *Environmental and Resource Economics*, **17**: 89-108.

Persson, T.H. 2002.'Welfare calculations in models of the demand for sanitation', *Applied Economics*, **34**: 1509-18.

Phaneuf, D. 1997. 'Generalized Corner Solution Models in Recreation Demand'. Ph.D. dissertation, Department of Economics, Iowa State University.

Phaneuf, D.J., J.A. Herriges and C.L. Kling. 2000. 'Estimation and welfare calculations in a generalized corner solution model with an application to recreation demand', *Review of Economics and Statistics,* **82**: 83-92.

Pimentel, D., C. Wilson, C. McCullum, R. Huang, P. Dwen, J. Flack, Q. Tran, T. Saltman and B. Cliff. 1997. 'Economic and environmental benefits of biodiversity', *BioScience*, **47**: 747-757.

Polinsky, A.M. and S. Shavell. 1976. 'Amenities and property values in a model of an urban area', *J. of Public Economics*, **5**: 119-29.

Pitt, M. and M. Rosenzweig. 1986. 'Agricultural prices, food consumption and the health and productivity of Indonesian farmers.' in Singh, I., L. Squire, and J. Strauss, eds., *Agricultural Household Models: Extensions, Applications and Policy*. Baltimore: Johns Hopkins University Press, pp. 153-182.

Pollak, R. and M. Wachter. 1975. 'The relevance of the household production function and its implications for the allocation of time', *J. of Political Economy*, **83**: 255-77.

Provencher, W., K. Baerenklau and R. Bishop. 2002. 'A finite mixture logit model of recreational angling with serially correlated random utility', *American J. of Agricultural Economics*, **84**: 1066-1075.

Provencher, W. and R. Bishop. 1997. 'An estimable dynamic model of recreation behavior with an application to Great Lakes angling'; *J. of Environmental Economics and Management*, **33**: 107-27.

Quiggin, J. 1992. 'Risk, self-protection and ex ante economic value–some positive results', *J. of Environmental Economics and Management*, **23**: 40-53.

Quiggin, J. and J.K. Horowitz. 1999. 'The impact of global warming on agriculture: a Ricardian analysis: comment', *American Economic Review*, **89**: 1044-1045.

Quigley, J.M. 1985. 'Consumer choice of dwelling, neighborhood and public services', *Regional Science and Urban Economics*, **15**: 41-63.

Randall, A. 1991. 'Total and nonuse values' in J. Braden and C. Kolstad, eds., *Measuring the Demand for Environmental Quality*. Amsterdam, Netherlands: North-Holland.

Randall, A. and J. Stoll. 1980. 'Consumer surplus in commodity space', *American Economic Review*, **70**: 449-455.

1983. 'Existence value in a total valuation framework' in R. Rowe and L. Chestnut, eds., *Managing Air Quality and Scenic Resources at National Parks and Wilderness Areas*, Boulder, CO: Westview Press.

Rappaport, J. 2004. 'Why are population flows so persistent', *J. of Urban Economics*, **56**: 554-580.

Roback, J. 1982. 'Wages, rents and the quality of life', *J. Political Economy*, **90**: 1257-1278.

1988. 'Wages, rents, and amenities: differences among workers and regions', *Economic Inquiry*, **26**:23-41.

Rosen, S. 1974. 'Hedonic prices and implicit markets: product differentiation in price competition', *J. of Political Economy*, **82**: 34-55.

1979. 'Wage-based indexes of urban quality of life' in P.Mieszkowski and M. Straszheim,eds., *Current Issues in Urban Economics*, Baltimore: Johns Hopkins Press.

Samuelson, P.A. 1942. 'Constancy of the marginal utility of income' in O. Lange, F. McIntyre and T. Yntema, eds., *Studies in Mathematical Economics and Econometrics in Memory of Henry Schultz*, Chicago: University of Chicago Press.

1956. 'Social indifference curves', *Quarterly J. of Economics*, **70**: 1-22.

Samuelson, W. and R. Zeckhauser. 1988. 'Status quo bias in decision making', *J. of Risk and Uncertainty*, **1**: 7-59.

Schelling, T. C. 1968. 'The life you save may be your own' in S.B. Chase, Jr., ed., *Problems in Public Expenditure Analysis*, Washington DC: Brookings Institution.

Scitovsky, T. 1941. 'A note on welfare propositions', *Review of Economic Studies*, **8**: 77-88.

Scotton, C. and L. O. Taylor. 2003. 'Mortality risk by occupation within an industry, inter-industry wage differentials and the value of a statistical life', working paper, Department of Economics, Andrew Young School of Policy Studies, Georgia State University.

2004. 'Do labor markets respond to different risks of mortality?', working paper, Department of Economics, Andrew Young School of Policy Studies, Georgia State University.

Shaikh, S. and D.M. Larson. 2003. 'A two-constraint almost ideal demand model of recreation and donations', *Review of Economics and Statistics*, **85**: 953-961.

Shaw, D.W. and P. Feather. 1999. 'Possibilities for including the opportunity cost of time in recreation demand systems', *Land Economics*, **75**: 592-602.

Shepard, D. S. and R. J. Zeckhauser. 1984. 'Survival vs. Consumption', *Management Science*, **30**: 423-439.

Shogren, J. and T. Crocker. 1991. 'Risk, self-protection and ex ante economic value', *J. of Environmental Economics and Management*, **20**: 1-15.

Sieg, H., V.K. Smith, H.S. Banzhaf and R. Walsh. 2004. 'Estimating the general equilibrium benefits of large changes in spatially delineated goods', *International Economic Review*, **45**: 1047-1077.

Simpson, R.D., R.A. Sedjo and J.W. Reid. 1996. 'Valuing biodiversity for use in pharmaceutical research', *J. of Political Economy*, **104**: 163-185.

Singh, I., L. Squire and J. Strauss. 1986. 'The basic model: theory, empirical results and policy conclusions' in I. Singh, L. Squire and J. Strauss, eds., *Agricultural Household Models*, Baltimore, MD: Johns Hopkins Press.

Slovic, P.B. 1987. 'Perception of risk', *Science*, **236**: 280-285.

Smith, R. 1979. 'Compensating wage differentials and public policy: a review', *Industrial and Labor Relations Review*, **32:** 339-352.

Smith, V.K. 1997a. 'Combining discrete choice and count models: a comment', Xerox.

1997b. 'Mispriced planet', *Regulation* **3**: 16-17.

Smith, V.K. and H.S. Banzhaf. 2004. 'A diagrammatic exposition of weak complementarity and the Willig condition', *American J. of Agricultural Economics*, **86**: 455-66.

Smith, V.K. and W.H. Desvousges. 1985. 'The generalized travel cost model and water quality benefits: a reconsideration', *Southern Economic J.*, **52**: 371-81.

1987. 'An empirical analysis of the economic value of changes in risks', *J. of Political Economy*, **95**: 89-114.

Smith, V.K., W.H. Desvousges and M.P. McGivney. 1983. 'The opportunity cost of travel time in recreation demand models', *Land Economics*, **59**: 259-277.

Smith, V.K., M. Evans, S. Banzhaf and C. Poulos. 2004. 'Rehabilitating weak substitution', Department of Agricultural Economics, North Carolina State University, working paper.

Smith, V.K., M.F. Evans, H. Kim and D.H. Taylor. 2004. 'Do the near-elderly value mortality risks differently?', *Review of Economics and Statistics*, **86**: 423-429.

Smith, V.K. and Y. Kaoru. 1987. 'The hedonic travel cost method: a view from the trenches', *Land Economics*, **63**: 179-192.

Smith, V.K., R. Palmquist and P. Jakus. 1991. 'Combining Farrell frontier and hedonic travel cost models for valuing estuarine quality', *Review of Economics and Statistics*, **73**: 694-99.

Smith, V.K., S.K. Pattanayak and G.L. Van Houtven. 2003. 'VSL reconsidered: what do labor supply estimates reveal about risk preferences?', *Economics Letters*, **80**: 147-153.

Smith, V.K., H. Sieg, S. Banzhaf and R.Walsh. 2004. 'General equilibrium benefits for environmental improvements: projected ozone reductions under EPA's Prospective Analysis for the Los Angeles air basin', *J. of Environmental Economics and Management*, **47**: 559-584.

Smith, V.K. and G. Van Houtven. 2004. 'Recovering Hicksian consumer surplus within a collective model: Hausman's method for the household', *Environmental and Resource Economics*, **28**: 153-167.

Stevens, J.B. 1966. 'Recreation benefits from water pollution control', *Water Resources Research*, **2:** 167-182.

Swallow, S. 1994. 'Resource economic theory, wetlands and fisheries', *Marine Resource Economics*, **9**: 291-310.

Thaler, R. 1989. 'Anomalies: interindustry wage differentials', *J. of Economic Perspectives*, **3**: 181-193.

Thaler, R. and S. Rosen. 1976. 'The value of saving a life: evidence from the labor market' in N. Terleckyj, ed., *Household Production and Consumption*, New York: Columbia University Press (for NBER).

Thornton, J. and B.K. Eakin. 1992. 'Virtual prices and a general theory of the owner operated firm', *Southern Economic J.*, **58**: 1015-1029.

Toman, M. 1998. 'Why not to calculate the value of the world's ecosystem services and natural capital', *Ecological Economics*, 25: 57-60.

Topel, R.H. 1986. 'Local labor markets', *J. of Political Economy,* **94**: S111-S143.

Train, K.E. 1998. 'Recreation demand models with taste difference over people', *Land Economics,* **74**: 230-39.

1999. 'Mixed logit models for recreation demand' in J.A. Herriges and C.L. Kling, eds., *Valuing Recreation and the Environment: Revealed Preference Methods in Theory and Practice*, Northampton, MA: Edward Elgar, 121-140.

2003. *Discrete Choice Methods with Simulation*, Cambridge: Cambridge University Press.

Treyz G.I., D.S. Rickman, G.L. Hunt and M.J. Greenwood. 1993. 'The dynamics of US internal migration', *Review of Economics and Statistics,* **75**: 209-214.

US Environmental Protection Agency. 1997. *The Benefits and Costs of the Clean Air Act, 1970 to 1990.* Office of Policy, Planning and Evaluation and Office of Air and Radiation. Report # EE-0295.

US Environmental Protection Agency. 1999. *The Benefits and Costs of the Clean Air Act 1990 to 2010: EPA Report to Congress.* Office of Policy and Office of Air and Radiation. Report # EPA-410-R-99-001.

Vartia, Y.O. 1983. 'Efficient methods of measuring welfare change and compensated income in terms of ordinary demand functions', *Econometrica,* **51**: 79-98.

Viscusi, W.K. 1980. 'Union, labor market structure and the welfare implications of the quality of work', *Journal of Labor Research,* **1:** 175-92.

1981. 'Occupational safety and health regulation: its impact and policy implications', *Research in Public Policy Analysis and Management,* **2**: 218-299.

1993. 'The value of risks to life and health'. *J. of Economic Literature,* **31**: 1912-1946.

2003. 'Racial differences in labor market values of a statistical life', *J. of Risk and Uncertainty,* **27**: 239-256.

2004. 'The value of life: estimates with risks by occupation and industry', *Economic Inquiry,* **42**: 29-48.

Viscusi, W.K. and J. Aldy. 2003. 'The value of a statistical life: a critical review of market estimates throughout the world', *J. of Risk and Uncertainty,* **27**: 5-76.

Viscusi, W.K. and M. Moore. 1987. 'Workers' compensation: wage effects, benefit inadequacies and the value of health losses', *Review of Economics and Statistics,* **69**: 249-261.

Viscusi, W.K and R.J. Zeckhauser. 2004. 'The denominator blindness effect: accident frequencies and the misjudement of recklessness', *American Law and Economics Review*, **6**: 72-94.

von Haefen, R. 2004. 'Empirical strategies for incorporating weak complementarity into continuous demand system models', working paper, Department of Agricultural and Resource Economics, North Carolina State University.

von Haefen, R. and D.J. Phaneuf. 'Estimating preferences for outdoor recreation: a comparison of continuous and count data demand systems', forthcoming, *J. of Environmental Economics and Management.*

von Haefen, R., D.J. Phaneuf and G. Parsons. 2004. 'Estimation and welfare analysis with large demand systems', *J. of Business and Economic Statistics,* **22**: 194-205.

Wales, T. and A. Woodland. 1983. 'Estimation of consumer demand systems with binding nonnegativity constraints', *J. of Econometrics*, **21**: 263-85.

Weitzman, M. 1976. 'On the welfare significance of national product in a dynamic economy', *Quarterly J. of Economics,* **90**: 156-162.

Willig, R.D. 1976. 'Consumer surplus without apology', *American Economic Review*, **66**: 589-597.

1978. 'Incremental consumer's surplus and hedonic price adjustment', *J. of Economic Theory*, **17**: 227-53.

Young, D. and S. Aidun. 1993. 'Ozone and wheat farming in Alberta: a micro-study of the effects of environmental change', *Canadian J. of Agricultural Economics,* **41**: 27-43.

Index

THE ECONOMICS OF NON-MARKET GOODS AND RESOURCES

1. J. Loomis and G. Helfand: *Environmental Policy Analysis for Decision Making*. 2001
   ISBN 0-7923-6500-3
2. L. Fernandez and R.T Carson (eds.): *Both Sides of the Border*. Transboundary Environmental Management Issues Facing Mexico and the United States. 2002
   ISBN 1-4020-7126-4
3. P.A. Champ, K.J. Boyle and T.C. Brown (eds.): *A Primer on Nonmarket Valuation*. 2003
   ISBN 0-7923-6498-8
4. P. Dasgupta and K.-G. Mäler (eds.): *The Economics of Non-Convex Ecosystems*. 2004
   ISBN Hb 1-4020-1945-9; ISBN Pb 1-4020-1864-9
5. R.T. Carson, M.B., Conaway, W.M. Hanemann, J.A. Krosnick, R.C. Mitchell and S. Presser: *Valuing Oil Spill Prevention*. A Case Study of California's Central Coast. 2004
   ISBN 0-7923-6497-X
6. R. Scarpa, A.A. Alberini: *Applications of Simulation Methods in Environmental and Resource Economics*. 2005
   ISBN 1-4020-3683-3
7. N.E. Bockstael and K.E. McConnell: *Environmental and Resource Valuation with Revealed Preferences*. A Theoretical Guide to Empirical Models. 2007
   ISBN 0-7923-6501-1
8. B.J. Kanninen (ed.): *Valuing Environmental Amenities Using Choice Experiments*. A Common Sense Approach to Theory and Practice. 2006
   ISBN 1-4020-4064-4

SPRINGER NATURE

GPSR Compliance

The European Union's (EU) General Product Safety Regulation (GPSR) is a set of rules that requires consumer products to be safe and our obligations to ensure this.

If you have any concerns about our products, you can contact us on ProductSafety@springernature.com

In case Publisher is established outside the EU, the EU authorized representative is:

Springer Nature Customer Service Center GmbH
Europaplatz 3
69115 Heidelberg, Germany

FSC
www.fsc.org
MIX
Papier | Fördert
gute Waldnutzung
FSC® C083411

Kolibri 360 GmbH
Ferdinand-Jühlke-Straße 7
99095 Erfurt, Deutschland
produktsicherheit@kolibri360.de